HUMAN CONSEQUENCES OF CROWDING

NATO CONFERENCE SERIES

I Ecology
II Systems Science
III Human Factors
IV Marine Sciences
V Air–Sea Interactions
VI Materials Science

III HUMAN FACTORS

HUMAN CONSEQUENCES OF CROWDING

Edited by
Mehmet R. Gürkaynak
Middle East Technical University
Ankara, Turkey

and
W. Ayhan LeCompte
Hacettepe University
Ankara, Turkey

Published in coordination with NATO Scientific Affairs Division by
PLENUM PRESS · NEW YORK AND LONDON

Library of Congress Cataloging in Publication Data

Symposium on Human Consequences of Crowding, Antalya, Turkey, 1977.
Human consequences of crowding.

(NATO conference series: III, Human factors; 10)
Selected lectures presented at the Symposium on Human Consequences of Crowding, held in Antalya, Turkey, Nov. 6–11, 1977.
Bibliography: p.
Includes indexes.
1. Crowding stress–Congresses. 2. Personal space–Congresses. I. Gürkaynak, Mehmet R. II. LeCompte, William Ayhan. III. Title. IV. Series.
HM291.S94 1977 301.1 79-19152
ISBN 0-306-40298-X

Lectures presented at the Symposium on Human Consequences of Crowding, held in Antalya, Turkey, November 6–11, 1977.

A Division of Plenum Publishing Corporation
227 West 17th Street, New York, N.Y. 10011

Printed in the United States of America

PREFACE

This volume contains papers selected from among those submitted to the Symposium on "Human Consequences of Crowding", held in Antalya, Turkey, 6-11 November, 1977.

Realizing an international symposium of this scope, and preparing the manuscript for publication afterwards, necessitated the assistance and support of so many people that it is impossible to name all but a few of them. First of all, we are particularly grateful to the Scientific Affairs Division of NATO (Special Programme Panel on Human Factors), and the Middle East Technical University in Ankara, Turkey, the co-sponsors of the Symposium. Dr. Robert B. Bechtel of the Environmental Research and Development Foundation, Tucson, Arizona, U.S.A., joined the editors of the present volume in planning the Symposium, and acted as a "point of contact" for the Americas and the Pacific Region. An advisory board consisting of Mithat Çoruh, M.D. (Hacettepe University, Ankara, Turkey), Dr. Vacit İmamoğlu, and Dr. Mete Turan (both of the Middle East Technical University) helped the directors of the Symposium in the initial selection of the papers to be included at the meeting. We are indebted to these persons.

We would also like to thank Aynur Başkan, who provided the initial secreterial assistance. Saide Çerezci, Bülent Tulun and Veysel Batmaz worked as assistants in the planning phases of the Symposium. During the Symposium, Dr. Güney LeCompte and Dr. Olcay İmamoğlu helped in organizing the sessions, and Meral Apaydın, Abdurrahman Eke and Bülent Tulun provided administrative assistance.

We are grateful to Dr. Hüsnü Arıcı and Dr. Bülent A. Bayraktar for their helpful suggestions and unfailing support during all phases of this undertaking. Similarly, Dr. İpek Gürkaynak's efforts were exceptional in making this conference a success. For eight months before the Symposium, the Middle East Technical University was closed because of political turmoil and we were unable to find the needed manpower to carry on the work. Dr. İ. Gürkaynak undertook the tasks of an administrative assistant, secretary and organizer. This involvement obliged her not only to attend the Symposium as a participant but also to work as a de facto "behind the scenes" administrative director in charge of registration and preparation of press releases.

A note of special thanks is due to the Mayor of Antalya, the Antalya Chamber of Commerce, the Institute for Population Studies of Hacettepe University, and the Ministry of Tourism of Turkey for providing generous contributions in the form of sight-seeing tours and entertainment. Also gratefully acknowledged are the efficient and friendly services of the personnel and managers of the Hotel Antalya where the Symposium was held.

Finally, with regard to the present publication, we owe a debt of gratitude to Ahmet T. Altiner, the owner of the Kelaynak Printing House, Ankara, who undertook the preparation of the manuscripts for publication, and to Dorothy A. Pedtke who did the tedious job of correcting the galley proofs and with outstanding competence modified the inadequacies of English which escaped the attention of the editors.

Mehmet R. Gürkaynak, Ph. D.
W. Ayhan LeCompte, Ph. D.

CONTENTS

INTRODUCTION

Crowding is good; crowding is bad; crowding is both good and bad; crowding is neither good nor bad. All of the possible variations in the effects of this phenomenon have been claimed in one context or another and most of them can be found in the collection of papers included in the present volume. To quote the writer of the first chapter of the book, "there are now more than one billion people on this planet and at times at least half of them seem to be doing research on crowding." Perhaps one attraction of this topic is precisely the apparently inexhaustible variety of perspectives that can be applied to study its consequences.

This book contains a selection of papers that were presented during a memorable week from the 6th to the 11th of November, 1977, in Antalya, Turkey, at a conference sponsored jointly by NATO and the Middle East Technical University on the topic of "the human consequences of crowding". The participants were from many different disciplines and countries, the setting was congenial, and the discussions often ran well into the evening hours, despite a very heavy schedule of papers during the days. The results may not be measurable entirely in terms of the contents of this book; by the demands of the space and cost, it must be only a selection of the many papers that were presented. In addition, many of the papers had been prepared well in advance of the meetings and could not be reformulated to fit the authors' new ideas. One must look to the future to see what new models and research might spring from the ideas planted at this conference. Nevertheless, the collection of papers in this volume represents as good a survey of the contemporary state of the crowding area as can be found. The following paragraphs will serve to highlight some of the remarkable variety to be found herein.

1. Crowding is good if you need it and haven't got it.

Ecologists maintain that the sperm whale is doomed, even if they are no longer hunted by man. Their number has been reduced too far, and their territorial range is too large to allow a sufficient number of them to find and mate with each other to maintain their population. They will presumably dwindle to extinction. Clearly, crowding is a preferable alternative for the sperm whale. In the United

States today, with well over 80 per cent of the population living in large metropolitan centers, the small, rural-situated town is an endangered species as well. Barker (1961) has documented the need of some of these small towns for people, characterizing this need for people as a condition of "undermanning". Two students of Barker's, Robert Bechtel (Part I) and Raj Srivastava (Part I), have contributed papers to the present volume, in which they have attempted to bridge the theoretical gap from "undermanning" to "overmanning".

2. Crowding is bad if you can't get away from it.

This important generalization is supported by many studies in the present volume, perhaps most dramatically by Chalsa Loo (Part II), with 5- and 10-year-old children; Marilyn Rall (Part II), with university students; and Susan Saegert (Part I), in naturally-occurring commercial settings. In addition, the labels of the dimensions emerging from the multivariate scaling studies of John Schopler, et al. (Part III), and Janet Stockdale, et al. (Part III) have a distinctly unpleasant flavor.

3. Crowding can be both good and bad, depending on X.

This conclusion is reached by many contributors to the present volume, but the definition of the moderating variable is nearly as varied as the number of writers. X may be identified at a philosophical level, as the "social versus the individual nature of man" (Christensen, Part I), as the organism's current state of physiological arousal in relation to its level of adaptation (Freedman, Part I), as a condition of a particular group of behavior settings in the ecological environment (Bechtel, Part I), as a critical level of "perceived control" of the environment (Greenberg, Part II), or as a cognition by an actor that a violation of their personal space occurs because it has been intended by another person, rather than as an impersonal characteristic of the environment (Nerenz, et al., Part II). This list by no means exhausts all of the theoretical possibilities but it does give some notion of the variety in hypothesized moderating variables.

4. Finally, crowding can be neither good nor bad.

This state of affairs may occur for a number of reasons. One's level of density chosen for a particular experiment may be subcritical or one's subjects, despite reported experiences of crowding, may remain intractable and refuse to display any particular consequences (Schultz-Gambard, Part II). On the other hand, the investigator may place more weight on another factor rather than perceived or actual crowding, as when Mikawa (Part II) defines one's reaction to actual crowding, as a personality variable. In the present collection of papers this is definitely a rarely chosen alternative, but the contributors to the present volume certainly do not exhaust the universe of theoretical alternatives on this issue.

Clearly, the foregoing variety suggests that the issue of crowding and its consequences will continue to generate excitement and controversy among behavioral scientists and environmental professionals for some time to come. One of the major reasons for such profound lack of agreement among the investigators who study crowding consequences is their lack of agreement on a common definition of the phenomenon itself. Among the chapters in this volume, crowding has been defined as population density, or the number of people per unit of space; as social density, or the number of people co-present in a defined space; and as household density, or the number of people per room. Turning from such "physicalistic" definitions to more "behavioral" ones, many investigators have defined crowding in terms of the activities engaged in by inhabitants of a given space. An activity requiring fewer people than the number actually present, for example, may be classified as an "over-manned" activity. Thus, the same level of density may be regarded as crowded or uncrowded, depending on the amount of role-related activity in the setting. Finally, there are many "psychological" definitions, that is, those infinitions in which a person's perceptual and/or cognitive apparatus is involved in the judgement of crowding.

Speaking generally, there seems to be a tendency for writers with the same definition of crowding to interact more frequently with each other than with others. Thus, for example, the people favoring one or the other type of "physicalistic" definition cite each other, but tend to ignore other work in the area. One effect of this tendency is to give the superficial appearance of consensus among writers until one stumbles across research done with a different definition. A second result of this lack of interaction is to reduce the number of methodological studies in which one crowding definition is systematically compared with another in the hope of finding commonality. Such studies, unfortunately, are as rare as vacant seats in a final match of the World Soccer Cup playoffs. A slightly different strategy, but one with some promise, is found in studies in which the independent variable is defined by some variant of the "physicalistic" criterion and the dependent variable contains variance from perceptual and/or cognitive measures of crowding. The paper by Derek Hall (Part IV), an English geographer, contrasts three communities in England on the basis of both ecological structure (e.g., community area) and such questionnaire variables as socio-economic status and ability to conceptualize "an area round here where you feel at home." In a similar vein, Edgüer and LeCompte (Part III) compared different ecologically defined sectors of Turkish communities by presenting slides of behavior settings to respondents to whom the specific settings were familiar, and having them rate their perceptions on semantic differential scales.

The controversy, the sudden flowering into productivity of researchable topics, together with their increasing relevance in today's overpopulated world, probably guarantees continued interest in the human consequences of crowding despite definitional disagreements. After all, the lack of a systematic definition of life has not done fatal damage to the discipline of biology. In the words that have been attributed (perhaps apocryphaly) to the psychologist, R.B. Cattell, "If we could precisely define our phenomena, it would be pointless to investigate it."

In addition to variation in physicalistic, behavioral and psychological definitions of human crowding, a second important dimension along which the chapters in this book distribute themselves is that of method. More specifically, in terms of techniques of data collection, the present group of studies runs all the way from non-interfering, systematic observation in naturally-occurring behavior settings to high-control, manipulative laboratory experiments. Rarely is an investigator working at more than a single point along this continuum (although Susan Saegert is a welcome exception to that rule) with the result that the same lack of coordination occurs that was previously noted among the definitions of crowding. Peter Stringer (Part I) has been especially sensitive to this methodological anarchy, and proposes a conceptual scheme for relating at least some of the various levels of analysis intelligibly together.

This is not to deny that creative and fascinating techniques for generating data on consequences of crowding are to be found among the studies reported in this book. On the contrary, brilliant innovations abound! Unfortunately however, the method variance is often so high that even working with the same conceptual level and definition of crowding, two investigators may not be dealing with the same phenomenon. The answer seems to be, as in the case with the definitional variance, for more studies to be designed with the explicit goal of convergent validation of two or more methods in mind. Thus, survey studies could canvass the same populations as are studied in experimental research, and/or systematic observation methods could be applied to groups on which a data base from demographic survey work also exists, etc.

Recognizing the imperfect nature of this world and the regrettable fact that the payoff matrix for crowding research, an in other areas of western science, is loaded for specialization and competition, rather than generality and cooperation, we nevertheless cannot resist the temptation to conclude this introduction with a suggestion for researchers of crowding phenomena. Let us consider the following principle: No empirical generalization in the human crowding area will be accepted as such until it has been confirmed by at least two studies with different methods or with different definitions of the phenomenon. Thus, for example, sex differences in reaction to crowding could be tested both with laboratory experimental work and with differential responses to questionnaire items. Alternatively, sex differences could be shown to exist following both density manipulations and variation in activity levels in the same setting. Insisting on confirmation across both method and definitional variance before accepting an empirical generalization strikes us as unrealistically conservative. However, asking the researcher for some evidence of generality would seem to be a feasible goal.

The careful reader may recognize in this principle a weak application of the multitrait-multimethod matrix notion (Campbell and Fiske, 1959). Admittedly, such ideas have been proposed before, although not, we think, in the crowding area. Following the accepted terminology, the rule stated above could be labeled as a "convergent validation" principle. It, or something like it, would seem to be worth thinking about in the area of Human Consequences of Crowding.

In any case, the amount of interest in the area of the consequences of crowding that has been evidenced during the preparation of this volume, bespeaks an extremely rich problem area for interdisciplinary studies and one that seems increasingly relevant to the conditions of today's world. During the regrettably long period in which the authors were at work organizing material, inquiries about the proceedings of the conference were received from scholars and institutions in many different parts of the world, representing a wide variety of disciplines.

The effect of this stimulation was to strengthen the conviction that a basic source-book with broad coverage of both conceptualization and research strategies on the phenomenon of crowding would be a true contribution. It is hoped that readers will judge this aim to be fulfilled in the contents of the present volume. If one result of the labor put into this volume is to stimulate readers to consider other perspectives and other levels of analysis in addition to those with which they may be already familiar, perhaps it would be a sign that this objective has been partially achieved. Beyond that, if a few would be motivated to take such differential levels of analysis seriously and would attempt a synthesis of some of them into a single model, it may be possible to label this effort as a success.

PART I

CONCEPTUAL DIMENSIONS OF CROWDING

This first division of the book contains nine papers focussing on a wide variety of theoretical models for the crowding process. It is perhaps representative of the present state of ferment and disarray of this topic that such a rich offering of different perspectives can be collected. Patient consideration of the ideas that are so persuasively presented by these various writers may lead the reader to agree with the editors of this volume that many of the models that are presented are not so much conflicting with each other as they are focussed on non-overlapping aspects of the crowding phenomenon.

The chapters vary not only in their views of the crowding phenomenon, but also in the degree of narrowness and breadth of scope used by the various authors. The first paper, by Jonathan Freedman, provides the text of his keynote address to the NATO symposium on Human Crowding. Freedman argues that the time has come to broaden the research scope as well as to integrate results through the use of more complex, interactive research designs. Peter Stringer and Dan Christensen follow, with papers placing the phenomenon of crowding within a political context. Next come two papers by Robert Bechtel, the second keynote speaker to the conference, and his colleague, Raj Srivastava, both of which regard crowding within the context of psychological ecology. Papers six and seven, contributed by Larry Severy and Susan Saegert, relate crowding to its demographic antecedents and its behavioral consequences. Finally, two chapters by environmental professionals in contrast to the previous group of social and behavioral scientists, round out this conceptual potpourri: Özcan Esmer focuses on the consequences of population density within the context of urban planning and the built environment.

There is a story that is told about the famous 15th century Turkish character, Nasreddin Hodja. It seems that Hodja was called upon one day to settle an argument that had grown to violent proportions between two neighbors. Hodja listened patiently to the story as presented by his first neighbor, remarking at the end that he was "absolutely in the right." This observation, however, started the second neighbor off, and his story was even longer and more detailed than the first. When he had finally run down, Hodja opined that he was "absolutely in the right" about the matter at hand. A third man, hearing the whole episode from beginning to end, turned to the Hodja and remarked, "Hodja, how can you say that each of them is right?" Whereupon the Hodja turned to him and said, "You are absolutely right!"

Perhaps the only safe position to maintain regarding the preference of one or the other theoretical model for the crowding process is that of the Hodja in this ancient Sufi yarn. Certainly, each of the points of view has a high degree of plausibility, coupled with a low probability of disconfirmability. To the extent that any perspective can lead to increased sensitivity to some aspect of this complex, multidisciplinary phenomenon, it probably cannot be safely ignored.

THE HUMAN CONSEQUENCES OF CROWDING: WHERE WE STAND, WHERE WE SHOULD BE GOING

Jonathan L.Freedman

Columbia University in the City of New York, Dept. of Psychology , New York, N.Y., U.S.A.

It is quite a privilege to have been asked to give the keynote address to a conference on the human consequences of crowding, especially since even a few years ago such a meeting would have been impossible. When I started my own work, this was a lonely field. Very few people were doing research on crowding and not many seemed to be interested in what we were doing. The situation has certainly changed. Now everyone is concerned with the population explosion, there are about one billion more people on earth, and at times you would think that a large percentage of them are studying crowding. When I started my research, we could hardly fill my small living room with people doing research on crowding; now a large group of us are gathered here in Turkey in this auditorium, and we represent only a fraction of those working on this issue.

Not only are more people doing research, but there has been real progress in the field. Of course, we know that progress in any scientific field tends to be slow. Our knowledge accumulates gradually, a bit here, a bit there. Often the bits are isolated for a while and do not add up to anything until more research has been done and we can see the connections. So, it is a mistake to be impatient and we should freely admit that we still know very little about crowding. Most of the questions are unanswered. Yet, I think that it is fair to say that we are past the first stage of work in this area. Although we do not know the answers, we are beginning to understand what questions we should be asking; and that is a big step forward.

To begin with, we know that the original question –Is crowding bad for people?– with which most of us started, was naive and simplistic. It was based on the research on non-humans that showed generally negative effects of crowding; and our work also perhaps grew out of the Chicago school of sociology that assumed crowding to be evil and harmful. By now most psychologists and sociologists who

----

Note: The preparation of this paper, some of the research reported in it and participation in the conference were supported in part by the National Science Foundation.

study crowding know that this was a foolish assumption and that even the question was misleading. The research, in fact, has shown quite convincingly that crowding sometimes has negative (harmful) effects, sometimes positive effects, and often no overall effect at all. Although for a while there was some controversy about this, I think that now virtually everyone who is at all familiar with the literature agrees that crowding is not simply bad –like pneumonia or higher taxes. We may still disagree on how often it is harmful, but not on the fact that sometimes it has positive effects and that its harmful effects depend in part on the situation.

Although there is general agreement about this, let me cite just a few of the studies to show why we have come to this conclusion. First, regarding simple performance on tasks, a small number of studies have found that crowding interfered with performance (Paulus et al, 1976, for males only; Saegert, Mackintosh and West, 1975; Sherrod, 1974; and perhaps one or two others). Other work has shown no effect of crowding on performance (Freedman, Klevansky and Ehrlich, 1971; Griffitt and Veitch, 1971; Stokols et al, 1973; and at this conference, Bharucha-Reid and Kiyak, 1977). Given the attitude of journal editors toward research that finds no effects, I would imagine that there are many more unpublished studies that have failed to demonstrate an effect of crowding on performance. But even from the published work, it is clear that we have gotten some harmful effects and some non-effects.

Second, there has been a great deal of work on social and mental effects of crowding. Again we find some studies reporting overall bad effects (Schmitt, 1957; Levy and Herzog, 1974; Hutt and Vaizey, 1966; Griffitt and Veitch, 1971; Aiello, Epstein and Karlin, 1975; McCain, Cox and Paulus, 1976). These studies vary greatly in quality and the meaningfulness of their findings, but I am citing them all without criticism, taking them at face value. In contrast, a larger body of studies, generally of higher quality and dealing with more substantial dependent variables, has found no harmful effects of crowding (Ross et al, 1973; Stokols at al, 1973; Freedman, Heshka and Levy, 1975; Freedman et al, 1972; Freedman, 1975; Price, 1971; Mitchell, 1971; Winsborough, 1965; Galle, Gove and McPherson, 1972; McPherson, 1975; Booth, 1975; and many more). The point is not to count the number of studies on each side, though I think that is of interest, but rather to show that both kinds of effects have been found and those studies getting null effects are at least as impressive as those getting harmful effects. Thus, it seems apparent that crowding is not one of those rare factors that invariably has harmful consequences; but instead can have a wide variety of effects, bad, good, and neutral, depending on the circumstances.

Indeed, the second major result of work over the past ten years or so is that we have shifted our focus to the real question, which is what determines what effect crowding has on people. For in contrast to the relative lack of main effects and especially to the inconsistency among these main effects, most of the studies that involve social behavior have found some effects of crowding, but they tend to be interactive effects, with their direction and magnitude dependent on other factors. This is not the place to summarize this complex literature. However, let me mention a few of these interactions. The most frequent finding has been that males and females differ in their reactions to crowding, with males usually responding negatively and females positively (Freedman et al, 1972; Stokols et al, 1973; Ross et al, 1973; Marshall and Heslin, 1975). Others have provided some evidence that the effect of crowding is mitigated or eliminated by a feeling of control over the situation

(Sherrod, 1974; Langer and Saegert, 1977). At least one study has made the obvious but important point that crowding is likely to have negative effects primarily when it interferes with ongoing behavior (Heller, Groff and Solomon, 1977); while the Sherrod experiment cited above (1974) found that effects showed up only after people were removed from the crowded situation, not while they were still in it. For the moment, the specific factors that have been found to affect crowding are probably less important than the general point that the effect of crowding depends on the specific conditions. Presumably, future research will determine what other factors play a role and may eventually allow us to explain the diverse effects have discovered.

A third bit of progress in work on crowding is that we have, I hope, gotten past our early confusion and dispute about a definition. We now know that there are two levels of distinctions that must be made. At the most general level, we distinguish between crowding as a physical state and crowding as a psychological emotion or feeling (Stokols, 1972). That is, we can study the effects of various aspects of the physical situation and see how they affect behavior, emotions, and so on (crowding as a physical state) or we can study how people perceive crowding (Schopler and Stockdale,1977; Stockdale, Jones and Wittman, 1977) or what affects that perception (Schmidt, Goldman and Feimer, 1976; Stockdale and Schopler, 1976; Stokols, Smith and Prostor, 1975). When we study crowding as a physical state, we must then make the additional distinction among several ways in which the physical state can vary. I have chosen to concentrate on density, the amount of space per person, and I think this has been the most common focus of the research. However, it is also possible to conceptualize crowding in terms of the number of people present (sometimes called social density - McGrew, 1970; Griffitt and Veitch, 1971; Saegert, Mackintosh and West, 1975; Paulus et al, 1976, and so on). The only difficulty here, one that I shall return to later, is to be certain that one holds physical density constant while varying number —otherwise the two are hopelessly confounded. And there are those who want to study the effects of privacy or lack of it; the volume of space available rather than floor space; and probably various other factors as well. My own feeling is that physical and social density are the most interesting variables, at least in regards to what we commonly think of as crowding; but anyone is free to study what interests them as long as they specify just what variable they are concerned with and try to avoid confusing the issue.

Finally, we come to the current state of theory in this field. Here I think we have progressed little. We have made the important step of discarding the notion that crowding is simply a stressor. If anything fits that definition (perhaps physical pain from any source does), surely crowding, however defined, does not. It does, under some circumstances, produce physiological arousal (as do all sorts of things not considered stressors such as erotic stimuli, funny movies, exciting football games, etc.); but it does not generally have harmful effects, is not associated with higher rates of illness or ulcers, and is obviously not a stressor as we ordinarily use the term. Having dropped that idea, we have not replaced it with any equally general notion though quite a few ideas have been proposed.

It has been suggested (Milgram, 1970) that crowding operates by producing stimulus overload so that people in crowded conditions respond with restricted behavior or with anxiety and stress. While some crowded conditions may produce too much stimulation and the accompanying negative effects, as we have seen, most crowded situations do not produce negative effects, and there is no evidence that, in general, people feel overloaded in crowded conditions. In short, there is virtually no evidence to support this idea and there is considerable evidence that directly contradicts it. Exactly why it continues to be the basis of research in its present form is not clear. However, I think the notion of overload may be helpful if we accept that fact that some people like high levels of stimulation while others prefer low levels. If crowding is one source of stimulation, it would follow that under crowded conditions, some people would be at their preferred level of stimulation and others would be overstimulated. The former would be comfortable and would thrive; the latter would be uncomfortable and would suffer. By distinguishing between the two types of people, we may be able to understand more about reactions to crowding, especially why some people do seem to respond more positively than others.

Another theoretical approach views crowding as an intensifier, which makes reactions stronger (Freedman, 1975). The idea is that under high density the other people who are present become more important stimulus objects and therefore whatever one's response to them, it is more powerful than under low density conditions. If the individual is afraid of the other people under low density, he will be more afraid under high density; if he likes them under low density, he will like them more under high density. More generally, if his response is positive under low density, it will be more positive under high density; if it is negative, it will be more negative. This has received some experimental support and is consistent with much, though not all of the results summarized previously. However, despite the fact that I proposed this intensification notion, I do not think that it provides an adequate account of the effects of crowding. It explains some of the effects quite nicely but one or more other principles are needed to provide a total explanation.

There are many other theories that have been proposed —overmanning, personal space as a mediator, attributional explanations, and so on— but at the moment, there is little empirical support for any of them and there is substantial evidence that contradicts them so they are not very helpful. My own guess is that the effects of crowding are quite complex and will require a combination of theories to account for them. In any case, despite the large number of theories that have been mentioned at one time or another, we need much more research before we will be in a position to construct a detailed theory that provides a general explanation of the effects.

Having summarized very briefly where I think we stand in our investigation of the effects of crowding on people, let me make some proposals as to what we should and should not be doing in the future. I realize this is somewhat presumptuous of me, but I hope I will be forgiven for giving in to the temptation posed by being a keynote speaker.

First, what should we avoid? 1. Let us stop debating whether crowding is bad or good. Surely by now we know that it is neither, but rather has complex effects that go in both directions depending on the situation. Let us get to the real work of finding out what determines the effects instead of arguing about whether they are harmful to us.

2. I think it is time to turn from research designed to demonstrate trivial, obvious effects. Several studies have shown that crowding has negative effects when

people are trying to perform activities that require more space than is available. Yes, I believe it -we all believe it. If you are trying to play tennis, having ten people on the court interferes; if ten people ore practicing their tennis swings, having only four square feet per person makes it difficult; if you are shopping in a grocery or a shoe store, the presence of many other people can make shopping harder. Perhaps these effects needed to be demonstrated. I think they tell us almost nothing about the psychological effects of crowding and, in any case, they can now be taken as established.

3. Let us try not to confuse the variables with which we are dealing. As I said earlier, crowding as a physical state can be manipulated by varying the amount of space per person, the number of people, and probably other factors. That is fine, but in talking about one factor, we should not confuse it with another. Adding ten people to a room that already has five present, increases crowding by varying both number and space per person, and hopelessly entangles the results. We should try to keep our manipulations pure —either number or space— and not talk about one as if it were the other. Even more critically, let us avoid confusing other factors with crowding. Several studies have looked at the effect of having three students in a dormitory room designed for two. This does involve crowding —more people and less space per person— but it also involves a host of other variables such as lack of facilities, irritation, and so on. Similarly, some naive studies of the effect of crowding on children's behavior have reduced the size of a room without maintaining the same level of facilities. More careful work has shown that it is the amount of facilities that is crucial, not space or number; so we must be careful to keep other factors, such as facilities and all other resources, constant because otherwise any "crowding" effects may be due to a reduction in resources. Similarly, there is a great deal of work on housing design —high rise vs. low rise, long corridors vs. suites, and so on. This is sometimes discussed in terms of crowding and that is a mistake. A high rise house need not be more crowded than a low rise one; a corridor dormitoryneed not be more crowded than a suite design one. Architectural design in a fascinating variable; sometimes it may interact with crowding; but let us be careful not to confuse the two.

Finally, and in the long run perhaps most important, let us not overstate our findings. We are just beginning work on crowding; it is an important social issue; and we should not make pronouncements unless we are very certain we are right and that we are taking into account not only our own work but that of others. As a minor example, a recent study that I will not mention by name, stated categorically that crowding does affect task performance even though many previous studies had found the opposite and even in this paper, only one of three studies found any effect, and that was only for males with females performing better under high density conditions. Yet, the authors blandly titled their paper "Density Does Affect Task Performance", and throughout the discussion declared that they had found crowding to affect task performance. Yes, they had, once; but certainly the bulk of the evidence from many different laboratories was that in general crowding had little or no effect on task performance. This is a trivial example of a relatively unimportant issue, but others have made similarly imprecise and grandiose statements without sufficient evidence and in the face of other, contradictory evidence.

As another example, a paper to be presented at this conference summarizes the inconsistent results of survey studies of crowding and then states that it is the author's impression that those studies that found no negative effects tended to use a narrower range of densities. Of course, anyone is entitled to his or her personal impressions, but there is no way that anyone familiar with the literature could have gotten this particular impression. The studies finding no effects were done on Hong Kong, New York City, Chicago, and Toronto, the first three being among the most densely populated areas of the world and the first two, at least, dealing with an enormous range of densities from the very low to the exceedingly high. Against this we have studies done in Honolulu and Holland, both of which have intermediate densities. This is an example of describing the evidence to suit one's purpose. It is especially damaging because the actual state of affairs is exactly contrary to that described; and this is not a question of interpretation but of facts. If our work is to have any scientific rigor and to have any positive effect, we should be careful to consider not only our own work that of others and to give balanced accounts of what we and others have found.

Now, what should we be doing? 1. Let us continue the very good trend of using both laboratory and field studies to investigate our complex problem. My own preference at the moment is field studies, with laboratory work used mainly to check on specific theories. But when we do field studies, we must be particularly careful to separate crowding from other factors such as lack of facilities, poor design, or whatever.

2. I think it is time to consider the possiblity that there may be major individual differences in responses to crowding. We should look at responses across different people, groups of people, and nationalities. I love cities such as New York and Istanbul, which I saw for the first time a few days ago; others hate this kind of crowded, teeming city. In contrast, some people love rural, country environments while others are bored and uncomfortable in them. The lack of main effects of crowding may be due to the fact that people differ greatly in their responses, cancelling out each others' responses.

3. Let us concentrate on discovering the factors that determine the effect of crowding. It seems clear now that almost all substantial effects will be interactions of one sort or another –interactions between degree of crowding and other factors such as individual differences, type of activity being performed, pleasantness or unpleasantness of the situation, type of situation, degree of control the individual feels over the situation, etc., etc.

4. Finally, and most generally, let us collect data, lots of it. We must accept the obvious fact that we are just beginning our investigations and we know very little thus far. Our data base is still very tiny, especially considering the great complexity of our topic. So, although new theories and ideas may help eventually what is needed most right now is careful, solid research designed to collect as much data as possible. Some years from now, when we have much more good data than we do now, perhaps we will be able to contruct a theory or theories that will explain it and then we will have some real understanding of the human consequences of crowding, which is the question posed by this conference.

ENVIRONMENTAL PLANNING IN A DEMOCRATIC SOCIETY

Dan Christensen, architect/planner

Greater Copenhagen Council

Denmark

ABSTRACT: The first part of this paper discussed environmental planning in the abstract, whereas the second part of the paper describes environmental planning as it is being practiced in Denmark at the national, the regional and the local level.

The main hypothesis of the paper is that the good environment must satisfy man's need for privacy as well as his need for social intercourse. The means and goals of planning in a given society show how this society defines the ideal balance between man's individuality and man's sociality. Successful environmental planning depends more on public participation than on ideological concepts.

I. MAN'S EXISTENTIAL NATURE AND MAN'S IDEAL SOCIETY

The Individual Man and the Social Man

Man has a dual nature. He is in an individual being as well as a social being. This duality makes man a difficult client for the environmental planner.

The individual man needs privacy. He must be allowed to do things alone and to think his own thoughts. But the individual man also needs to express his feelings and to communicate his thoughts. He needs to share some of his happiness and most of his burdens.

Man's social nature demands that he lives in a society. Only in a social setting can man fully use his talents for spiritual and material development. Isolation may impare man's sanity, and exclusion from co-existence with one's fellow men is generally considered to be the ultimate penalty for social misbehaviour.

Throughout history man has been striving to balance his needs for privacy with his needs for social intercourse. History shows us that an extreme emphasis on individual freedom ignores man's social needs, whereas an extreme emphasis on social responsibility diminishes man's individual value and his dignity.

An ideal society should be conceived of not only as a philosophical abstraction but also as a concrete environment. Man will always depend on his environment because it contitutes the physical framework for his activities. Environmental planningis concerned not only with man's material needs but also with his spiritual needs as man has a body as well as a soul. To provide shelter for man is not enough. The human habitat must be as complex as man, otherwise it can not support his creativity but will gradually deform his aspirations.

Man plans his environment in order to create a still better future for himself and his offspring. Planning cannot be only a follow-up process based on trends from the past. Planning must be based on defined goals. But as man's imagination is usually rather limited in scope the goals he formulates for tomorrow clearly reflect his situation today.

The planner is a servant of his own society and the way in which his present society is organised politically will determine the planner's frame of reference. Only means that are politically acceptable will be put at the planner's disposal.

The goals of planning as well as the means of planning are thus reflections of man's present society. Goals and means show a great deal of variation from country to country because the view of man varies from country to country. In some countries man's individuality is rated higher than his sociality. In other countries the opposite is the case. It is this very rating that determines the goals and means of planning in a given society.

Ideology and Environmental Planning

Many labels are used when a society demonstrates how it evaluates man's nature and man's need. Very familiar labels are "capitalism", "communism", "democracy" and "dictatorship". We must try to understand these familiar words not as campaign slogans but as interpretations of the ideal society as seen by ideological eyes.

In decribing man's material needs and his relationship to property the ideologists offer two roads. One road is for the individual man and leads to individual ownership of property –this is the "capitalistic" road. The other road is for the social man and leads to collective ownership of property– this is the "communistic" road.

In decribing man's spiritual needs and his relationship to politics the ideologists likewise offer two roads. One road is for the individual man and leads to a multitude of political parties (in extremis: one man his own party) –this is the "democratic" road. The other road is for the social man and leads to only one all-encompassing party which embodies man's longing for universal harmony– this is the "dictatorial" road.

Man's relationship to property and politics thus demonstrates ideological consequences of his dual nature, but so does man's attitude towards environmental planning.

A society that idealises man's individuality will see the ideal society as the one in which everybody is his own master, owns his own property, etc. Environmental planning in this society must be kept at a minimum as planning by its very nature restricts uninhibited individual development. For ideological reasons then, planning is not held in high esteem in the individualistic society.

A political system which could be described as "democratic" would confront the planner with man's aspirations in all its complexity. When the planner asks for a simple set of guidelines for his planning and powers to execute his plans, he will be met with suspicion. No simple goal can be provided because all of his fellow men want to influence the direction in which society should be going. The process of formulating the goals of planning will be a very tedious one and probably not lead towards any one goal but towards many that should preferably be reached at the same time.

Likewise the planner will not be encouraged with great powers of execution as the people –through their elected representatives– want to preserve for themselves the option of changing goals if they happen to change their minds along the way. This is not an ideal situation for a planner who wants to achieve striking results.

The very opposite situation may occur in a society that idealises man's sociality and which will see the ideal society as the one in which everybody works for the common good disregarding individual ambitions. Environmental planning in this society must be extended to a maximum as planning by its very nature promotes public interests. For ideological reasons then, planning is held in high esteem in the socialistic society.

A political system which could be described as "dictatorial" would provide the planner with a simple set of guidelines and probably also give him the means to execute his plans. So the planner has an easy time provided he is not overly concerned about the contradiction between his simple task and the complexity of man's aspirations when these are not restricted by ideology.

"Process" and "Results" in Environmental Planning

In his famous book "Understanding Media", Marshall McLuhan discusses the relationship between the way we do things and the means with which we do them. McLuhan points out that the means we possess very strongly influence the goals we choose. The author summarises this relationship in a striking announcement: "The medium is the message".

In environmental planning this philosophy could lead to the dictum that the process is as important as the results. This dictum then would be a modern alternative to the old thesis that "the end justifies the means". The cynicism of Macchiavelli might give way for the pragmatism of McLuhan.

When the planner ceases to be an absolute sovereign he can no longer try to achieve his ends by all means. Ends and means in planning can not be separated but must suit each other. To benefit from the results of planning man must participate in the process of planning. Man in this sense is the end as well as the means. The man we talk about here is the whole man with his individualistic and his socialistic nature recognised and respected. A pure ideological interpretation of man must be abandoned.

It should be accepted that the goals of planning cannot satisfy everybody but must reflect the majority view after careful consideration of the minority views. "Democracy" in this case does not mean endless discussions about ends and means. The "democratic" process of goal formulation has to lead to the process of goal achievement. Preparation cannot replace achievement. Process must lead to results. The execution of planning goals may sometimes require the power to overrule individual objections in which case the protester might see the achievement of the "democratically" formulated goals as "dictatorial".

Individual objections to certain planning goals may be of a philosophical nature, but could also be based on personal economical interests like land ownership. Therefore "capitalism" must be modified to an extent where private property is no longer sacrosanct. Achieving planning goals that clearly benefit all citizens –such as clean air and good housing– does not per se exclude private ownership or individual expression within limits. Complete public ownership –"communism" is on the other hand not a necessary requisite in successful environmental planning. It may even be counter –productive as it would eliminate a great deal of the vigour which is so important in any social enterprise.

So where are we? In my optimistic estimation we are in the middle of the road between the extremes of the individual and the social Utopia. We are on the road towards an unspectacular Utopia, where man is accepted as he is, individualistic and socialistic at the same time. The name "Utopia" is then justified in the sense that this society will be built to fit man because man could not be manipulated to fit a spectacular Utopia.

Should we relax now that the pure utopian ambition in planning has been abandoned? Not at all! On the contrary. Instead of planning for the imaginary man we can plan for the real man. We do not have to wait until tomorrow, but must start today.

Denmark - an Unspectacular Utopia?

In fact this planning for today's man is well on its way in the little country I come from, Denmark. I shall not propose that Denmark is the unspectacular Utopia just mentioned, because that surely would be a chauvinistic proposition. But I shall dare the statement that Denmark in its environmental planning has placed itself right in the middle of the road! Extreme interpretations of man's individuality and man's sociality have been abandoned. Consequently the economical system is "mixed" and so is the planning legislation and practice which puts great emphasis on "the process" but also stresses "results" in planning. The goal is not to make everybody a planner but to create a better environment for everybody.

II. ENVIRONMENTAL PLANNING IN DENMARK

Environmental Planning at the National Level

Environmental planning in Denmark is based on legislation passed by the national government. This legislation is the first step of the planning process and the most important one because it is in the laws that the planning goals are defined. Producing and realising schemes are of course necessary next steps if the goals are to be achieved. But the law tells you where you are going and also which means you may use to get there.

A number of environmental planning issues are decided on at parliamentary level and executed through the state bureaucracy. Other environmental planning issues are decided on at the regional or local district level.

National planning issues may be issues which concern the whole nation, but could also be issues which primarily concern only one region or an even smaller geographical area. In this latter case, the issue has become national because it involves investments that only the state can provide. The national transportation systems –railways and motorways– are financed and planned nationally. The location of airports and nuclear power stations is a national planning issue too, as are bridges between the Danish islands.

To co-ordinate all these major investments the government attempts to formulate a national planning strategy. When Parliament reconvenes every autumn the Minister of the Environment presents to the legislators a "national planning statement" in which the government's environmental policy intentions are outlined. The Minister's statement usually is vague enough to allow for a consensus in Parliament.

National planning legislation by consensus is necessary in Denmark as no single party ever controls a majority in Parliament. Any government is either a minority government or a coalition government. In both cases the government must act cautiously so as not to challenge its own existence. Consensus is thus a requirement of a stable parliamentary government but it is at the same time the hallmark of a democracy in which you promote your own interests, not by force but by persuasion.

The environmental legislation passed by consensus in recent years has followed two different trends. On the one hand, the planning authorities have been given much greater power to plan the environment, but on the other hand, public participation in the planning process has been strongly encouraged and is now a mandatory element in the process.

Two planning laws illustrate the two different trends. One is the zoning act of 1970, the other is the community planning act of 1977. According to the zoning act, general building activities can take place only within the urban zone whereas building activities directly related to agriculture will be permitted in the rural zone. According to the community planning act every local district in Denmark must have a development plan that defines the limits of the urban zone and specifies the land uses within this zone.

The national government no longer executes a quality control of the environmental planning at the local district level. The planning goals are formulated by the central government for all of Denmark but the goals are to be reached at the local level in a way that suits the people who happen to live at this local level.

To facilitate this decentralisation of power Denmark's 1400 local districts were reorganised into only 270. Each of the new districts should have a sufficient number of taxpayers to support an effective local government. (My own local district has 15,000 inhabitants and a local public administration of about 160 civil servants).

In the light of our previous discussion about the "individual man" versus the "social man" one should say that stronger planning laws favour the "social man" because environmental planning per se aims at a common goal –a good environment– disregarding individual interests. The new planning legislation does not only define the goals of planning in general and specific terms, but also provides the legal means necessary to achieve these goals. The public authorities' use of compulsory land acquisition has thus been markedly extended.

This trend in environmental planning legislation is not a result of a sudden "communistic" or "dictatorial" current in the country but stems from the legislators' reaction to the excesses of past individualistic use of property rights.

The legislation for a more powerful public participation in environmental planning could –in the light of our previous discussion about the individual man versus the social man– be interpreted as favouring the "individual man" because it enables the citizens, individually or organised in groups, to express objections to any one environmental planning project which either a public or a private builder wants to realise.

This trend in environmental planning legislation is not a result of a sudden "capitalistic" or "democratic" current in the country but stems from the legislator's reaction to a markedly increasing alienation of the common man in his relationship to "big" government. The common man felt that the ever more effective nursing by the "welfare state" had reduced him from being an individual person to being merely an anonymous citizen –a number to be serviced according to the law.

When man ceases to regard his government as a servant but begins to view it as an oppressor he will disregard the actual services that this government provides for him. He will seek to destroy the present social order and may even be ready for an extremist form of government.

The Danish mentality is rather sensitive to the dangers of extreme positions. The Danes live in a small country and have for centuries been accustomed to a foreign policy of balancing between major powers. Perhaps this is the reason for the balance the Danes try to achieve in all matters - including environmental planning legislation. It is therefore a typical Danish phenomenon to follow two trends at the same time: to provide for more efficient planning means and concurrently secure a more effective public participation in the use of these means.

Environmental Planning at the Regional Level

Halfway between the "big" national government and the "small" local governmet (the district council) we find the regional governments. The regions have, like the local districts, been given new powers and new obligations to promote environmental planning and to involve the public in this planning.

The regions, though, are not the link between the state and the local districts but the third corner of a triangle. This is the state's own interpretation which is promoted by a triangular logo which decorates the front cover of all official reports produced by the Ministry of the Environment concerning regional planning. The state has never specified who is at the top of the triangle but it is hardly the regions.

A regional plan in Denmark must be approved by the Minister of the Environment who evaluates the plan on the basis of the government's "national planning statement". A regional plan must be based on the expressed wishes of the local districts within the region. The plan therefore has to comply with national as well as local planning guidelines. This in itself makes regional planning a difficult technical and political exercise.

The situation is further complicated by the fact that the state has no direct quality control of the environmental planning at the local level, but can execute an indirect control through the regional plans which give guidelines for the planning at district level. Suppose the state has a special interest in a local district's planning affairs (that local district might be the capital, Copenhagen), in full accordance with the letter of the law the state then can provide the regional planning authority with detailed guidelines for the future development in this specific local district as a condition for the state's approval of the regional plan.

Whether the state will use the regional plans to control the future development in the local districts has yet to be seen as no regional plan has so far been approved by the state. The purpose of a regional plan of course is not primarily to function as a counterbalance to the decentralisation of planning power, but to secure that the planning powers are not used at the local level in a way that disregards the relationship between the local districts themselves.

In the Copenhagen region the planning process has been in progress since 1968. The regional plan is now in a last phase of confirmation by the state. To illustrate the extent to which public participation has played a role in shaping "Regional Plan 1973 for the Greater Copenhagen Region" the production of this plan should be described in very general terms.

The regional planning council initiated the planning process in 1968. By the end of 1971 the council published two planning reports. No. 1 describes the planning problems and no.2 presents four proposals as to how the problems should be solved, i.e. four draft regional plans. The two reports were sent to all 50 local district councils and to the Minstry of the Environment.

Through out 1972 the four proposals were debated and so were the problems as defined by the regional council. The course of events showed that different eyes see different problems and that different authorities prefer different solutions.

This first experiment in public participation in regional planning did not result in a consensus nor did it lead to chaos. The regional planning council digested several hundred sets of commentary and produced a new simplified plan, the "Structure Plan 1972" (planning report no. 3). The council had managed to accumulate all the basic planning suggestions of the four draft plans against which no overwhelming objections had been raised during the public debate. This minimum plan though was far more substantial than many had feared. The process could proceed.

A year later the council published a new plan "Regional Plan 1973" (planning report no. 4) which gives detailed guidelines for the future urban growth within the region. Again a round of public meetings and consultations with local authorities was held. The sensation of 1971 had subsided and been replaced by a more relaxed planning discussion. By Midsummer 1975 the council could confirm "Regional Plan 1973" as the region's wish for future development.

The plan was submitted to the Minister of the Environment for approval. Lengthy negotiations led to a temporary compromise in 1976 which enabled the regional council to produce alternative proposals for the implementation of the plan. A third public debate took place and finally by the end of 1977 the regional council will finalise its policy for future regional development before meeting the Minister again for a new confrontation.

While all this debating takes place, the regional council continues to do business as usual with the local districts as well as with the state. The local district councils are encouraged to press ahead with their local environmental planning, whereas changing Ministers of the Environment are invited to sales promotion sessions by the chairman of the regional council.

Our previous discussion about "process" and "results" is not based on theory only but reflects daily practice in Denmark. Everybody is busy debating and these discussions produce a greater awareness in the public opinion of the importance of environmental planning. Awareness is the basis for any constructive action. Not only the ordinary citizens who are the consumers of the regional plan but also the regional council members who are the producers of the regional plan benefit from this procedure of indefinite debate. As consumer and producer become still more particular, the product —the plan— is steadily improved.

A regional plan deals with 15 to 25 years development. We discuss the environment of our children. We are setting goals to be achieved by them. Most people though are rather less patient than regional planners. Local planning interests them more because it shapes the view from their own windows. Environmental planning at the district level deals with tomorrow, not years from today. Let us visit a local district in the midst of development.

Environmental Planning at the Local District Level

The local district I shall now describe is Karlebo Kommune north of

Copenhagen. I moved to Karlebo in 1972 with my family from Copenhagen because the district was in rapid development and modern housing was available in great variety sharply contrasting with the housing situation in the inner city area where we lived before.

The urban development in my local district takes place in two seperate townships around two stations from which commuter trains take you to the city of Copenhagen in 1/2 hour. The open countryside between the two townships will be developed into public open space. The rest of the local district will be kept in its present state of agriculture. This then is the district plan: two townships, one park and a rural hinterland.

The two towns though are not "real" towns but only "dormitory" towns. Industrialisation has transformed man's living pattern such that work has been separated from habitat. Home is where I sleep, not where I work. This fate I share with all other Danes who happen not to be farmers. In accordance with the regional plan there is made no land use allocations for major employment in Karlebo. The two townships will house 20,000 people but most of these people will work somewhere else in the region.

Most Danish families change residence now and then. In the Greater Copenhagen region people move from the older urban areas to newer ones where they find modern housing and good environmental planning. If a local district council in a development area initiates residential development within the urban zone it can be fairly sure that housing construction will take place and new citizens (taxpayers) will be moving in a few years later.

Any urban development in the district must be preceded by environmental planning. The local planning authority is the district council to which the Minister of the Environment has delegated his previous planning control powers. The district council now must approve its own plan for the district as a whole and for every development scheme within the framework of the district plan.

When the national legislators gave the elected district councilmen more planningpower they simultaneously gave the electorate a more clearly defined role in the planning process. Every planning proposal has to be presented to the citizens of the district to facilitate a general and public debate about the desirability and merits of the actual scheme. If the public opinion is strongly opposed to the proposal the city council would be ill advised to approve it, remembering that the councilmen are up for re-election every four years. The city council therefore in its execution of planningpowers usually reflects the general political inclination of its constituency.

A district council with a conservative flavour will interpret its planning role rather narrowly and leave most of the environmental planning to private developers. The developers will acquire the land and build on it according to the national building code –no more, no less. This "minimum" quality control by a public authority could be said to reflect the "individual man" philosophy.

Another district council with a more progressive flavour will interpret its planning role rather more broadly and leave as little environmental planning as possible to the developer. The council may acquire the land –at market value– and prepare a very detailed masterplan for its use. Alternatively the council may negotiate with the landowner a development layout that satisfies not only the quantative aims of the national building code but also more ambitious qualitative aims as defined by the district council and its local planning department. This "maximum" quality control by a public authority could be said to reflect the "social man" philosophy.

In my district both philosophies have been applied as development has now taken place for 15 years under different local councils with changing political flavours.

The neighborhood I live in has been blessed with "maximum" quality control. Although it was built in phases, one can hardly see where one phase stops and a new developer took over. The plan for the development was so rigid that it defined not only the exact layout of the neighborhood, but also the building materials to be used, the size of the houses, the shape of the windows, etc. The last building contractor confessed to me one day that this to him was "dictatorship". To me and to my happy neighbors, it is "quality control".

The neighborhood consists solely of patio houses (140 dwellings). It exemplifies quality control in environmental planning but also how the physical framework could be formed to satisfy both individual man and social man. In this low-rise, high-density neighborhood every resident can develop both his individual and his social nature. The individual man stays behind his garden wall in complete privacy whereas the social man joins his neighbors on the green or on the many small squares between the houses. In accordance with man's moods he can remain on his private territory or move freely outside on the common territory. Man's individuality and sociality is not restricted by the physical environment in this case.

The houses are individually owned but the common amenities such as heating plant and the public open spaces are owned and administrated collectively. The house owners elect a board of administrators to function as a mini local council. I was chairman of this council for a while and I can assure you that democracy at this very local level demands energy, deafness and compassion.

III. CONCLUSION

Environmental planning is a long winding process. The result of this process is the physical environment that takes shape around us and which is the actual framework for all of man's spiritual and material activities. But man's environment also has an indirect impact on man's frame of mind.

If man finds his physical environment inhuman he will distrust the validity of the society which has provided him with this environment. But if the physical environment respects the dignity of man and man's participation is expected in the planning of his own community then man will find his environment meaningful. "Meaningful", though, is an existential question. Society and environment are inseparable answers to this question.

Society lives by order, it cannot live without it. A democratic society lives by a democratic order. This order must be learned and practiced at all levels of society. Democracy means to possess and to share. In a democratic society man possesses an individuality and shares it with his fellow men. This sharing reflects man's sociality.

Environmental planning in a democratic society therefore must respect man's individual and social nature. The purpose of public participation in the planning process is not only to guarantee that the citizen's wishes are known to the planners. The purpose is also to make the very same citizen ever more conscious of his democratic rights, in fact convince him that democracy is not an abstract ideal but a living and demanding reality.

Therefore environmental planning as described in this paper is probably one of the most effective defense systems a democratic society has.

A POLITICO-PSYCHOLOGICAL PERSPECTIVE ON CROWDING

Peter Stringer

Department of Psychology, University of Surrey

Guildford, Surrey, UK.

Some nineteen hundred years ago Seneca was complaining bitterly in one of his surviving letters, of the crowded, noisy and polluted condition of Rome. The phenomena and experiences referred to by words like 'crowd', 'crowding', 'crowded' are not new. The words have various connotations and denotations. In ancient Greece and modern times the crowd, 'hoi polloi', could be used as a term of political abuse. But 'our crowd', the 'in crowd' has a positive, warm sound to it. No single analysis, whether experimental or conceptual, can do justice to these complexities. For this reason it is important that any study of crowding should be semantically defined.

Our interest in this week's symposium, which reflects the much wider contemporary concern with the experience of human crowding, is principally motivated by practical considerations. It is largely because politicians and laypersons, as well as scientists, believe that crowding constitutes a problem, that so much effort is being devoted to its study today. The expectation is that people such as ourselves should find solutions to the problem, or at least explain convincingly why it is not a problem. The difficulties of carrying out research, the results of which can inform a feasible practical solution to an everyday problem cannot be over-emphasised. Studies not only need to be relevant to everyday life in a general sense; their formulation, and therefore their output, should be directly congruent with the context of their application.

Crowding is not necessarily an urban phenomenon, but it is urban density and scale which is often associated with the problem (Freedman, 1975). Between the middle of the nineteenth and twentieth centuries there was a thirty-five fold increase in the number of cities with more than one million inhabitants(Toffler, 1970). The doubling time of the world's population is now some thirty-seven years, and much of that will occur in cities(Ehrlich and Ehrlich, 1970). Population growth and urbanisation go hand-in-hand, and lead in their train fears of the over-use of resources and of social disorder and breakdown. The four factors which Calhoun(1970) sees as contributing to crowdedness indicate why it is so often seen as a peculiarly urban

Table 1 A Multi-Dimensional Scheme for Considering Crowding Studies

MODE	Dimension	Technical	Social	Personal
ANALYSIS	Politics	Bureaucracy	Representation	Participation
	Social psychology	Individuals	Roles	Relations
PROBLEM	Crowding	Neutral cons- straint ; overload	Overmanning	Personal constraint, alienation
SOLUTION	Planning	Rational	Consensual	Participatory

feature. Apart from high unit densities, crowding can entail a lack of harmony in the population's value systems, competition for resources, and a high rate of contact between individuals.

Baldassare and Fischer(1977) have argued that experiments on crowding have had very little relevance to the study of urban life. They urge that theoretical connections be made between the experiments and urbanism by way of urban theories. The theme of this paper is similar. I want to suggest that for crowding experiments to provide the information for feasible solutions of urban problems they have to be situated in relating to the political and planning context within which action will occur. The meaning that one attaches to 'crowding' and the kind of analysis and solution one conceives of should be intimately related to one's view of urban politics.

It is impossible to do justice here to the complexities of this context and of the research endeavour. I shall simply draw attention in a schematic way to some of the dimensions involved and hint at their relations to one another. In doing so I shall rely implicitly on research I have been engaged on into local government in Britain and participatory modes of planning (Cf. Boaden et al., in press; Boaden et al., forthcoming; Stringer, forthcoming). The institutions and practices which attempt to deal with the consequences of crowding vary from one country to another. But I hope that the dimensions below, although by no means an exhaustive set, can provide a framework for comparing a range of strategies in many different contexts. The argument which follows is best appreciated by interrelating at all times the categories in each column of Table I.

There are four dimensions, non-orthogonal as you see, and each with three points indicated on them. The first two are basic, political and social psychological dimensions; the third deals with our focus of interest, crowding; the fourth describes one aspect of the institutions and practices which might attempt to alleviate the consequences of crowding. Studies of crowding may be described by the position they occupy, often in an implicit sense only, on each of the four dimensions. Ideally the attributes of a study would place it at a corresponding point vertically on all the dimensions; that would constitute the congruence which I believe promotes informative and practical research. More often there will be a disparity which reflects the hidden nature of the assumptions which are present in any piece of research.

Even if the assumptions are clarified, in many circumstances reasons will be found for the study having noncorresponding characteristics. I will leave it to your ingenuity to provide examples of crowding studies which lie at each of the three different points or which show heterogeneous profiles.

The Political Dimension

One of the principal functions of government is the provision of services and resources to localities and to individuals. The manner of distribution of resources is a political matter and its organs of administration are usually democratically elected bodies. The competition for resources is likely to be acute in large, densely populated cities and inequalities in distribution seem to be particularly hard to tolerate. Urbanisation and industrialisation both demand and put heavy strains on services such as sewage and refuse treatment, transportation, energy supplies, health and educational provision, employment opportunities, housing, shopping and recreational facilities.

In most countries the provision of these governmental services is decentralised to some extent. But the form of local government can vary between the large-scale bureaucratic organisation and the more localised participatory community. In Britain, local autonomy exists, though it is often subject to centralised control. Employment opportunities, for example, are administered through decentralised offices of central government. The police are variously the responsibility of the Home Secretary, in London, and of local government elsewhere. Some services are catered for through the appointment of a special board, such as the British Water Authorities. Increasingly, financial control for locally organised services, particularly in urban areas, is closely maintained at the centre.

The route by which the administrators arrive at their positions can vary. The bureaucratic end of the dimension is characterised by officials who are central appointees. At an intermediary point are the elected representatives and the officers who may be appointed by the elected members or be themselves elected. Representative democracy does not necessarily allow for wide public involvement. Mass participation is at the further end of the continuum. Advocates of participatory democracy take a different view in claiming that it is the people who have the right directly to make decisions affecting their own activities.

The scale, complexity and degree of autonomy of local government are relevant parameters when considering crowding as a problem. Cities tend to tackle their problems by large-scale, complex and distant forms of administration. Few political roles are open to their citizens and the individual is given no encouragement to govern his own affairs. Crowding and its consequences is interpreted as a technical issue. Following the sentiments of a recent British committee the objective is to make efficiency democratic, rather than democracy efficient (HMSO, 1972, para. 4. 33).

The Social-Psychological Dimension

We can distinguish broadly three models of man which have variously been assumed in social psychological theories: man as an individual, an organism; man as the occupant of social roles; and man as person. These stipulations underlie three types of theory: the behaviouristic approach, role theory, and relational theories. A comparable analysis could be made of psychological theories. But the desirability of treating crowding within a social psychological perspective is central to this paper.

Within social psychology, behaviouristic theories are oriented primarily to the individual, and in a passive or mechanistic mood. For example, social learning theory (e.g. Bandura, 1971), attempts to explain the processes by which an individual acquires behaviour patterns and attitudes in the context of interaction with other people. Studies of socialisation focus on the end-state of the learning process in the developing individual. In exchange theory (e.g. Thibaut and Kelley, 1959; Blau, 1964) social interaction is explained through concepts of 'reward', 'cost', 'outcome' and 'comparison level'. The former two terms are familiar behaviouristic notions. Two-person interaction in a dyad is examined. But it is the individual's behavior and the outcomes for him which is of prime interest rather than the dyad as a system. Although trait theories of personality (e.g. Cattell, 1950) have a mixed parentage —psychoanalysis and behaviourism— they can be considered in the same light. They have biological overtones, personality being defined in terms of an inner, essential aspect of man. Traits, like the acquired behaviour and attitudes of social learning theory and the given dyadic situations of exchange theory, have a positivist air of stability and finality.

Behaviouristic theories deal essentially with the past of psychological phenomena rather than the present or future. They can predict repeated events. But their emphasis on static conditions or equilibrium contains nothing within it which suggests effective action for change. Change in behaviouristic terms tends to be more of something rather similar.

The second type of social psychological theory, role theory, needs only a brief description. "Man has certain positions within the social system and related to these positions are normative expectations concerning the individual's behaviour and concerning relevant attributes. Positions are independent of a specific occupant. The same is true of the expectations directed towards a position; they are defined as the role of the incumbent of a position" (Sarbin and Allen, 1969).

Role theory has a strong social orientation and subordinates or excludes the individual's active interests. He is seen as passively submitting to the influence of social or political institutions. Prospects for change are dependent on change in the environment. But there is little that the individual can do to determine that change.

A relational model of Man is more appropriate to a properly social psychological theory. It concentrates neither on the individual nor on social processes, but on the person, the individual-in-society. An example of a relational theory is that of G.H. Mead, in his assumption that "A self can arise only where there is a social process within which this self had its iniation. It arises within that process" (Mead, 1956). Man is not conceived of as a bundle of traits or other properties, but _as_ his social relationships. Man is the sum of his social interactions. Throughout his constant interaction with others, his self is continuously changing.

Interaction is fully reciprocal.

The development or education of the individual and of the social group are common, interdependent processes. The social environment is the setting for all individual acts; but the acts cannot thereby be purely individual because they implicate other members of the group. Identities and institutions develop simultaneously and integratively. Process or change is crucially involved in the relational model; and it emphasises man's active role in his development.

The Crowding Dimension

It is convenient to divide theories of crowding into three parallel categories (Schopler and Stokols, 1976; Fisher et al.,). The lack of theoretical specification makes this an uncertain task. Definitions are very varied; and variables are introduced in a somewhat arbitrary fashion. The list cannot be exhaustive. Proponents of the theories may resist firm categorisation. But the points on the continuum are intended to represent the centre of gravity of a theory, rather than its tentacles.

A preponderance of approaches to crowding are either biological or psychological; just as so much social psychology is organismic or individualistic. Biological theories refer to innate mechanisms of population control, for example; territorial instinct and conflict; and limited capacities for physiological stimulation. The models of personal space and psychic overload rely on traditional accounts of cognitive and learning processes. Most 'social psychological' theories which define crowding in terms of interpersonal events are in fact essentially individualistic. Typically the interest is in the individual's inability to control interaction with others and his resultant behaviour, rather than in the relational process under such circumstances.

An emphasis on roles is evident in the theory of undermanning of Barker and his colleagues (Barker, 1968). A central assumption is that all behaviour settings have particular tasks or functions that are associated with specific manning requirements. Barker's theory has been extrapolated(Wicker, 1973) to provide an account of human crowding. The experience of crowding is a function of the number of social roles available in a setting in proportion to the number of eligible and willing participants.

It is difficult to find a theory of crowding which has an explicitly relational basis. There are hints in certain uses of the overload framework. The exposure of city residents to very high levels of social stimulation may impede the more usual forms of relationship and replace them by apathy and isolation (cf. Milgram, 1970). Situations which involve coacting strangers are marked by their lack of relation. Stokols' distinction (1975) between 'neutral' and 'personal' crowding could be extended and moved from its individualistic foundation to a relational base. The emphasis is on the individual's control of situations, which is variously impeded by unintentional or intentional interference imposed by the physical or social environment. (The introduction by Stokols of intentionality is an unusual but welcome feature of discussions of crowding). One might alternatively place emphasis on the inter-subjective process which constructs a crowded relation between individuals or groups.

The Planning Dimension

Whether crowding is considered at the level of housing, neighbourhood, city or region, the planning profession will be drawn into the search for and implementation of solutions. The interests represented on the planning dimension are, like the political positions above, evolutionary. The gradual movement away from a rational model of planning, with its high technical content and reliance upon professionalised officers, and the greater acceptance of consensual and participatory models has accompanied a growing awareness of the essentially political nature of planning. The distinction between the three models has been elaborated by Smith (1973). Rational planning is concerned with achieving specified goals by the most efficient means. The goals are unquestioned and may be defined in general terms, the "good life". Planning is seen as a technical exercise with little or no political content, a matter of professional responsibility. This view of planning dominated until the early 1960's at least, and can still be found in practice. It leaves little opportunity for public involvement.

Consensual planning (Simmie, 1974) allows a somewhat greater role for the public. Planners search for support or for consensus, both about the means to be used and the ends to be achieved. The political content of planning is accepted, although the technical component remains considerable. Those individuals or groups whose interests are directly and manifestly affected by planning proposals will be given an opportunity to be involved, in order that the consensus may be promoted. But the representation of general public interests is limited, indirect and partial.

Finally, participatory planning engages the individual citizen in the planning process to the fullest extent(cf. Cole, 1973; Coppock and Sewell, 1975). Both the policies to be pursued and the preferred strategy for achieving them may be decided by those for whom the plan is being prepared. The people are actively involved. The planner's role is technical, advisory and subsidiary.

Each move on the continuum toward participatory planning (Stringer, 1977) subsumes the earlier positions. Participatory planning has its technical and consensual, as well as participatory modes. The rational aspects of participatory planning arise from the growing complexity of the planning process and the need for a continuing flow of relevant information. Individuals and groups can provide this information since they are in direct contact with the scene of planned change.

The Contexts of Research

Just as a participatory approach to planning does not exclude technical and consensual elements, so also can there be a variety of approaches to the study of crowding. Man is undoubtedly, in some sense, at once an individual organism, an occupier of social roles and a person defined by his relations with others. His reflexive nature and his capacity for constructivist thought offer him considerable flexibility to decide in what situations and for what purposes he will stress one stipulation rather than another. We find individuals and cultures who are thoroughgoing behaviourists; others who take a relational position; and others who vacillate between the extremes.

There is a similar variety of political institutions, in different countries and even within a single country. Basically most governmental systems have adopted or are adopting a middle position on the continuum. A multi-tiered representative system is common, with differing degrees of central government control or influence and of wider public involvement. The residues of bureaucracy and the infusion of participatory characteristics are the result of the critical nature of service provision today. Simultaneously it breeds an ever-heightened desire for efficiency and a desire for protest and self-determination.

The argument here is that there are political components in the antecedents and consequences of crowding. Their interpretation will vary according to our model of Man in society. The way we treat them will be dependent on the ideologies and methods of designated experts. As a practical and human problem, crowding should be discussed in its full and varied contexts.

Crowding as a Technical Problem

The processes of industrial and post-industrial urbanisation manifest themselves on all the four dimensions and toward the left-hand end of the continuum. Crowding is an aspect of insufficient service provision, experienced in the shape of weak land-use policy, housing shortage and inadequate educational, health and recreational facilities. These shortcomings are in some respects a function of the scale of cities and of their rate of growth. Attempts to correct them have introduced administrative systems which depend heavily on bureaucratisation, appointed expertise and direction or control from the centre. The scale of local government has grown at the same time. An earlier organic growth in cities has been replaced by a positivist engineering, under the belief that their form can now be rationally and efficiently planned so as to promote a generalised physical and cultural well-being.

Members of urbanised societies not only suffer from inadequate services. Crowding takes on its metaphorical guises. Large scale bureaucratised administrations place clear constraints on people. Government becomes distant and hard of access to them (HMSO, 1967; HMSO, 1969; HMSO, 1973). Its complexities are a baffling overload. People find they have no control over what is provided for them. They have no path by which to intersect the administrative process, which they variously experience as interfering, constraining or withholding. Density on a large scale breeds institutions which both constrain and overload. These features may be more fundamental and more aggravating than crowding in its more common sense.

The individualist bias fits a technical interpretation of crowding. It is convenient to treat people as a uniform aggregate of units. Anything other than one-way communication becomes impossible. Differentiation is unnecessary. Goals are set by human needs, independently of values or existing inequalities. The technical solution to the problem relies on individuals being ultimately predictable both in their requirements and their satisfactions. But it is this very interpretation which constitutes a large measure of the problem for those who experience it directly, An essential aspect of feeling constrained, distant, anonymous is the knowledge that one is treated as an individual unit, identical to and undifferentiated from other individuals. To conceive of crowding purely as a technical problem is to risk replacing one of its manifestations by another more fundamental one.

Crowding as a Social Problem

Efficient service provision is undoubtedly desirable. Whether or not crowding in itself is noxious, its physical concomitants in the state of housing, education, social service facilities and so on are distressing and contrary to the value system prevailing in most Westernised societies. However, attention to problems of scale and differentiation enables provision to be made in a less bureaucratised and bewildering fashion. Smaller administrative units are possible for many services, if not for exceptional services such as water supply or trunk road planning. The size and boundaries of the administrative area can be related to local needs and traditions and coincide with a natural focus of civic and social concern.

The aim of the 'social' approach to crowding is to identify the societal groupings which are particularly disadvantaged in terms of crowding, and which at the same time can form the setting within which solutions are applied. The distress and potential involvement of these groups will be related to their perceptions of their areas or communities and to particular services at particular times.

The way in which the dimensions of crowding analysis and solution are vested in one another can be illustrated by pointing to three possible approaches to group involvement at the local level. They correspond to points to the rational-consensual-participatory continuum. Firstly, groups and organisations are recognized by government as a valid source of opinion and information on local matters affecting policy and proposals (Stringer and Plumridge, 1974; HMSO, 1975). Their contribution is treated as technical, an extension of the data base used by rational decision-makers. Second, local government can decentralise some of its activities through area committees (cf. Stewart, 1974). Local groups may be represented together with elected council members and their officers. An advantage of these committees is that they consider problems on an area-wide basis rather than in terms of the separate service functions around which local government committees are normally organised. This corporate strategy is basically consensual, a pluralistic extension of existing representation. Thirdly, a new institution is created to represent people at a neighbourhood level (cf, Dixey, 1975). The idea of formal neighbourhood councils has been widely discussed in Britain in recent years, and has been adopted in some parts. Non-statutory councils have also been introduced. Representation on these bodies and the scale at which they operate gives fresh opportunities for participation in local government. Which of these three forms of local group involvement is pursued in relation to the consequences of crowding, will affect both the analysis of the problem and the manner of solution attempted. The corporate approach will take a systematic viewpoint. The participatory neighbourhood council will look for opportunities for self-help and self-determination by the council.

Crowding as a Personal Problem

At a personal, as opposed to individual or social level, crowding is concerned with the relation, as such, between an individual and a social group or institution. (The person is defined here as the sum of his relations with others.) Psychologically, a crowded relation is characterised in terms of unwanted constraint or interference by another; or by circumstances in which it is difficult or impossible to progress beyond simple co-prescence with another toward a relation which has a wider social meaning. The intersubjective process is either almost entirely absent or is transformed into a subject-object relation.

Politically, the relational failure is between the individual and others, in their political institutions. This also has been referred to in terms of deliberate control and constraint, exercised by the institutions and with very limited possibilities for reciprocal control. The absence of representation or its interpretation as an unrestricted mandate by the elected, render citizens relatively powerless (Murgatroyd et al., 1977). Political institutions at a geographical scale and with a focus of concern which corresponds to people's daily lives (cf. Lee, 1968; Hampton, 1970) will give the individual an opportunity for participation in them and for political integration.

Participation enables the citizen to become actively involved in the determination of his circumstances. Ideally there should be provision for his civic education (HMSO, 1977) so that he may acquire the knowledge and confidence to take responsibilty for decisions which intimately affect his life. By this means he will come to appreciate his relation to his fellows both in practical activity and in self-awareness. In formulating and implementing measures to relieve the consequences of crowding in its material manifestations, he will also find relief from its more generalised forms – constraint and lack of perceived control.

Conclusion

In focussing upon crowding in this conference we are in danger of investing it with the character of a disease. In reality it is only a symptom of the state of political and psychological life in urbanised, industrialised societies. The symptom is uncomfortable. But treating it in isolation may cause other troubles to erupt in its place.

Crowding is both about the material provision of spatially organised facilities and about the relations between people. It is about who gets what, and about who does what to whom. In this sense it is both a political and a psychological issue. I have attempted to suggest in this paper that neither individualism nor pluralism in politics and psychology gives a fully adequate analysis of the issue. The adoption of a participatory and relational approach may seem idealistic, even obscurantist. Progress and understanding in these directions may be gradual and lengthy. But initiatives are already being made.

We can continue to follow the other two sets of assumptions. However, if solutions are attempted in such terms, the problem will merely be aggravated. The individualist and pluralist assumptions and procedures actually constitute a large measure of the problem itself. There needs to be a will to change and a readiness to cast aside more familiar models. The potential adaptivity of society as a system can be tapped only through facilitating the richness and varied forms of direct, personal involvement in society. Solutions can be attempted from within the process of formulation of personal identity, constructed in relations between people in the context of their social and political institutions.

I began by saying that crowding was not a new problem, and that the term had several senses and shades of meaning. What is new is the social and political context within which crowding is manifested. Its meaning for us today as a practical problem is defined by that context and we must reflect it both in our analyses and in the solutions we suggest.

THE TASK INTERFERENCE MODEL OF CROWDING

Robert B.Bechtel

Environmental Research and Development Foundation, Tucson,

Arizona. U.S.A.

ABSTRACT: The recent problems of defining crowding are discussed with the suggestion that a more workable definition can be derived from the early animal studies. A task interference model is proposed that can be used in either animal or human studies. This model is compared with three other models used in crowding studies and has some overlap. The main problem with the model is to empirically define the seriousness of crowding effects on the various tasks of human behavior. Crowding that interferes with reproductive tasks is potentially the most serious. This is followed by crowding that affects sustenance tasks.

INTRODUCTION

The Problem of Definition

The topic of crowding is currently in vogue. The recent books by Freedman (1975), Altman (1975), and Saegert (1976) are representative of a burgeoning literature on the subject. Yet, despite this proliferation there seems to be a basic problem of definition. There is some agreement among researchers (Esser, 1971; Stokols, 1974; Rapoport, 1976) that simple density alone will not serve as an adequate definition of crowding, but there is little agreement as to what can take the place of density as a more proper definition. Rapoport (1976), for example, wants to modify density by partitioning it into perceptual, associational, temporal, physical and socio-cultural aspects in order to explain its contradictory effects. Esser (1971) wants to define crowding as a "mental state in which stimuli are experienced as inappropriate and stressful." The problem with many of these definitions is that they tend to confuse crowding as a cause with crowding as an effect. The problem arises when one wants to try to explain why simple density does not have a uniform effect on human beings. The animal studies do not have such knotty problems. In Calhoun's experiments (Calhoun, 1962, 1973) density serves as a perfectly adequate definition of crowding and it can be shown quite clearly that the dependent variables (the effects) are directly related to measures of density.

When one attempts to translate the simplicity of the animal experiments to human studies, however, he is confronted with the famous "Hong-Kong" dilemma; that people in Hong Kong can live apparently without ill effect in 43 square feet per person, while places in the United States are considered crowded which have under 340 square feet per person. (Rapoport, 1976). To explain this dilemma researchers on human subjects do something the animal researchers seldom do, they go inside the subject. Thus, we hear about "perceived" aspects of crowding (Stokols, 1974) and the "experience" of crowding (Esser, 1971). From the simple denotative definition of density with its easily observable effects in the animal experiments we move to the complexities of human perception and experience which mediate the denotative aspects of density to render them useless as a definition. Hence, we seek to define human crowding in terms of internal, hypothetical constructs. One might take the posture that since humans are more complicated, the concept of human crowding should therefore be more complicated than animal crowding. It does not necessarily follow, however, that the waters need to be so muddy.

A New Definition

The chief problem with internal models of crowding is the tendency to confuse cause and effect. If crowding is an internal experience, then what causes it? Crowding? We are left with a neat tautology that the cause and effect are the same. And this is not a mere play on words. A useful definition of crowding must have a measurable, denotative quality located in the environment, not in the head. Without this requirement "crowding" becomes an internal state independent of environmental influence. In short, crowding has to be located external to the subject as a cause, and whatever effects crowding may have should be objectively measurable and easily separated from the cause.

The proper place to begin in searching for a better definition of crowding is in the earlier animal studies. Calhoun (1962) was quite clear in his definitions because he increased population while keeping the available space the same. But even in these first two experiments, some important and subtle distinctions were made. In the first experiment a behavioral sink developed while in the second experiment, the behavioral sink was prevented because the food was given to the animals in a different way. "A powdered food was set out in an open hopper. Since it took the animals only a little while to eat, the probability that two animals would be eating simultaneously was considerably reduced," (Calhoun, 1962, page 145). It is this reduction in number of contacts that is the important clue. Does this mean that the critical variable is mere number of contacts? Not quite. Calhoun's observation of what causes the behavioral pathology is precise. He describes why the female rat fails to make a proper nest: "In the midst of transporting a bit of material she would drop it to engage in some other activity occasioned by contact and interaction with other individuals met on the way." Thus, it was not the mere contact but the interfering quality of the contact that made the crucial difference.

Let us summarize the important features of these two experiments reported by Calhoun. First, it was discovered that the pathological effect of crowding, as defined by density, could be reduced by restructuring the environment to reduce the number of contacts between animals. Second, it was observed that the interfering quality of contacts was directly responsible for the pathological behavior.

We have, then, an operational definition of crowding that is located in the environment, external to the animal and easily separated from the effects of crowding. It can be seen that this definition has potential for explaining the "Hong Kong" dilemma in the same way that it explained why there were no behavioral sinks in the second experiment, i.e., that the chances for interfering contacts were not the same. New York may seem crowded at 340 square feet per person because there are more interfering contacts than in Hong Kong with only 43 square feet per person. Yet, the differences may be due to cultural norms but the cultural norms define at what point the number of human contacts will become interfering.

At this point "interfering" needs wider definition. To say that crowding can be defined by the number of interfering contacts is to raise the question, interfering with what? In the case of the maternal rat, it was clear that the interference was with nest building. There are a host of other behaviors such as drinking, eating, sleeping, urinating, mating, etc. Crowding can be defined and measured as the number of interfering contacts with each of these behaviors, yet it soon becomes obvious that the quality of these interferences can be different, and that the effects of these interferences can be even more drastically different.

The Quality of Interference

An interesting fact is observed when one tries to determine how animal contacts in the behavioral sink interfere with behavior as opposed to merely interfering contacts outside the behavioral sink phenomenon. The behavioral sink begins when a number of animals are forced to eat and/or drink together for a period of time so that eating and drinking become conditioned to large groups. It then becomes impossible for the animal so conditioned to eat or drink alone. If left by the group, it will stop drinking or eating to rejoin the group. Thus, we have the seeming paradox that a behavior is interfered with by lack of group as opposed to behavior that is interfered with by presence of a group. This phenomenon alone could account for many of the contradictory differences in crowding studies. However, there are certainly limits to the crowd facilitation concept and it is perfectly obvious that not too many individuals could eat or drink at the same time, so that the population breaks up into different groups and we have the "changing of the guard" phenomenon when one group stops eating or drinking and another takes over the feeding area.

There is the possibility, then, that some behaviors become conditioned to large groups, and perhaps even facilitated by them. Researchers like Freedman (1975), Whyte (1968), and Walter (via Solomon, 1972), in fact, do report positive effects of crowding which may, indeed be due to a phenomenon similar to the behavioral sink. But, as Patterson (1976) points out, some of these positive effects may be due to the way that crowding is defined. Certainly, crowding was not defined in these cited research reports as the number of interfering contacts.

The Quality of Effects

The problem of defining the object of interference remains. If crowding is to be defined as interference, what it interferes with is a second and critical part of the definition. One connot merely say that crowding has the potential for interfering with any human or animal behavior because this leaves the quality of interference at random, and as has been shown, some of these qualities can contradict one another. It is the kinds of behaviors interfered with that determine the seriousness or the degree of quality of crowding. If eating and drinking are interfered with, this will have a more immediately serious effect than if playing and sleeping are interfered with. Yet, if mating is interfered with, this could eventually have the most disastrous consequences of all.

One can order these behaviors in the degree of seriousness of consequences to the individual animals and to the species as a whole. For individual animals, interference with drinking would be most serious, followed by eating, sleeping, and other basic needs. For humans one could possibly follow Maslow's hierarchy of needs yet, such an ordering remains to be empirically determined both for animals and humans. A way to generalize these possible effects is to characterize them as tasks necessary for the maintenance of the individual animal. Crowding, then, is interference with these basic maintenance tasks.

And there are tasks necessary for species maintenance. If individual animals fail to reproduce in sufficient numbers the species dies. It was Calhoun (1973), again who showed that it was possible through crowding to interfere with species maintenance tasks while maintaining the individual animals. He calls this "death of the spirit."

In sum, crowding is interference, caused by frequent contacts, with the basic tasks necessary to sustain individual animals and species. At the minimum, this interference is an annoyance, at the maximum it can result in the deaths of individual animals and/or the entire species.

APPLICATIONS TO HUMAN CROWDING

The question might be raised whether this definition applies to humans. After all cannot such "non-essential" things as recreation be interfered with, too, and is this not an example of crowding? Certainly it could be, and with the greater complexity of human behavior one would expect that the opportunity for interference through human contacts would be very great across the range of many behaviors. However, the essential ingredient remains the same, and the definition is equally applicable across both human and animal studies once the life habits are seen as a whole and the degree of interference is measured by the interruption of these habits.

What about the comparison of this definition with other models of crowding? Stokols (1974) describes three basic models which encompass the range of crowding studies fairly well: The Overload Models, Behavioral Constraint Models, and Ecological Models.

Overload models stem from a concept of the human being as an electrical circuit which has a given and limited carrying capacity. If too many stimuli are provided, the organism must either select out certain stimuli or suffer from the "stress of overload." In fact, the process of having to select from so many stimuli may itself be stressful. Milgram (1970) is the researcher who more recently introduced such a model and Stokols describes Desor (1972), Esser (1972), Baum and Valins (1973) and Zlutnick and Altman (1972) as doing research under such a concept.

Behavioral Constraint models evolve around either a "perceived limitation of behavioral freedom" (Stokols, 1974, page 4) a la Brehm (1966) or an invasion of Sommer's (1969) personal space bubble. In any case, the organism will react to restore either the lost behavioral freedom or the violated space bubble. Work by Proshansky, Ittelson, and Rivlin (1970) and Stokols (1972) himself fit under this model.

Ecological models are derived from Barker's (1968) ecological psychology concepts, particularly undermanning theory. It was Wicker (Wicker, McGrath and Armstrong 1972, Wicker, 1973) that developed the concept of overmanning as a definition of crowding. Barker's original undermanning concept was that when there are fewer people than are ordinarily needed for a given task, then a condition called undermanning exists. This situation has certain positive benefits for all present in the situation, among which are a greater feeling of being needed, greater work output, greater satisfaction and other positive effects. Overmanning, by contrast, is the same situation but with more than enough people present to do the task. Presumably, the positive benefits of undermanning are reversed and their negative counterparts take over. Persons in an overmanned situation feel less needed, have a smaller work output, less satisfaction and other negative experiences. Barker and Gump (1964) showed these relationships held for large vs. small schools, Wicker (1969) showed these relationships held for churches.

The overload models are not really models of crowding at all. They are stress models which may or may not show that if people are crowded, they feel stress. The burden of defining crowding remains.

The behavioral constraint models, however, show more promise in being able to define crowding. If, for example, people do carry about them Sommer's "emotionally charged bubble" and this is violated, then crowding has taken place. Sommer's (1969) experiments operationally define crowding by having stooges move closer to subjects until there is a reaction. However, there are conditions under which personal space does not operate. People have intimate relations, they move closer together in conversation than with strangers in public, and there are a whole host of conditions under which personal space is either inoperable or modified. In short, we have returned to the basic concept again. Whether personal space operates depends on whether some behavioral task is being interfered with. The same may be said for Brehm's "perceived limitation of behavioral freedom."

However, in defining undermanned and overmanned settings Barker's methods have not yet evolved to the point where task interference can be pinpointed, or where overmanning can be defined. Overmanning deals mainly with the principal actors in behavior settings. It is easy to see when a lawyer's office becomes overmanned when there are too many lawyers for the clients. But it is more difficult to talk about overmanning when settings such as theaters are being studied. At what point the audience becomes overmanned is not clear. A crowded theater is often seen by both management and audience as a positive stimulus. Wicker (Wicker, McGrath and Armstrong, 1972) has to introduce the concept of capacity to account for "physical" crowding. But capacities are defined by the particular task at hand. A movie house requires so many seats to be profitable. A truck cab must have sufficient room to allow the driver to shift gears but not so much as to add too much extra weight to the load. These capacities are determined by culture and culture defines how a task shall be done and when it is being interfered with.

Exactly how to compare the concept of overmanning as a definition of crowding and the task interference model is not immediately clear. It is conceivable that overmanning effects could occur without observable task interference. Dropouts to an overmanning setting could occur without the participants being visibly impeded in performance, merely because they perceive no chance to achieve responsible positions. Thus, they would leave the setting without any observable interference with any task. Yet, a case might be made that the reason they left was they had been prevented from taking part either in any task or not enough tasks. Indeed, overmanning literature (Barker and Gump, 1964) shows that one of the main effects of overmanning is to decrease active participation levels. This would mean that settings which, in effect, turned participants away because of exceeded capacity, or few chances of "advancement" were also interfering with tasks in a subtler way. These would not be the tasks of the leaders or performers, as such, but the tasks that potential leaders would perform if allowed.

There might be a further dimension to task interference that occurs in the above situation. It is possible that the tasks of the leaders or performers would be interfered with if pressures to enter the setting became too severe. Then, the leaders or performers would need to spend a certain amount of time dealing with these applicants to turn them away. And time here becomes an important measure of interference. How much time must be taken away from the prescribed tasks to deal with interference? These amounts of time could be empirically manipulated in order to determine what constituted serious interference.

It would be remiss not to point out that the undermanning concept adds an entirely new dimension to the concept of crowding. Whatever definition has been applied, with the exception of Freedman's (1975) and Whyte's (1968) definitions, crowding has usually been seen as a negative effect on human behavior. Undermanning, however, describes a positive effect of "undercrowding," if you will. In other words, crowding may be seen as the negative end of a continuum with undermanning at the opposite extreme. Undermanning posits a continuous dimension of allotment of persons for any given task such that "too few" is actually the most beneficial assignment while "adequate" has already shifted toward the crowded end of this dimension. This is an entirely new way of looking at crowding. In the undermanning concept, crowding begins long before most researchers considered it possible. The relationship is shown in Figure 1.

The problem is that under extremely undermanned conditions there are so few persons, the tasks cannot be performed. Thus, we have a paradox that on the right hand side of the undermanned situation conditions exist prior to "normal" crowding

Number of Tasks per Person Decreasing by this Dimension →

Crowded	Not Crowded	Crowded	
Extremely Undermanned	Undermanned	Adequately Manned	Overmanned

Manpower supply increasing in this dimension →

Figure 1

Undermanning and Crowding

that are only slightly negative, and on the left hand side conditions exist which are even more debilitating but which are not usually considered crowding –in fact, its antithesis.

Again, task interference can rescue us from this dilemma. But it can only be done by adding the dimension of tasks per person at the top of the figure. The conclusion is that having too many tasks per person can be just as interfering and can be seen as "crowding" even though there appear to be few people present.

We have then, three dimensions of task interference. 1) Too many people per task which leads to interfering contacts, 2) Too many tasks per person which does not permit completion of tasks, and finally, 3) Wicker's capacity dimension which does not permit task performance in a given space. And to each of these we use interference over a given time as a measure of degree of interference.

But the paradox of adequately manned crowding remains. If undermanning is the truly "uncrowded" state, then why should adequately manned situations be seen as "crowded?" The clue is seen in Wicker's (1968) school study. In this study Wicker found that the positive benefits usually attributed to undermanning also occurred in overmanned settings but only in the few persons in responsible positions. Thus, the undermanned setting achieves its positive benefits by distributing responsibilities more evenly throughout the population. This is, of course, easier when there are fewer people. Therefore, the adequately manned situation is "crowded" in the sense that its participants do not have a sufficient share of the responsible tasks to receive the benefits of undermanning. In short, the ratio of tasks to persons is still too low. There must be either fewer people or more tasks to achieve the undermanned state.

The task interference model of crowding overlaps with the behavioral constraint and ecological models but it attempts to be more precise than either of them in denoting crowding as a cause of certain human behavior. Is there any independent evidence besides that already cited that task interference is the critical defining factor? An interesting study was reported recently by Heller, Groff and Solomon (1977). They attempted to show that density does have some effect on task performance. Their results are worth quoting (page 188):

"It seems clear that high density adversely affects performance only when individuals must physically interact to perform the task. Neither high density nor physical interaction alone is sufficient to produce these effects."

As so often happens with attempts at conceptual clarification, more questions are raised than are answered. Yet, that is sufficient goal in itself, to move forward in the questions being asked.

CROWDING: AN ECOLOGICAL APPROACH

Rajendra Kumar Srivastava

Environmental Research and Development Foundation

Tucson, Arizona, USA

ABSTRACT: This paper treats crowding as an ecological phenomenon and as an objective and behavioral construct. According to this approach crowding is overmanning which refers to the larger size of population than is needed for the number of tasks within a given behavior setting or environment. An equation for the quantative measure of overmanning (crowding) has been provided. Overmanning is shown to be associated with a variety of negative human-behavioral consequences. It is suggested that overmanning and its negative consequences can be controlled by keeping the population size small and expanding its enterprise scope and behavior repertoire.

CROWDING: AN ELUSIVE CONCEPT

In spite of its apparent simplicity crowding is an elusive concept. We do not know what it means, testified by the fact that we use the term crowding for a number of totally different phenomena. We also do not reliably know what its effects are which is testified by the fact that the results of crowding studies are found to be contradictory. Consider Schmitt (1966) finding density to be related to crime, only to be contradicted by Galle, McCarthy and Gove (1974) who found no negative relationship between the two. The literature on crowding is full of such inconsistent findings (Wicker, 1973; Saegert, et al., 1976, for example), which could be because of differences in the population samples used for crowding studies, or the kind of research designs employed, or lack of adequate and appropriate controls in experimental studies, or attempts at cross species generalizations from mice to men, or differences between laboratory and field studies, or the kind of variables selected for the study, or how crowding is defined, or a combination of these and other reasons. Instead of investigating all these reasons, I will limit my discussion to the definition of crowding which, I feel, is at the core of the problem of inconsistent results.

All the existing approaches to the definition of crowding can be placed into five broad categories. According to them crowding is

a) a population phenomenon or
b) a physical phenomenon or
c) a perceptual phenomenon or
d) a bio-social phenomenon or
e) an ecological phenomenon.

CROWDING AS A POPULATION PHENOMENON

When crowding is looked at as a population phenomenon it is primarily a matter of number of bodies in the context of essential resources for their survival and without much regard to the specific area of the space containing them. Another more appropriate term for this phenomenon is overpopulation. The suicidal march of the lemmings is a good example. These rodents in Norway when over-populated move toward the mountain cliffs in large numbers and fall off into the ocean and die (Clough, 1968). What happens is that with the steady increase in their population the lemmings exhaust their available food supply and then start to wander off in search of new grazing grounds, plundering and stripping the land in their way, eventually come to the mountain cliffs and accidently fall off them into the ocean and drown, thereby reducing their population. This allows the plundered land time to regain its vegetation making it possible to sustain another increase in the lemming population followed by another suicidal march, and the population cycle of increase and decrease continues. Clearly, this phenomenon is related to the amount of food supply and is a classic example of overpopulation and not that of density as some (i.e., Freedman, 1975) assert it to be because with increase in population the lemmings also increase their area of grazing and wandering keeping the density, i.e., the ratio of the number of bodies to the physical area, low.

Thus, the reference points for "overpopulation" of a land are its natural and manmade resources that sustain the population and neither its area nor anything else. In this context, China and India are overpopulated, not because they have the largest number of people of any other country in the world, but primarily because many of their essential natural and manmade resources are not sufficient to sustain their populations. In the same sense this earth is overpopulated whose limited resources are fast disappering while the population keeps increasing at a rate of 1.9% per year with a projection of 7 billion by the end of the century according to the United Nations Demographic Yearbook of 1977. So, we are beginning to look into the possibility of settling in space or on other planets (Johnson, 1977) without really knowing what we are headed towards. Are we the lemmings headed toward the ocean?

What this means is that overpopulation is an issue of grave importance for many individual countries and even for the earth. This problem should be studied and ways must be found to solve it. But, let us not confuse it with density and other crowding phenomena.

CROWDING AS A PHYSICAL PHENOMENON

When crowding is considered as a physical phenomenon it is essentially a case of density, defined as number of bodies per unit area. Freedman (1975) and Saegert (1976) define crowding in terms of density. Most animal studies of crowding have used density as the independent variable but have frequently called it crowding (Calhoun, 1971, 1973, Marsden, 1972, Southwick, 1955, 1967, for example). In everyday language we use the word crowding to mean density. When we say a house or a bus or an auditorium or any other place is crowded we are referring to the number of bodies within a given area of space.

Density is an important phenomenon with far reaching consequences for mankind particularly in the world of today where the physical area per person is fast decreasing. It has been shown to increase the mortality rates and decrease fertility rates among rats (Calhoun, 1973), promote mental deficit in old age (Mitchell, 1971), relate to infant deaths, age-standardized deaths, and tuberculosis among humans (Winsborough, 1965), influence nursery school children's use of space, and activities (Preiser, 1972), affect children's aggressiveness, social interaction and time spent in group involvement (Loo, 1972).

This suggests that density is a legitimate subject of study and an understanding of its consequences for human and animal behavior may have far reaching impact on the solution of many of our behavioral and social problems. It is not synonymous, however, with other phenomena treated as crowding.

CROWDING AS A PERCEPTUAL PHENOMENON

According to this approach a condition of crowding exists only if the persons involved so perceive it. This perception is always negative. And since perceptions are always influenced by a host of personal, social, cultural, physical, temporal and other variables, crowding becomes a multivariate concept.

This position is suggested by Rapoport (1976) according to whom a given density, if perceived as congruent with the perceivers' social, cultural and other norms, would lead to a situation of "OK" and if incongruent the result would be "isolation" or "crowding" depending upon which way the balance on incongruency tilts. In both "isolation" and "crowding" the affect is negative or unpleasant and only in "OK" is it positive or pleasant.

Stokols (1974) also presents crowding as a perceptual or experiential counterpart of density. He proposes an equilibrium model which says that density in interaction with numerous personality, social, cultural and physical variables produces physical and/or psychological stress in the individuals leading to all sorts of responses on the part of these individuals designed to overcome the stress and strain. Thus, when stress is felt in the context of spatial restriction it is interpreted as crowding.

The literature is full of research studies that have studied the perception of density in a variety of situations as a function of any kind of intervening variable fancied, such as body-buffer zones (Kinze, 1971), personal space requirements

(Dooley, 1974), architectural features (Desor, 1972), dorm design (Baum and Valins, 1973), competitive or cooperative game conditions (Stokols, et al., 1973), dyadic or similar interaction settings (Fisher, 1973), internal locus of control (Schopler and Walton, 1974), and number of people, amount of space, lighting, type of activity, social factors, etc. (McClelland and Auslander, 1976).

These studies are important in telling us about how different density situations are perceived by different people as a function of a variety of intervening variables. Yet, they are studies of perceptions, not of crowding, in the same manner as studies of sound perception, color perception, depth perception, etc., are the studies of perception and not the studies of sound, color and depth.

CROWDING AS A BIO-SOCIAL PHENOMENON

The main proponent of crowding as a bio-social phenomenon is Esser. According to him crowding is "conceptualized as a mental state in which stimuli are experienced as inappropriate and stressful" (Esser, 1971, p.2). He posits brain functions as a necessary construct in the experience or perception of crowding. Esser states, "Man will feel crowded if in his attempts at cognitive constructs he finds that his human environment (including himself) cannot be completely experienced (perceptual overload) or does not consensually validate his images (differences in level of brain-environment transaction)" (1973, p. 210).

Because of its reliance on experience through brain functions for an explanation of crowding this approach is able to explain why people feel crowded. It can also explain individual differences and differences between normal population and the psychiatric patients in their perception of crowding within the same physical, social, spatial, and temporal parameters. It is also able to explain why people feel crowded even though they are physically alone (because experience of crowding can be attained through the real or imagined presence of others) or not crowded in the midst of a large number of people.

In spite of these positive features one major deficiency of this approach is its treatment of crowding as a subjective experience. This makes it more amenable to the study of personality, individual differences and perception and less directed toward the study of crowding. Experience of crowding and crowding are not the same. And since the bio-social approach focuses on the experience of crowding it cannot be accepted as an approach which has crowding as its legitimate subject of study.

CROWDING AS AN ECOLOGICAL PHENOMENON

The notion that crowding is an ecological phenomenon was first propounded by Wicker (1973, 1974), and is an outgrowth of the ecological psychology movement started by Barker (1968). In this paper the ecology concept is further refined, and suggested to be superior to other approaches of crowding.

Concepts

There are four ecological psychology concepts which are critical to the understanding of crowding.

1. Behavior Settings: Simply stated, behavior settings are environment-behavior units in synomorphic relationship. Thus, the stadium (environment) and ball game (behavior) together constitute a behavior setting "baseball game." Behavior settings are the basic unit of analyses.

There may be one or more behavior settings within a larger environment. Crowding, therefore, has to be measured, studied and understood first at the behavior setting level. The crowding situation in all the behavior settings together within an environment would provide an understanding of the crowding in that environment.

2. Task: The term "task" refers to the specific behavior or activities that take place within a behavior setting. The number of tasks within a behavior setting provides a measure of the extent of its behavioral requirements against which the level of crowding is determined.

3. Population: The existence of tasks depends upon the existence of people who perform them. It is these task performing people for whom the term "population" is being used here.

The tasks vary on many dimensions, the goals, the type, the complexity, etc., and, therefore, they also vary on the requirement of the size of the population for their accomplishment. The actual size of the population in a behavior setting, however, may be lower or higher than what is required by the tasks. The size of population with reference to the number of tasks within a behavior setting is an important criterion for a measure of crowding.

4. Manning Level: The manning level refers to the differences between the size of population in existence (Pe) and the optimal size of population (Po) to get a given task (T) accomplished. This may also be stated as

$$ML_T = (Pe\text{-}Po)$$

In the context of a behavior setting (BS) which may have more than one task the manning level would be represented as a ratio of the difference between Pe and Po to the total number of tasks, i.e.,

$$ML_{BS} = \frac{Pe - Po}{T}$$

When all the behavior settings within a larger environment are considered together the equation to provide the manning level of the environment (Env.) would read as follows

$$ML_{Env.} = \frac{\left(\frac{Pe - Po}{T}\right)_1 + \left(\frac{Pe - Po}{T}\right)_2 + \cdots\cdots + \left(\frac{Pe - Po}{T}\right)_n}{BS}$$

In other words the manning level of a larger environment would be determined as the mean of the manning level ratio of all the behavior settings within that environment.

The manning level scores may be positive (+) indicating that there are more people than needed (overmanning), or negative (-), indicating that there are fewer people than needed (undermanning), or a zero (0) indicating that there are as many people as needed (adequately manned). The size of positive and negative scores indicates the extent of overmanning and undermanning. The crowding is measured in terms of manning levels.

Ecological Definition of Crowding

Simply stated, crowding is overmanning. It refers to the larger size of population than is needed for the number of tasks within a given behavior setting and can be measured by the equation given in the preceding section.

This definition of crowding has several special characteristics which set it apart from other definitions of crowding and establish its superiority over them.

1. According to the definition of crowding as overmanning, the only relevant variables are the number of people and the number of tasks. All other variables such as space (area) available, individual characteristics of the people, social, cultural, physical and other variables do not affect the objective fact of crowding. If a behavior setting isovermannedit is crowded and it will have predicted consequences irrespective of the existence or nonexistence of other variables which seem to be so critical for other definitions of crowding.

2. By its emphasis on the tasks and people this definition treats crowding as a behavioral construct. It is, thus, an improvement over the population and density approaches, both of which treat crowding as a nonbehavioral construct. The population approach emphasizes the number of people in relation to the extent of resources and the density approach emphasizes the number of people in relation to the amount of physical space, none of which has anything to do with behavior. Of course both population and density have consequences for behaviors, but the two approaches themselves are not qualified by any behaviors.

3. Crowding as overmanning is an objective fact and can be measured objectively, It is not in the head of the people. Because of this characteristic, it is an improvement over the perceptual and bio-social approaches both of which treat crowding as a subjective perceptual and experiential construct, making its existence entirely dependent upon the perceiving individual.

4. Crowding or overmanning as an objective fact is subject to perception. Thus, a given situation, even though overmanned, may or may not be perceived as crowded, but this will not change the objective fact that the situation is overmanned and therefore crowded. In spite of this, it is expected that overmanned settings generally would be perceived as crowded and the correlation between the two would be positive and high.

5. Crowding as overmanning considers the population size as a critical variable. In thes respect it is similar to all other four approaches. However, the five approaches differ in their treatment of the concept of population itself. For population approach the term population refers to the consumers of resources, for density approach it refers to the occupiers of spaces, for perceptual approach it refers to the people who are perceiving the presence of other people in a given situation in the context of their physical, cultural, social, temporal and other variables, for bio-social approach it refers to all sorts of real or imaginary people in a given situation and for ecological approach it refers to people whose behaviors are tied to their tasks. The differences in these approaches are also reflected in the research design and methods used by them. Thus, the five approaches are different from each other to such an extent that they are not dealing with the same phenomena even though they use the same term "crowding" for the phenomena they do study.

CONSEQUENCES OF OVERMANNING

The direct research evidence on the consequences of overmanning is lacking primarily because it is a new concept and has not had a chance to be properly investigated. The knowledge that we do have in this regard is largely inferential and derived from the studies of undermanning.

The undermanning theory was originally proposed by Barker (1960) at the Nebraska Symposium on Motivation. According to it when the number of inhabitants in a behavior setting falls below an optimum number needed for the given number of tasks a condition of undermanning exists which exerts a motivational force on the inhabitants and a number of positive consequences ensue. The studies that followed, conducted in communities and schools, revealed that the inhabitants of undermanned settings in comparison to optimally manned and presumably overmanned (crowded) setting,

(1) engage in more program actions,
(2) engage in more varied program actions,
(3) engage in more maintenance actions,
(4) engage in more varied maintenance actions,
(5) engage in stronger maintenance actions,
(6) engage in more deviation countering maintenance actions,
(7) engage in fewer vetoing maintenance actions,
(8) engage in more induced actions,
(9) enter more frequently into the central zones of behavior settings,
(10) engage in difficult actions more frequently,
(11) engage in important actions more frequently,
(12) behave in response to important actions more frequently,
(13) have less sensitivity to and are less evaluative of individual differences,
(14) have greater functional importance,
(15) have more responsibility,
(16) have greater functional identity,
(17) experience greater insecurity.

This list indicates that there are identifiable and measurable differences in human consequences of differentially manned behavior settings and that the identified consequences of undermanned settings generally with some exceptions are positive and desirable in terms of the goals and the primary functions of the settings.

The studies in schools (Barker and Gump, 1964) and more recent studies in churches (Wicker and Kauma, 1974), have also supported this conclusion by pointing out that compared to large settings people in small settings have greater feelings of responsibility, involvement, obligation to participate, high level of participation, etc. The problem with these studies is that they have typically used settings consisting of different size populations and on the basis of their comparative population sizes have assumed the small population settings to be undermanned and the large population settings to be overmanned. No precise measures of manning levels have been used.

There are some laboratory studies, however, which have used precise measures of manning level. For example, Petty and Wicker (1971) and later Petty (1974) used a slot car task with designated number of obstacles. The object was to run the slot car on the track as fast as possible and one of the things that needed to be done was to remove the obstacles. So the number of obstacles defined the optimum number of people needed to remove them to keep the car running. A reduction in this optimum number created the condition of undermanning. Consistent with the predictions of the theory the people in the undermanned settings compared to optimally manned settings felt more needed and felt that their roles were more important. The problem with these studies is that they have not even used overmanning as an independent variable. They only leave us to imply that if optimally manned settings are producing negative or less positive consequences than undermanned settings the overmanned settings would certainly do so.

In spite of the deficiencies of these studies, lack of precision in the measurement of manning levels in field studies, and lack of direct focus on overmanning as a variable in laboratory studies, the implications are clear that overmanning is generally associated with negative consequences.

CONTINUUM OF MANNING LEVELS

The research treatment of the manning levels seems to suggest a trichotomy of undermanning, adequate manning and overmanning. Wicker (1973) at a conceptual level differentiated between these three levels and proposed a continuity between them. Actually, if we use the quantitative measure of manning levels suggested in this paper, many more levels of manning are possible. The following diagram is presented to illustrate the continuum of manning levels and their relationship to crowding.

According to this diagram positive consequences may be expected only at level C "slightly undermanned" and as we move up and down the scale from that point we can expect increasingly negative consequences of varied types.

This diagram shatters two beliefs.

1. It contradicts the notion that manning levels are trichotomous. It demonstrates that many finer levels can be conceptualized and empirically

Manning Levels		Maintenance Minimum → Capacity	Consequences	Crowding
Undermanned	A	Extremely Undermanned	Negative	Uncrowded
	B	Seriously Undermanned		
	C	Slightly Undermanned	Positive	
Adequately Manned	D	Poorly Manned	Negative	Neutral
	E	Richly Manned		
Overmanned	F	Seriously Overmanned		Crowded
	G	Extremely Overmanned		

Figure 1
Continuum of Manning Levels

discriminated between the extremes of maintenance minimum (a condition of population deprivation to the level that the setting cannot exist such as lack of quorum leads to the cancellation of a board meeting) and capacity (a condition of such population overload that the setting is incapable of containing any more increase in number, such as a movie theater can accommodate no more people than there is room for).

2. It dispels the myth that one can attain positive consequences just by achieving undermanning. It shows that the undermanning can go too far and end up yielding negative results. The phenomenon of extreme undermanning was discovered while analyzing some data collected in mental health care settings (Srivastava, 1974), and it was found that not only the overmanned but also the extremely undermanned settings were associated with negative consequences. The fact of extreme undermanning further emphasizes the need for the consideration of multiple manning levels and their precise measurement and empirical testing.

CRITICAL AREAS OF ENQUIRY

This conceptualization of crowding as overmanning leaves a lot of questions unanswered. It even raises a lot of new questions. They define the critical areas of enquiry needing our immediate attention. Some of them are mentioned here.

1. The manning level equation proposed here is a beginning toward the quantitative discrimination between broad categories of overmanning, adequate manning and undermanning. However, we still need to know at what score variations within each broad category a particular manning level ceases to be and the adjacent one comes to be. In other words we need empirical data to set critical quantitative manning level limits.

2. While manning level remains the sole critical defining variable of crowding it is possible that the anticipated and observed positive and negative consequences may be mediated by some other variables. We need to know what those variables might be and how they influence the consequences of manning levels.

3. The negative consequences of manning levels below and above the slightly undermanned level cannot be assumed to be the same. In what ways they differ needs to be empirically determined.

4. Not all consequences of slightly undermanned level would be positive and not all those of other levels negative. The consequences that deviate from the norms need to be investigated.

5. We need to examine overmanning in its own right and need to determine by direct evidence what its consequences are. We might be able to uncover many negative consequences that we were neither able to speculate or infer. Or for that matter, we might even find that there are some positive consequences of overmanning in certain task situations.

HOW TO CONTROL OVERMANNING?

This discussion of overmanning or crowding will not be complete without pondering, at least briefly, over what to do about it. There are at least two possibilities: One, think small, and two, increase task repertoire.

Think small (Schumacher, 1973) is an antithesis of the current way of thinking. We, the homo sapiens, seem to be obsessed with the idea of more, more people, more things, more resources, So, we are multiplying at an alarming rate and instead of conquering the world, we are preparing it to be fossilized, devoid of any life. And we are acquiring more things, collecting things that have little purpose and no use. More things require more resources. We have already depleted many natural resources and we are working hard at eliminating the rest from the face of the earth. The earth has been plundered enough and even before its demise we are beginning to engage in monstrous thinking of how to plunder the moon and other planets. Worse still, the "more" syndrome is one opposed to activity. Don't we see advertisement every day of products that claim to save us time and work. We are beginning to be a society which does not want to work. No wonder an average person in the industrialized society is bored with his life yet keeps acquiring more things which promise still lesser behavior on his part. If we have to reverse the onslaught of this madness there is no other way but to think small. Gandhi had the wisdom to make a call for it several decades ago. He visualized small units of villages instead of megalopolis, the great moments of insanity. These small units were also visualized to be self-sufficient depending upon the natural resources that are in cycle so there was little fear of running out of them. The concept of the small unit consisting of a few hundred or at the most a few thousand people was the criticial idea. If we quickly glance through what Barker found in undermanned settings we can easily see that the small village unit will have people who will have a lot to do, will have a deep involvement in what they do, will have major decision making powers, will feel that they are needed, will have a sense of accomplishment and many other things that this life is all about. Think small, therefore, is not "do nothing" or "do little." Actually, small population units force us to do more; it is a behavior rich situation. Do we want a living and behaving or a dead planet? The choice is not hard to make.

"Increase task repertoire" cannot be separated from "think small". Its emphasis is on tasks, activities. By this term I do not mean all sorts of molecular behaviors. I am referring to molar behaviors which are specific, limited, and consistent with the goals and purposes of a given environmental setting. For example, in a school teaching, examination, student record keeping, discussing student progress, are examples of tasks but not blinking the eye, shuffling the feet, taking a drink at the fountain, all of which also surely take place. Increase in the tasks not only provides more things for people to do creating undermanning but it actually forces them to engage in them. But we must remember that increases in task repertoire should not be used as an excuse to increase the population. Surely, more tasks can accommodate more people and at times increase in population in a setting may be necessary in order to get all the things done that need to be done. But the emphasis should still be on increasing the tasks and not on the negative approach of increasing the population. The big governments, the big corporations, the big anything are examples of this negative approach, where new positions are constantly created, new employees are hired, yet the actual output of product or services remains the same or decreases.

Actually, for the most effective outcome "think small" meaning keeping population size small and "increase task repertoire" meaning expanding enterprise scope and behavior, should be considered together. Care must also be taken that this activity does not go so far as to create extreme undermanning and yield a new set of negative consequences. Thus, the solution lies in small overall populations with task intensive life style and certainly not in large populations and leisure living as currently practiced. If my solution sounds primitive, it also sounds sane.

INDIVIDUAL VARIATION IN PERCEPTION, ADAPTATION, AND CONSEQUENT CROWDING

Lawrence J.Severy

University of Florida, USA and Behavioral Research Institute, Boulder, Colorado, USA

ABSTRACT: A predictive model of response to differential density is developed. It is suggested that individual variation in the magnitude of density - consequent social behavior relationships must be researched. The model incorporates individual variation in: the perception of crowding, the stress that perception creates, and adaptation. Crowding results when the density of an environment deviates from appropriate levels. Stress is influenced by situational control and cognitive style. Further, stress increases in proportion to the extent that crowding is perceived as a status rather than a state. Adaptation to status is imperative.

—·—·—

Two of the basic concerns of populationists are population growth and population distribution. As population size increases in an uncontrolled fashion - and as distribution of that population occurs in a systematic rather than a random fashion, concern for the consequences of high density and clustering of persons obtains. On a general level, we at this conference give proof to the claim that there are behavioral consequences of these population processes. Specifically, questions have arisen, theories have been developed, idiosyncratic positions have been argued - and finally, this symposium is being held to determine: (1) If there is cause for concern; (2) What we know so far; and (3) If there is reason for worry, how to appropriately plan for the alleviation of potential problems. Given the expenditure of energy and resources associated with our meeting, one would have to conclude that this is certainly an observable consequence of human crowding.

Recognizing that we as social scientists and planners are responding to potential problems with population distribution does not, however, in and of itself, help to advance our efforts. What is needed is an individual or person based understanding of behavior generated in response to living in – or being exposed to– a variety of population distribution patterns. It is, therefore, the intent of this paper to delineate a social-psychological model of individual variation in the perception, adaptation, and consequent response to human crowding. The position taken here is that there is an exponentially expanding amount of information that can be synthesized into a predictive model of such behavior.

The point of my introductory comments is that: (1) it appears that population growth is going to continue which will create higher densities; (2) people will probably continue to environmentally cluster their living arrangements rather than spread themselves evenly throughout the world; and (3) environmental planning in response to these two truisms must be based upon a conceptually sound, empirically based understanding of individual behavior in differentially dense environments.

EMPIRICAL FOUNDATIONS

Turning to previous empirical work, the importance of environment and environmental design in and of itself on behavior is well demonstrated by those such as Barker (1965). Further, the specific impact of density or crowding on human behavior can be noted in; laboratory research by social psychologists, the research of environmental and population geographers, migration and fertility theorists, designers and planners, and last, but most intriguingly, the investigations of the epidemiologists. (All these without mentioning work with lower species .)

For example, as regards social psychological research, Loo (1972) investigated effects of various methods of creating crowding on children's aggressive behavior; Desor (1972) noted the influence of partitions on crowding perceptions; Worchel and Teddlie (1976) attempted to further delineate determinants of such perception; and Sundstrom (1975) addressed the impact on nonverbal behavior, self disclosure, and selfreported stress. All this in the face of Freedman's (1975) conclusion, based on his own and others' research, that density's effects are not inherently negative. It should be noted that his density-intensity hypotheses do assume a behavioral consequence to high density. The problem is to develop a more thorough explication of the variables that, in conjunction with density, will allow for precise behavioral predictions.

Canter and Canter (1971) and Mitchell (1971), working with Asian cities, demonstrate complexities with housing characteristics. Chombart de Lauwe (1959) obtained dramatic results concerning the impact of urban housing design in France as regards social and physical behaviors. In a related vein, recent theorizing and empirical investigation by social demographers interested in household migration (residential mobility) has great relevance to an analysis of crowding. Both Lansing and Hendricks (1967)and Speare, Goldstein, and Frey (1975) obtained evidence suggesting density and crowding ratios play major determinant roles in predisposing a person to be dissatisfied with current housing arrangements and a desire to move. The Speare et al. conclusion is that, based on three large scale longitudinal investigations, results provide general support for a stress-threshold theory of human mobility.

This orientation derives from Wolpert's 1966 formulation suggesting that people are motivated to move (as an adjustment) when their current environment is stressful due to factors such as family size, small rooms, etc. In another investigation of migration based upon 4000 couples, Chevan (1971) finds that "if the persons per-room ratio of non-movers represents an acceptable household density, moving can be shown to be an adjustment mechanism whereby housing space is brought in balance with family needs. When couples who move within a given three-year period, in terms of marriage duration, are compared with couples who did not move during that period, movers are found to have higher initial densities, would have had substantially higher densities had they not moved, but have similar terminal (after move) densities (p. 451)." Going one step further, Severy and Atkins (1976) delineated the impact that such migratory motivation would have on fertility intentions and behavior. This work clearly demonstrates the important influence that density considerations have on such critical life behaviors as the determination of family size and household migration.

Lastly, the epidemiological investigations of Galle, Gove, and McPherson (1972), Booth and Johnson (1975), and Levy and Herzog (1974) tempt with ideas but incorporate enough methodological drawbacks to leave large question marks. Many (for example Roncek, 1975) have critiqued the inherent problem of excessive aggregation of data which might produce artifactual results as well as making inferences to specific individuals untenable. Goldstein, in his presidential address to the Population Association of America, argued a similar point when he claimed that "it is difficult to relate aggregate models, which are useful for predicting the volume of migration streams, to individual models, which attempt to explain why people move" (1976, p. 427). Simply put, these studies based on aggregate data present extremely interesting data but are not appropriate for, and may obscure, the effects of population density on individual behavior.

As a consequence of the state of the art, what follows is an attempt to provide a careful conceptual analysis of the existent possibilities when individuals find themselves in settings of differential density. (In essence, the title of this paper is incorrect in that I am really addressing a model of response to differential density rather than only crowding.) In identifying this work as a conceptual analysis, it should be emphasized that the complete formulation has yet to stand the test of either empirical investigation or application. On the other hand, whenever components of the analysis can be substantiated by prior empiricism, such evidence will be discussed. With these thoughts in mind, I turn to the concern for how an individual perceives, adapts, and consequently, responds to differentially dense settings.

PERCEPTION, STRESS AND ADAPTATION

The conceptual analysis of individual response to differential density settings involves three distinct processes that also serve to structure much of the remainder of this paper. First, in order to respond to any stimulus input, the organism must, on some level, actually <u>perceive</u> that stimuli as an input. Consequently, an analysis of individual variation in the perception of density and crowding is necessary. Second, the input has to be <u>interpreted</u>. Our target person may experience crowding and its accompanying stress, or, alternatively that person may label the input in another way that would not produce stress. Hence, an analysis of individual variation in the generation of stress is necessary. Third, based on the interpretation of stimulus information, people search their cognitive and behavioral repertoires for potential alternative responses designed to adapt to the environmental configuration. The analysis of this step would then complete the delineation of the conceptual analysis of individual response to high density settings.

Perception and Stress

The analysis of individual variation in the perception of crowding and consequent stress necessarily entails examination of three concepts. First, density needs to be considered in conjunction with previous adaptation and social learning regarding perceptions of appropriateness. Second, the concept of situational control appears crucial and requires delineation. Third, the distinction between state and status impacts upon the analysis of the severity of stress that is generated.

The first task of such an analysis is to capture as precisely as is possible what is meant by the term crowding. Up to this point I have been deliberately loose in phraseology and have attempted to employ terminology such as population distribution, differentially dense setting, etc. However, it is of critical importance that one analyze the distinction between the concepts density and crowding. My analysis is similar to that of Stokols (1972), If I were a physicist, I would be content to utilize the theoretically precise unit of analysis termed population density. This objective index certainly provides description regarding the number of persons per unit area - but offers little more towards predicting human response. The cornerstone of the social sciences is that what is expensive for one person or group of people may or may not be for another; and the same obtains for other concepts such as "pretty", "evil", etc. In the same way, the physical description of 3.1 persons-per-room has different connotations for different people. What is crowded for one person may not be for another. Further, just as the description "a beautiful blue sky" loses something when presented in terms of characteristic light wavelength, to a social scientist, the concept crowded has meaning above and beyond density that is lost when only density is known.

This formulation derives from that of Severy, Brigham, and Schlenker (1976) and Wicker (1973). These writers propose that there is individual variation with regard to the perceived appropriate levels of manning for social environments. Further crowding can be thought of as the perception or attribution of an inappropriate number of people for the setting by virtue of having too many people present. Clearly, "too many" people can be 3, 4, 50 or any number depending upon the situation and what one is accustomed to. Consequently, population density simply is not sufficient (although necessary) to capture a human's perception or attribution of crowding. In other words, crowding is taken by this researcher to be an individual experiential state rather than a physical description. This distinction between density and the perception of crowding has important implications for any model of response to high density. And, it appears to this writer that most researchers recognize and share this distinction between density as a physical condition of limited space and crowding as an experiential state.

When one is raised in a rural farm environment, average inter-personal distance is quite high. This person, being used to having a great deal of space might believe even a group of five people in a room is quite high density. Alternatively, if a person has spent most of his or her life in one of the many truly large urban centers in this world, they are probably used to not having nor expecting much space. The thrust of this digression is that humans tend to adapt over time. One form of such adaptation is that of personal space requirements. Recall that much research has supported the general proposition that people vary with regard to the amount of space that they wish to have around them. Further, evidence suggests cultural, ethnicity, class, and rural-urban differences on this dimension.

At the same time adaptation to the two described environments involves the process of socially learning both what is normative and appropriate as regards the number of people involved in different social settings. If you spend most of your working life surrounded by many others, working in a large room with few others would not seem to be high density. Alternatively, if you expect to work alone, job settings with others present seem non-normative, inappropriate, overmanned, etc.

Finally, these double processes of developing expected or desired personal space and manning levels in social settings impacts upon the perception of crowding. As Stokols suggests, "crowding exists and is perceived as such by an individual when the individual's demand for space exceeds the available supply of such space" (1972, p. 75). Since this analysis delineates two ways in which demand can be understood to be an individual difference parameter, the perception of crowding in any situation must be highly idiosyncratic. Another way of coming to the same conclusion stems from Sundstrom's analysis of crowding. He posits that "crowding represents a sequence of interpersonal events in which high room density gives rise to either intrusions or interference" (1975, p. 138). Clearly, just how high the density has to be to constitute intrusion or interference also depends upon an individual's past history and his or her expectations about the manning of the settings involved. To summarize this first point, the perception of crowding is some function of density and the function varies for each individual based on past adaptation regarding both desired and appropriate personal demands for space or lack of intrusion.

The second concept to be delineated is that of situational control. Individuals' need, or perceived need, to feel that they are controlling, or will have an opportunity to control their destiny, their fertility behavior, their migratory behavior, or their input to social occasions is a theme which consistently surfaces in analyses of crowding. As mentioned earlier, Booth's epidemiological research led him to posit control as an important mediating variable. He particularly was concerned with child development as he viewed youngsters as having very little control. Mitchell observed little in the way of density related problems in Hong Kong except "when the individual's movement is restricted" (1971, p. 24); in other words, when the individual suffers a reduction in control. In another vein, Sundstrom's (1975) analysis of crowding involved intrusion and interference. Clearly, intrusion demonstrates a lack of control by way of an inability to prevent such interference.

On a psychologically oriented level, the concept of internal-external control (not the operationalization of it) has enjoyed much attention for a number of years and has been theoretically and empirically linked with most every social behavior imaginable. Included in this grand claim or indictment is research on control and fertility, control and personal space, etc. However, instead of conceptualizing control in its typical global fashion, the analysis of crowding would seem to suggest a second, more specific or situational version. An analogous formulation exists in the attitude-behavior linkage. Some find attitudes to be related to behavior while others do not. Those who do suggest that it is foolish to attempt to utilize global orientations in predicting specific behaviors only abstractly linked one to another. Instead, the more promising approach is to measure specific attitudes and attempt to predict the proximal dependent variable of behavioral intentions (e.g., Fishbein and Ajzen, 1975). For crowding, this means that attention should be paid to the concept of control as it connotates a perceived ability or inability to manipulate the amount of space available in a social situation. For example, does he or doesn't he feel that he could decrease the density in a social setting if he so desired? As argued throughout this paper, although dependent to a large extent upon the realities of the situation, this type of perceived situational control, as well as more global personal control, can easily be seen to vary among individuals.

The importance of the ability to feel that you can control or alter the amount of space available is that it influences the creation of stress. It is this writer's position that stress is created by the perception of crowding compounded by the perception of an inability to control or alter the amount of space available. This formulation delineates the way stress is generated in a number of other conceptualizations of crowding. For example, Sundstrom posits that "an individual's experience of stress depends upon (a) whether he or she can successfully reduce the effects of intrusion... through coping behaviors, and (b) how much effort is required" (1975, p. 138). Others such as Schopler or Worchel also suggest formulations of crowding in which stress is generated when interference by others prohibits one from doing whatever is desired. Such interference again demonstrates lack of situational control. Worchel and Teddlie (1976) buttress the above interpretation via an investigation wherein stress is avoided by virtue of requiring subjects to engage in cognitive activity which would serve to divert one's concern for lack of control - this even though objective density remains high.

The generation of stress is enhanced by three other aspects of the crowding situation. First, the perception of crowding implies an attribution of inappropriateness (either in the general sense or in the sense of Wicker's manning formulation). Inappropriate environments, particularly ones wherein the further perception is that one can not remedy the situation, are stressful in and of themselves. Second, a related phenomenon is that of a general tendency on the part of humans (at least as assumed in many theories of social behavior) to strive for simple structure, balance, or symmetry. The absence of such is hypothesized to be stressful and, as in the case of attitude change models, sets into motion a chain of events designed to dissipate the stress. Crowded settings are hard to conceptualize as being indicative of symmetry, balance, etc. Third, Brehm's (1966) theory of reactance has application to this analysis of crowding. If people strive for situational control, but the situation both frustrates that attempt as well as continues to offer less than desired space, Brehm would suggest that the needs and desires would become even more salient than they were in the first place. (If you cannot have something that you want, you want it even more.) In the case of crowding, such a process could only create a spiraling effect with stress increasing concommitantly - particularly so for those characteristically high in personal control in the global sense.

Returning to the concept of control itself, several last points are relevant. Control implies that people allow certain conditions to exist even though they may not be optimal. For example, even though you perceive a crowded situation, if you like or are attracted to the people involved, the generation of stress is diminished. The point is that control is not synonymous with action, rather the option to act. The other point will be delineated in a more comprehensive fashion later, but also has relevance to the concept of situational control. Simply put, if one cannot conceive of options, it is hard to feel that one is in control of the situation. Consequently, understanding the situation for what it is should increase the perception of situational control thereby reducing the creation of stress.

Turning to the third aspect of the conceptualization of individual variation in the perception of crowding and the generation of stress, state should be differentiated from status. A state is conceptualized as a temporary and reversible condition. Examples of states might be: sick, drunk, infatuated, drugged, scared, etc. Alternatively, status is conceptualized as a more permanent and irreversible demarkation. Examples of status might be: terminally ill, married, mentally retarded, Anglo-Saxon, etc. Implication for the response to the perception of crowding is obvious. The potential for stress increases in proportion to the extent that crowding is perceived to be more than a state - more than temporary. Even though a state of crowding may exist in the laboratory, in elevators, etc., not as much stress would be created as would exist in the family home or other more permanent environments. Sundstrom's (1975) analysis of short and long term stress in response to crowding has relevance here. Clearly, meaningful adaptation so as to avoid long term stress seems of paramount importance if the density situation appears permanent.

Adaptation

After a density situation has been perceived, after it has been interpreted as crowding with accompanying stress beginning to develop, response in order to alter the situation or adapt to it is necessary. In fact, the position taken here is that the potential for socially pathological behavior is associated with density or crowding only when there is a failure to progress completely through the above three phase process concluding with a socially acceptable coping strategy. The very ability to generate a successful strategy, however, appears to be another individual difference parameter and appears to have both cognitive as well as economic and other real world constraints.

Several investigators, such as Langer and Saegert (1977), have demonstrated that by giving subjects information about the effects of crowding and of the situation involved, negative effects of crowding can be diminished. They suggest that such information is a form of cognitive control of the situation. The point is that such information allows people to select from a broader range of potential responses. The more information one has, the more alternatives for response that can be generated. In an analogous fashion, Severy (1976, 1977) has argued that some think in terms of the interrelationships among spatial behavior, fertility, and household migratory decisions. Specifically, it was suggested that there are individual differences with regard to the cognitive strategies with which different people conceptualize population and the environment. It was proposed that there are individuals that could be termed ecosystem types. Namely, individuals, who on an abstract level, see interrelationships between environment, population, resources, etc. as indispensible and inseparable aspects of ecology and ecosystems. Alternatively, there appear to be individuals who keep these areas concrete and separate - non-ecosystem types. Severy and Atkins (1976) have demonstrated that the ecosystem types are cognitively more abstract and complex. Further, it was found that ecosystem types evince a relationship between their fertility behavior and their household migratory behavior. These people seem to understand the potential for crowded life situations and are cognitively adopting strategies to avoid such problems. It was speculated that it is the ecosystem type who needs to move, but is trapped (either psychologically or economically) who displays the kind of social pathology reported in some of the previous literature. Further, it was suggested that non-ecosystem types, on the other hand, might not be able to generate successful strategies for coping due to the inability to understand the problem at hand. The thrust of this speculation, of course, is that there are individual differences with regard to conceptions of "ecological systems" (crowding-migratory-fertility relationships) and that these cognitive strategies will have differential impact upon adaptive behavior.

CONSEQUENT RESPONSE

The following model attempts to integrate the pices from the previous literature and conceptual analyses in order to delineate a model of the relationship between density and consequent social behavior. It suggest that old density-stress formulations are too simple and too general, and that although more complex, the propositions delineated here have the potential for better levels of explanation. Characteristics and assumptions of the model are as follows:

1. People differ with regard to which density settings they feel are crowded.
2. People differ with regard to the feelings of power/control/opportunity to respond to the situation in which they find themselves.
3. People differ with regard to the way they conceptualize (or understand) the family size, residence size, and other environmental high density settings relationships.
4. People differ with regard to the perceived psychological, economic, and normative constraints permitting or prohibiting adaptive environmental or personal change, such as migratory behavior.
5. As a consequence, people differ with regard to the amount of adaptation versus stress that is generated in response to different density settings.

Given these assumptions, a variety of predictions would be possible. For purposes of simplicity, I shall concentrate on variables such as family size, migration, health, and delinquency. (As social planners, we need be concerned with negative effects and with areas requiring intervention programs. This does not deny the possibility that there may be positive effects of density.) To start, a number of permutations exist. Consider those individuals who feel crowded and are ecosystem types:

1. For these activists (high control individuals not perceiving normative or economic constraints), the theory is essentially similar to Wolpert's migration theory dependent upon environmental stress. When the crowding stress becomes too great, a move will occur.
2. For those who feel economically trapped, yet perceive that a state of normlessness exists (including potential access to the illegitimate opportunity structure), serious delinquency will obtain. This type of instrumental deviance might provide the "big score" that could be perceived as a way out of the trap. It can be hypothesized

that the above delinquency occurs just prior to the "solution" of a move (which would account for some of the high crime in high migration areas).

3. For those individuals who feel trapped and bound by norms from deviant (as well as legitimate) behavior, a retreatist response is more likely. Retreatist responses deemed pathological would be family disturbance, drug usage and abuse, and deleterious physical and mental health.

4. Similarly consider those who feel crowded, but are unable to conceptualize, for example, the relationship between the family size and residential space.

5. The ability to even envision a move or limit the number of children as possible solutions would appear to be minimal. In this case frustration might lead to either: cathartic delinquency (thrashing out and certainly not "problem solving" delinquency); a projection that it is society's problem (which, in turn, may lead to deviance designed to "punish" that society); or, again if bound by norms, a retreatist response.

A number of other predictions are possible for those who do not currently perceive crowding although they are in high density settings.

6. For those who do not feel crowded there should be minimal press to move. Further, any engagement in pathological behavior would be due to explanatory schemes other than "density-stress" models.

7. For those who do not feel crowded, feel constrained from moving, but feel that additions to the family would overcrowd the residence, fertility intentions would be limited.

Consequently, the model proposes, in a logical fashion, that we should not expect density to result in social pathology for everyone. The model presented here accounts for individual behavior - not for the institutional response. In conclusion, I have argued that human response to high density, the perception of crowding, the generation of stress, and adaptation or social pathology can be conceptualized as influenced by previous adaptation and social learning regarding space desires and normative manning of settings, situational control, the potential permanence of the crowding, and the ability of the individual to conceptualize or understand the situation and generate and implement ameliorative strategies.

A SYSTEMATIC APPROACH TO HIGH DENSITY SETTINGS: PSYCHOLOGICAL, SOCIAL, AND PHYSICAL ENVIRONMENTAL FACTORS

Susan Saegert

City University of New York Graduate Center, U.S.A.

ABSTRACT: The difficulties high density are viewed in light of behavioral constraints encountered and the experience of cognitive overload. These concepts are explained and several field experiments are reported supporting the importance of each separately and in interaction in determining responses to density. Perceptions of crowding stress spontaneously reported in an interview study of urban stress are then described; different place-related patterns of crowding stress and coping emerge from this analysis that confirm and extend the previous concepts. Finally several studies of high density residential environments are presented that lead to a more strongly ecological theoretical perspective.

INTRODUCTION

The investigation of the human consequences of crowding continues to be a controversy-laden area of research, full of significant and insignificant findings and burgeoning with theories. The whole topic of crowds and crowding fits into a long tradition of concern among social thinkers from the time of the industrial revolution and the great political and social changes that accompanied it (c.f. Fischer, 1976). Early writers such as Le Bon (1895) and the sociologist Simmel (1905) and Wirth (1938) have viewed with alarm the expanding size of basic social institutions and agglomerations. With this background of emotional theorizing, Calhoun's research demonstrating the disastrous effects of crowding on caged laboratory rats (c.f. Calhoun, 1962) lead to the expectation on the part of many researchers that similarly negative consequences of high density experience would result for humans. However, studies of human crowding did not clearly support this expectation, nor did they clearly contradict it.

A substantial body of research has accumulated suggesting that some aspects of high density experiences and living conditions can provoke stress as well as anti-social behaviors and attitudes. Studies of individuals' physiological reactions to high density situations indicate that density is positively related to increased physiological arousal and even high blood pressure (D'Atri, 1975; Aiello, Epstein and Karlin, 1975; Saegert, 1974). Prisoners living in large groups are more prone to become ill (McCain, Cox and Paulus, 1976). Valins and Baum (1973) and Bickman and his colleagues (1973) have found that dormitory residents exposed to excessive interactions with others or living in higher density dormitories felt more withdrawn, less friendly and more crowded; further, they behaved less cooperatively and in a less socially responsible manner. Children in higher density play groups have been observed to become more aggressive and withdrawing (Hutt and Vaizey, 1966). Murray (1974) reports that children raised in higher density conditions tend to be seen as more aggressive, impulsive and extroverted by their classmates. In both laboratory and field studies, adult subjects have expressed more feelings af anxiety, aggressiveness and discomfort in high density settings than other subjects who were in lower density conditions (Griffitt and Veitch, 1971; Saegert, Mackintosh and West, 1975). Task performance has also been shown to be negatively affected by high densities (Paulus et al., 1976; Saegert, 1974; Saegert et al., 1975). Analyses of aggregate data have revealed strong relationships between area density and mortality, delinquency, illegitimate births and divorce (Levy and Herzog, 1974; Schmitt, 1966). Other investigators have suggested such relationships between interior dwelling unit density and signs of pathology (Galle, Gove and McPherson, 1971).

These studies and others like them are numerous enough and careful enough to prevent their dismissal as chance findings or methodological artifacts. Yet there are many contradictory findings that also cannot be ignored. Other researchers have found no task effects (Freedman, Klevansky and Ehrlich, 1971), no mood effects (Stokols, Rall, Pinner and Schopler, 1973), and no important independent correlates of area or dwelling unit density (e.g., Schmitt, 1963; Mitchell, 1971; Winsborough, 1965). It is my impression that fewer studies of individuals have failed to show differential reactions to high and low densities. It also appears that many aggregate data studies have focused on settings in which the range of densities investigated was narrow, with both "low" and "high" density conditions falling in the low range of absolute density (e.g. Welch and Booth, 1973; Michelson and Garland, 1974). In other studies high density and economic conditions have been inextricably intertwined (e.g., Winsborough, 1965). The strong independent effects of area density in Levy and Herzog's recent study of the Netherlands are particularly significant because the levels of density studied were rather high and were not confounded with poverty.

In summary, methodological criticism, of both studies reporting effects of high densities and those reporting none are possible. It appears that high density situations can and do exist that do not cause stress, discomfort or anti-social relationships among people and that in some situations, these responses are found. The divergence of results in studies of the effects of high density naturally leads to an attempt to delineate which aspects of the situation are associated with which outcomes. My own research has focussed on two possible problems a person in a high density setting is likely to encounter: behavioral constraint and information overload (Saegert, 1973; Saegert, 1978).

Behavioral Constraint

Some of the negative experiences people have in crowded settings are so obvious that they seem not to require comment. These include being pushed and shoved, not having enough room to move freely, being excluded from access to certain resources, foregoing some activities because the presence of others some how interferes with them and so on. These impediments can lead to frustration, negative affect and stress, especially if they are frequently encountered. In some ways the negative consequences of these experiences are so well known that they have been largely neglected by researchers and theorists, with some exceptions (c.f. Heller, Groff, and Solomon, 1977; Saegert, 1973; Stokols, 1972; Schopler and Stockdale, 1977). Yet in making policy and design decisions, such factors are crucial.

Because of the immediacy of behavioral difficulties, it is appealing to argue that behavioral constraint, in and of itself, is sufficient to explain any negative consequences that may be associated with high density settings. Yet this approach leaves open a wide range of questions related to which conditions people experience as constraining, which environmental parameters contribute to or ameliorate such experiences, and how behavioral goals and strategies develop in settings. The problems of cognitive overload are seen as related to these issues. In the following section, particular behaviors, goals, and constraints will be examined as they may influence cognitive overload. Unlike interference with or frustration of behavioral goals, it is my view that cognitive overload is not necessarily experienced as unpleasant or stressful. However, being in a state of overload limits one's range of behavioral options which makes one more vulnerable to interference and frustration.

Information Overload

The concept of information overload has implicitly or explicitly shaped speculation about the impact of high-density human settings, and even of large-scale organizations in which face to face contact does not occur. Le Bon (1865) believed that mere partication in large groups such as parliaments or mass political parties somehow induced people to take leave of their normal, individual common sense and become prone to simple-minded judgements, tyranny, propaganda and incitement to violence. Simmel (1905), while making somewhat less apocalyptic claims, held a similarly negative view of cities based on the social isolation, distrust, and anomie he thought arose from the dense aggregation of large numbers of diverse individuals. In his essay "The Metropolis and Mental Life, " he identified the essential concept of information overload when he stated that "The psychological basis of the metropolitan type of individuality consists in the intensification of nervous stimulation which results from the swift and uninterrupted change of inner and outer stimuli (Simmel, 1905, p.48)." This line of thought was revived by Milgram (1970) as a basis for understanding city life and has been extended to an analysis of the effects of density (c.f. Rapoport, 1975; Saegert, 1973, 1977) and of other potential environmental stressors (c.f. Cohen, 1977).

I have discussed the concept of information in more detail elsewhere (Saegert, 1977). Here I will briefly state what appear to be its most important aspects for understanding the consequences of high densities. First, the assumption is made that people normally attend to only a certain amount of information whether because of limited attentional capacity or because they employ limited-capacity attentional strategies. Data to be discussed later generally support the assumption of limited attentional capacity but also suggest a certain cognitive flexibility that can either reduce the experience of overload or minimize its effects (Langer and Saegert, 1977).

Excessive demands on attentional capacity are created by intense, upredictable, uncontrollable or simply extremely numerous environmental events. Thus large numbers of other people are seen as one potential source of overload: not only do they create an information field of numerous elements, but they also may often be unpredictable and uncontrollable.

Cognitive overload occasioned by high densities is not simply a matter of excessive stimulation. This may or may not be present, depending on the physical qualities and behavior of people in the setting. Furthermore, people usually habituate to sheer stimulation, making it therefore a rather minor contributor to information overload (Saegert, 1975). In contrast, a surprising pattern of human responses to high densities seems to be emerging in which some effects actually become more pronounced over time (D'Atri, 1975; Aiello, Epstein and Karlin, 1975; McCarthy and Saegert, in press; Saegert, 1974).

Secondly, other overload effects associated with high densities and/or large numbers of others do <u>not</u> seem to require the actual face to face experience of those others (c.f. Baum and Greenberg, 1975; Baum and Koman, 1976). Conversely, perhaps the lack of response to high densities found in some studies reflects the fact that participants' activities and expectations or the environmental configuration were such that little attention to others was required in the situation.

Maximal experience of cognitive overload is predicted when participants in the setting are actively engaged in seeking information about the environment, when they must physically move through and interact with the setting, and especially when the behavior of others in the situation is difficult to predict or control. Rapoport (1975) has provided a thoughtful and subtle analysis of the qualities of socio-physical environments that are likely to create such conditions or to give the impression that such conditions might occur. Because he has done such a thorough job of discussing the environmental conditions related to high densities in which people are likely to be presented with more information than they can handle, I will not further pursue this topic. However, I concur with his conclusion that "It is essential to consider in detail and to a high degree of specificity, the relationship of given socio-cultural groups to traditional density figures (people per unit area), the relationship of the particular area to the larger context, the specific activities taking place and their meaning, the detailed layout and design of the setting in terms of privacy..., the facilities available, the social characteristics of the area in terms of life style, homogeneity, the social rules available and used, and so on..." (Rapoport, 1975, p. 153).

One quality of high density settings that seems more likely to lead to overload than others is the presence of large numbers of other people because they are sources of more potential information and interference, as well as creating greater coordination demands (Saegert, 1973; 1978). Limited space availability is likely only to be extremely salient under conditions of confinement or when one desires to engage in activities that physically require more space. Limited space however may raise the impact of others present. When large numbers of others are occupying a rather limited space, problems of overload, interference and coordination are expected to be exacerbated. In most of the studies to be described, high density environments are characterized by relatively large numbers of people occupying bounded spaces in which high density conditions arise from an increase in number of people occupying the setting.

The following sections of this paper will begin by examining some research on the cognitive consequences of high density settings as well as some differential effects of the cognitions people bring to these settings. Then the question of more serious consequences will be addressed as well as the relative contribution to these of different conditions and processes.

CROWDING AND COGNITIVE PROCESSES

A fundamental question for the theoretical perspective being presented is whether or not people really do experience cognitive overload in high density settings. If so,how is this overload handled and what are its consequences? Research conducted by myself and my colleagues provides a number of different types of evidence that such overload does indeed occur, that people deal with it by attending to information most closely related to their tasks and to information that is most visually differentiated, as well as by the use of other coping strategies. Task demands and cognitive set also are found to influence the other consequences associated with such overload.

A study of people's cognitive performance in a department store under crowded and uncrowded conditions (Saegert, Mackintosh and West, 1975) revealed that performance and memory for descriptions of a specified number of objects (shoes in the shoe section of the store) were not affected by crowding. However, crowded subjects seemed unable to remember much about the rest of the physical environment. Their maps of the area they had been in, drawn soon after the completion of the focal task, were less accurate and complete than those of subjects who were exposed to less crowded conditions. In this study, high densities did not interfere with the task subjects thought to be their major responsibility. Those who were crowded and those who were not reported similar affective responses.

A second study was then carried out in a train station (Saegert et al, 1975) in which subjects were given a task that required more active transactions with the environment and which was thus more likely to be interfered with by both cognitive overload and behavioral constraint. In this situation, subjects performed more poorly on the focal task when they were crowded. In addition crowded subjects expressed more negative, anxious and anti-social feelings. Two interesting interactions between density condition and sex of subject also occured. As in a number of other studies (c.f. Freedman et al, 1972), crowded males expressed more aggressive feelings whereas crowded females did not. A post-experimental test (The Stroop Color-word Test) for task performanence decrements as an after-effect of crowding was administered. On this measure, females who had been in the high density condition performed more poorly than previously uncrowded females, but males who had been crowded did better than other males. There was a significant though modest positive correlation between the aggression scores and Stroop performance. These findings seem to suggest that the ways in which men and women handled the density situations were different, both affectively and cognitively.

We have recently replicated the results of the first study (this time in a supermarket), indicating that people can recall less about the environment after having experienced it in crowded conditions (Love and Saegert, Note 1). In addition, we found that our subjects were less able when crowded to calculate correctly which product would be most economical. In this study, the task was designed so that behavioral interference in the calculation task did not occur, but scanning of the environment was necessary. Even without behavioral interference, task decrements were accompanied by negative feelings and evaluations of the setting and its occupants by crowded subjects. Of course this does not rule out the possibility that our crowded subjects suffered from behavioral difficulties encountered in moving about the store between tasks. Our procedure of leading subjects to the location of the array of products to be evaluated did, however, minimize physical interference.

Not only did we find performance decrements for crowded subjects on both the main task and in incidental learning of the environment, but we also obtained data that suggests that the environmental information retained by crowded subjects differed in a systematic way from that recalled by uncrowded subjets. The environmental recall task consisted of asking subjects to draw a map of the grocery store and to locate certain products where they belonged in the store (we had made sure that the paths subjects followed passed by all products on the list). While crowded subjects made more mistakes in locating all types of items, they made disproportionately more errors on items displayed in the usual shelf-fashion than on those with distinctive visual displays, such as a special refrigerated case for milk. These data seem to indicate that crowded subjects construct a sketchier image of the environment in which only the more distinctive features tend to be retained, although even these are sometimes lost. These map data are particularly interesting because subjects in this study were regular users of the grocery store.

Subjects' estimates of the number of people present also give another indication of lack of cognitive clarity among crowded subjects. We found that high density subjects were vastly more inaccurate in estimating the numbers of people in the store, estimating an average of 30 more people than were actually present; low density subjects under estimated by, on the average, about four fewer people.

In another attempt to look at cognitive involvement in the effects of high densities, Ellen Langer and I (Langer and Saegert, 1977) provided some subjects with information about the fact that the environment (again a grocery store) might become crowded and that this sometimes made people feel anxious and aroused. This information was given to half the subjects in the high density condition and to half of those in the low density condition. They were then handed a grocery list and told to find each product, compare prices and quantities, and record the most

-----

Note 1: Saegert, S. and Love, K. manuscript in preparation concerning task effects and mental images of those in high-density settings. Environmental Psychology Program, C.U.N.Y. Graduate Center.

economical choice of brand and quantity for each item. More items were listed than could possibly be completed in the thirty-minute session. After the task time was up, subjects were then asked to rate the store, the degree of crowding, their own feelings, the difficulty of the task and how much they thought other shoppers interfered with them.

As in the previous study, high density subjects were less economical in their choices and more negative on all affective measures. There was also a main effect of information on task performance and on most of the ratings such that informed subjects in both density conditions did better than the uninformed subjects in that density condition.

Generally, low-density informed subjects performed better and were more positive than low-density, uninformed subjects who in turn did better and felt better than high-density, informed subjects with the high-density, uninformed group coming in last. Selected contrasts (McNemar, 1969) revealed that the experimental intervention brought the high-density subjects' responses to the same level as those of subjects in the low-density, uninformed condition. The same comparisons between uninformed subjects in the two density conditions were significant at the .01 level for all measures except satisfaction with the store and ease of finding items. Thus we can conclude that the information did significantly ameliorate the consequences of the high-density setting.

Two mysteries nonetheless remain in this study: why did low-density subjects also benefit from the intervention and how exactly is the information provided used by the subject to improve task performance and to make themselves feel better. With respect to the first question it is quite possible that people in New York City expect that a grocery store may become crowded at any time, even if it is currently not crowded, so they act somewhat like crowded subjects. Or it may be that obtaining potentially relevant information about the environment and one's relationships to it evokes a more efficient set of strategies for dealing with the environment, which also leads to a more pleasant experience.

It is much more difficult to plausibly explain how the information we gave subjects could be used by them to improve task performance and affective responses. The findings may bear some resemblance to those of studies of dissonance reduction. It is equally possible that the receipt of the information cues off in the subject a set of behaviors and cognitive strategies that are more suited to the environment and the task at hand. In this case the information would be seen as a warning signal. In either case a more detailed understanding of what subjects actually do and how their cognitive processes work would be desirable.

In summary then our research findings from these various studies are consistent with the hypothesis that high-density situations present a surfeit of potential information to the setting occupant. When transactions with the environment involve scanning, movement and coordination with others, this potential excess of information is experienced as cognitive overload and is accompanied by negative affective states. However, it also appears that people's motivation and their cognitive and behavioral strategies play an important role in determining the extent to which cognitive overload occurs and what its consequences will be. McClelland and

Auslander's findings (in press) that number of people in a setting more strongly predicts ratings of crowdedness than does space available and that crowdedness and pleasantness ratings are negatively correlated only in shopping and work settings further support the overload analysis. They also emphasize the importance of tasks in determining the consequences of crowding.

CROWDING STRESS AND SETTING

Thus far the studies reported, while they were conducted in field settings, required the participants to engage in tasks set by the experimenter, thus limiting the range of responses that could be observed and also perhaps inducing difficulties that people normally do not experience. Furthermore, we do not know whether the experience of crowding in very different settings leads to different consequences. Certainly the analysis thus far would suggest that cognitive overload would be most salient in task-oriented settings where the activities engaged in require relatively high levels of information processing and when crowding involves encountering a large number of other people.

An interview study of people's perceptions of urban stress helps us address these issues (Roberts and Saegert, note 2). Of the 80 residents of Manhattan questioned in an open-ended manner about urban stress, all but four mentioned crowding as stressful. Distinctive patterns of crowding experiences for different settings emerge from their accounts. They also reported a variety of definitions of crowding, responses to it, reasons for considering it stressful and coping strategies.

Crowding stress was most frequently reported in public settings, particularly on public transportation, in streets and traffic and in service facilities such as stores, agencies and entertainment places. In this sample only seven people mentioned crowding as a problem in their homes. This smaller number probably reflected sample characteristics. Most respondents were middle class and lived alone or with only one other person. A moderate number of responses involved a general feeling of crowding in the city as a whole.

Crowding in public transportation situations arose, of course, from both the presence of large numbers of others and the limited amount of space available. Predictably, major complaints centered around being pushed and shoved and being too close to other people. These situations, and other high-density settings characterized by many people in a small space, elicited the strongest emotions and the most mentions of physical reactions. Almost all reports of aggressive responses to crowding occurred for transportation settings as did the preponderance of descriptions of coping by fantisizing and ignoring the situation. Fewer coping strategies involving rational problem solving were mentioned for dealing with crowded public transportation than for coping with other crowded settings.

Service facilities were seen as giving rise to crowding stress due to the sheer

----

Note 2: Roberts, C. and Saegert, S. Crowding in the Big City: manuscript in final revision in fall 1977.

number of people who used them, as were stituations involving automobile traffic and parking. Being delayed and waiting in line were given as the reasons why these settings were stressful when crowded. In respondent's descriptions of crowding stress in which the large number of other people present was the salient dimension, people complained of feeling that extra demands were being made on them. Being with large numbers of people in any settings was also more likely to lead to feelings of alienation and loneliness. Most mentions of feeling confused, mentally overloaded and so on were also related to exposure to large numbers of people. Fewer people reported coping strategies for high number situations than for those involving both limited space and high numbers, perhaps becauce the former here were less stressful. Avoiding using service facilities that tended to be crowded was the most commonly described way to eliminate this kind of crowding stress, although some respondents did then resent the additional constraint this placed on their activities.

While patterns of crowding stress typical of different kinds of settings did appear, there was also some overlap in descriptions of crowding stress. The more common reasons for why crowding was stressful were given across many different settings. Some reasons also were given about equally in all settings. Table 1 lists these reasons and the frequency of mention for each one.

There are also some general ways in which respondents' reports of coping with crowding stress differed from those concerning other stresses such as noise, crime and so on. People were more likely to avoid crowded settings and less likely to report active coping strategies. Interestingly, respondents were more likely to state that there was nothing they could do about crowding stress than to say this about all other stresses combined.

The findings of this study indicate (a) that people do indeed experience crowding as stressful, (b) that large numbers of people are most often associated with stressful high-density situations, (c) that people are sometimes aware of cognitive overload as an aspect of crowding stress, especially when sheer number of people is a salient component of their experience. Furthermore, the settings I and my colleagues have employed in our studies are ones in which crowding stress frequently occurs. In these ways then the hypothesis and studies presented in the earlier part of this paper appear to be of relevance to people's everyday experience of crowding.

TABLE 1: REASONS FOR CROWDING STRESS REPORTED BY RESPONDENTS

	Frequency	Percent
Feeling blocked or delayed	89	.21
Experiencing extra demand	79	.19
Not liking to be pushed and shoved	59	.14
Being irritated by unpleasant strangers	40	.09
Being too close to others	36	.08
Disliking the physical environment	30	.07
Feeling alienated or alone	28	.07
Fear of physical harm	26	.06
Not liking own reactions	16	.04
Feelings of spatial confinement	15	.03
Needing to be alone	5	.01

Interference and behavioral constraint are clearly part of most of the respondents perceptions of why crowding is stressful. While there are some signs that they also experience cognitive overload, this is not frequently named as a cause of crowding stress. This in itself does not necessarily mean that it isn't. As Nisbett and Wilson (1977) have persuasively argued, cognitive processes are often not verdically available to introspection. But it does raise some questions about where cognitive overload fits in the total pattern of people's experiences of high-density environments.

CROWDING AND SOCIAL RELATIONS

In thinking about this question it is necessary to keep in mind the inherently social nature of density. High densities involve large numbers of people in some proximal relationship to each other, not boxes or visual stimuli or noises. Each person thus must be a part of the environment for others and at the same time subject to the effects of the high-density setting. Coping strategies of one person may affect the success or failure of different or similar strategies employed by others.

Because of this and because of the usually great amount of meaning attached to people as compared to other kinds of stimuli, information overload occasioned by high densities is likely to have some consequences that differ from those of, for example, noise. Much of the power of Milgram's (1970) overload analysis grew out of his linking the large number of people in a city to a tendency of urbanites to ignore people they did not know, to refrain from offering them aid, and to keep transactions with them as brief and focused as possible. This is interesting to us exactly because people seem to be responding to others as if they were not the compelling stimuli we normally think them to be. Simply ignoring a noise is considered to be habituation and apparently is a common, adaptive phenomenon when people are exposed to excessive noise. Ignoring other people is social withdrawal. Similarly, removing distracting or offensive visual stimuli when one is trying to concentrate is a reasonable thing to do, but forcibly removing distracting or offensive people can be a crime. Sources of cognitive overload other than people are primarily problematic if they lead to discomfort or stress. Crowding may result in negative consequences for people not only when it directly leads to cognitive interference, physiological upset, or negative feelings, but also when coping with it involves behaving in a way that violates important social and personal values or prevents the establishment of desired social relationships. The trade-offs required in this adaptation may help explain lack of habituation to crowding and the tendency sometimes found for reactions to become stronger over time.

Groups and institutions may formally or informally organize social life to mitigate the effects of high densities or numerous participants, but that in itself may demand that certain values be given precedence over others. For example, the need to attend to other people can perhaps be reduced by limiting their behavioral freedom. The value consensus and functional relationships mandated by these organizations can be expected to reflect the constraints on information processing capacity and to provide simplified means of coordinations and ways of limiting the affective and attentional demands of large numbers of people. Fischer's (1976) account of urban life emphasized the value differences commonly associated with population size. Elsewhere, I have also considered the value and functional implications of the overload problem for social organizations (Saegert, 1978).

The crucial significance of social values and relationships in determining the consequences of high density can be seen when one looks at the studies of Baum and Valins in university dormitories (cf. Baum, Valins and Harpin, 1975; Valins and Baum, 1973) and in several studies my colleagues and I have conducted of high-rise residential buildings (McCarthy and Saegert, in press; Mackintosh et al., note 3). In the settings studied, architectural features of the physical environment, management policies, and trust and friendship among occupants mutually shape the consequences the large number of setting occupants have for each other.

Baum and his colleagues compared perceptions of crowding, of unwanted social interaction, and of the quality of social relationships in two university dormitory settings, one designed so that residents on a corridor shared all facilities except sleeping quarters and the other in which four or six students shared facilities in a suite arrangement. Corridor residents were found to feel more crowded, to like their neighbors less, to know less about each other, to prefer individual ways of solving problems to group solutions, and generally to withdraw more from social contact (Baum et al 1975; Valins and Baum, 1973). When in the second year of the study some groups of students had arranged to live on corridors occupied by friends these effects did not occur for them, while they continued for other corridor residents. Thus it appears that the suite design provides a manageable group size for students and prevents the sense of being exposed to unpredictable and uncontrollable interactions that characterized the reports of corridor residents. The ameliorating effect of friendship groups in this environment suggests that these students may not have felt as burdened by the need to attend to the possible behaviors of their friends and probably less concern about their own ability to control these interactions.

The question of why friendship has these effects is theoretically a very interesting one. Baron and Rodin (1977) have provided a rational for this by stating that one has a greater sense of physchological control when among friends. This is, I believe, true, and they offer some interesting and useful ways of thinking about the relationship of crowding to control. But it begs the question in the same way that the cognitive overload analysis does when one hypothesizes that friends' behavior demands less attentive monitoring and easier coordination. Further, the word control in their analysis of the situation could as well be replaced by the word "trust." This replacement would serve to emphasize the important fact that people can be supported and enriched by the social and physical environment as well as stressed by it.

What seems to me significant about the presence of friends in a high-density environment is the greater freedom it seems to give people to psychologically extend themselves. As Fischer (1976) has pointed out, the aggregation of large numbers of people in a geographical locale tends to provide a much wider range of possibly satisfying relationships, even for people with statistically infrequent personal and social characteristics. Yet I would argue that for these possibilities to become realities, the immediate life space of the person, that is the environments in which he or she

----

Note 3: Mackintosh, E., Olsen, R. and Wenworth, W. The attitudes and experiences of the middle income family in an urban high-rise complex and the suburbs. Manuscript in preparation. Environmental Psychology Program, C.U.N.Y. Graduate Center.

lives, works, travels, shops, etc., must be physically and socially organized in such a way that the person can come to identify and in some way "know" a significant set of the people encountered daily. Otherwise one is in the position similar to someone needing water but having a leaky pail in a rainstorm. Two studies conducted in our reseacrh program will illustrate this point.

One of these investigations was conducted in a low-income, public housing project of the City of New York, located in the Bronx (McCarthy and Saegert, in press). The tenants were assigned to apartments on the basis of availability of units and family size. In one of the two types of buildings, either three-story walk-ups containing 12 families or fourteen-story towers with elevators housing about 110 families. Two thousand families lived in the project's 53 three-story buildings and 12 fourteen-story high-rises.

Because of several factors, this situation was expected to be one in which residents would have great difficulty in establishing secure and meaningful social relationships, and also to be a case in which such relationships would be particularly crucial for the conduct of life. Like most low-income people, those residents had little opoprtunity to escape from the environment which, as Rainwater has described (1966), tends to be pervaded by an atmosphere on uncertainty, fear and mistrust. Thus the need to monitor behavior, and to be aware of potential encounters is high. Furthermore, the random assignment of resident to apartments does not usually allow them to locate near others whom they already know or who are from similar racial and ethnic backgrounds.In this context, we found that the patterns of social relationships that developed in the smaller buildings were strikingly different from and much more benign than the social withdrawal and mistrust that characterized high-rise residents'relationships.

Our analysis showed that residents did not differ on any demographic variables other than family size. Somewhat larger families lived in the lower buildings because each walk-up had a couple of larger apartments. However when family-size was statistically controlled in our analysis our results remained the same.

Overall, comparisons of living experience, of high - and low-rise tenants revealed that residents of the fourteen-story buildings were more likely (1) to report experiences of social overload and crowding, (2) to feel a weaker sense of control, privacy and safety in various interior spaces of their buildings, (3) to experience greater difficulty in their social relations, and (4) to feel more alienated and detached from their own building and the project as a whole, as well as less satisfied and involved with them.

Cognitive overload seems to be one part of the high-rise residents'experience in that they report seeing more people generally in the elevators and lobbies and more people whom they did not know. These reports were strongly related to perceived building crowding and to reports of trouble among tenants, as well as to a feeling of not knowing people well enough to even say hello to them.

The measures of control, safety, and privacy were strongly correlated with each other and very different in the two types of buildings. While for residents in both groups this cluster of feelings was strongest in the apartment, moderate within the floor the resident lived on, less in the elevators, stairs and lobby, and least in the outdoor areas adjacent to the building. There were no differences between residents' feelings of safety, privacy and control within their apartments. However their average ratings on these dimensions of all other spaces did not even overlap; that is, the most public areas in the low-rise buildings were rated as more safe, private and controllable than were the least public areas in the high-rises.

In the light of continued construction of high-rise facilities for low-income families in many areas of the world, our most disturbing findings concern the quality of social relationships in the larger buildings as compared to the small ones.Low-rise tenants felt reasonably confident that they could count on others for small favors, help in emergencies and mutual aid in case of attack or vandalism. These expectations were much weaker among high-rise residents. Some of them explained that people only help those whom they know, and no one knew any one else in their building.

We also have strong indications that the social anomie experienced by residents of these high-rise buildings extended outward to their perceptions of and activities in the larger environment. High-rise tenants were likely to state that they felt they belonged just to their apartment, whereas a sizeable proportion of low-rise residents said they felt they belonged to the project as a whole. Tenants in the smaller buildings also thought that they could have more effect on management decision. In the large buildings, respondents not only reported visiting less with other project residents, but they also ventured beyond the project less frequently to socialize with friends and relatives outside the project. (There were no differences however in the reports of number of friends or relatives within the area). More low-rise respondents reported belonging to voluntary organizations such as churches, social clubs, unions or tenants groups. This difference was strongest when we looked only at membership in politically oriented groups such as the National Association for the Advancement of Colored People and the Tenant's Organization (25% in high-rises vs. 60% in low-rises).

These findings were so strong, and we think significant, that we are presently in the process of analyzing data we have collected from several other public housing projects including one containing 30-story buildings. We feel that the generality of findings like these must be empirically investigated.

Is the implication of this research, then, that high-rise buildings or other building designs that increase numbers of people encountered have negative consequences for their occupants? While I am often tempted to believe this is true, influenced probably by aesthetic criteria as much as by these data, other research we have conducted leads me to qualify this conclusion and to emphasize once again the systemic nature of high-density environments.

As part of the same research program, Elizabeth Mackintosh became interested in effects of high-rise housing on child-rearing. One way this problem was approached was to investigate a middle-income high-rise development that was considered by many to provide one of the best family residential facilities in New York City (Mackintosh, Olsen and Wentworth, note 3; Mackintosh, note 4). Management selec-

----

Note 4: As her dissertation research Elizabeth Mackintosh has extended the

tion of tenants has been notoriously stringent, favoring white, well-educated middle class families. The development has a record of excellent maintanence and responsiveness, as well as a history of racial exclusion and difficulty with the lower-income community that surrounds it. While the focus and method of this study were different from those of the previously described investigation, some interesting comparisons can be made.

This middle-income development is extremely similar in appearance to many public-housing projects in New York City, and to the high-rise sections of the project investigated in the Bronx. Most respondents interviewed commented negatively on this aspect and often remarked on the beehive-like sameness and massiveness of the physical structure. Yet the experiences of these residents were usually quite different from those in public housing. Most of them found the development to be a very satisfying place to live especially because of the social relationships that were available to them as well as the good maintenance and easy access to the amenities of city life. The respondents who all had young children found that they had much in common with the numerous other young parents in the development, and that the opportunities for their children to find playmates were an asset. The mothers of young children were especially pleased about the network that developed among themselves, which abounded in playgroups, babysitting arrangements, cordial encounters and significant friendships.

Their responses were compared with those of former residents who had moved to suburban settings. It was found that the women in these families often longed for the social relationships that they had previously experienced in the development. It appears that people, such as mothers of young children, are forced to spend a good deal of time in their residential environment, and when the others who occupyit are perceived as similar and trustworthy, then a high density of people in the immediate area provides both more opportunities for and diversity of social contacts. While the suburban families were very happy with the effects of the new home on children's freedom and both children and husband's opportunities for outdoor activities, a significant proportion of wives experienced loneliness,feelings of isolation, and lack of options for personal fulfillment.

Even from the women's perspective however, the situation was a complex one.In the city development, crowding in the apartment was seen as a serious problem resulting in family tension and friction. Again the wives were especially exposed to this difficulty since they were inside the apartment with the children for a large amount of time most days. Furthermore, if children were to be allowed to go outside, mothers accompanied them until they reached the age of eight or 10. In the suburbs, the greater living space in the house plus the ease of access to the outdoors meant that mothers did not constantly have young ones under foot and that making the children out ceased to be a demand on their time.

In looking at these studies of high density-residential environments the following conclusions seem warranted:

----

comparisons referred to in Note 3 to include additional cities with different building structures and sites properties.

(1) When groups of homogenous or familiar and trustworthy individuals exist in high-density environments, there may be few or no negative consequences of the situation.

(2) When people enter without the company of such people into environments that create large numbers of actual and potential encounters, people will tend to withdraw and may develop mistrust of their fellow residents.

(a) When residents have other reasons to distrust or be fearful of their fellow residents, this kind of social overload can lead to anomic, apathetic social atmosphere.

(b) When residents trust the management of the residential complex, and feel common bonds with the other people, or when other aspects of the context contribute to residents' perceiving the environment as benign, trustworthy and a potential source of satisfactions, the effects of high densities will tend to be ameliorated, and may even be positive.

THE ECOLOGY OF HIGH-DENSITY SETTINGS

The various studies discussed lead me to look at the original concepts that have organized my research on crowding as being overly focused on describing individual psychological processes and insufficiently rich in environmental concepts. While I consider the data presented as supportive of the importance of behavioral constraint and cognitive overload in determining the effects of density under various conditions, it does not begin to systematize the qualities and processes of the socio-physical environment that create such conditions. Hopefully, the recent studies of residential developments and of perceptions of urban stress will be a beginning step in this direction. Winkel (1978) has pointed out the need for such an approach and suggested different criteria that must be employed in selecting theoretical concepts, measurement tools, and analytic procedures.

Recent studies have served to reemphasize the transactional nature of the person-environment relationship (Ittelson, Proshansky, Rivlin and Winkel, 1972). If we are to understand and, within certain limits, predict the effects of high densities on people in different settings we must develop a more thorough analysis of the following characteristics of situations:

(1) tasks and activities
(2) goals, values, norms
(3) physical configuration
(4) social organization
(5) place of the setting in inhabitant's life space

This last problem has been approached somewhat by Stokols (1976) in his attempt to distinguish the impacts of crowding in primary and secondary environments. Our research on residential settings suggests that this distinction is important but that it must be expanded to address the duration of time spent in the environment as well as the range of primary and secondary environments that make up the person's lifespace.

In summary, it is my impression that the question of the human consequences of high densities should be rephrased to ask what are the most significant human ecological dependencies in different types of high-density settings.

SPATIAL DIMENSIONS OF HUMAN CROWDING

An Analysis by the Density Gradients for Ankara, 1965-70

Özcan Esmer

Middle East Technical University, Ankara, Turkey

ABSTRACT: Crowding, density and intensity are highly relative in their meaning and lack diagnostic values unless they are applied to their related spatial contexts. This paper proposes that these terms should be considered within the general concepts of population distribution and spatial organization. The pattern of population density for Ankara between 1965-70 is analysed by the gradients of the negative exponential function of C.Clark which is a good approximation to actual intra-urban population distributions. Results show that Ankara is not very different from the B.Berry-type Non-Western model with a high population concentration at the center and a sharp density gradient declining also by time.

INTRODUCTION: A PERSPECTIVE OF SOME DEFINITE PROBLEMS

Urban planners frequently encounter concepts like crowding or density which are ambiguous in their meaning. The ambiguity of these terms seems to have two technical sources: the kind of units involved in definition of these terms; second, the way these units are related to each other. Furthermore, for planning purposes, questions revolve about the prescription of the right or optimum level of crowding or density for a given urban area, and the determination of the causal relationship among the dependent and independent variables. All these questions have far-reaching implications in the urban planning field, as this paper aims to point out, and it is hoped that this symposium will at least raise some important questions for their solutions.

Density is a well-defined term for a chemist or physicist --it is the ratio of mass to volume. Although in the general approach adapted to urban-geographic analyses, it can be defined as "quantity divided by area or by space enveloping" this raises some other problems concerning the definition of quantity and the area and/or space to be taken at a time. Land area (LA), population (P), and the building bulk (BB) are the basic set of units by which the different definitions can be derived, by the process of combination of any two units. To introduce a spatial context to the definitions, settlement units can be considered in an hierarchy of micro, intermediary and macro spatial levels. Room, house and the residential area can constitute the micro-level; whereas megalopolis, region, national boundary, or some form of higher groups may describe the macro-level settlement areas. Neighborhoods, small cities and urban-metropolitan centers may have an intermediary location in such an hierarchical system of settlements. Criteria for delimitation techniques of the metropolitan areas are discussed and given by J. Gibbs (1961) and Berry, et al. (1968). Delimitation of the neighborhood units or other spatial patterns of similar socio-economic characteristics within the metropolitan areas cannot be achieved by mono-criterion techniques, but requires factor-analytic approaches because of the large number of variables involved (Berry and Horton, 1970).

At the micro-level, Occupancy Rate (OR), i.e. the crowding index, and the Floor Area Rate are frequently used to describe the housing conditions. OR is defined as the number of persons per habitable room; and floor area rate as the ratio of floor area to the family size. In such definitions, "habitable rooms" require more clarification, and OR has an obvious inadequacy from the fact that room sizes differ from each other. Moreover, because of the wide differences in room sizes, the threshold of OR=1.51 persons/room may in fact represent either overcrowding if the rooms are relatively small; or ample living space if the rooms are relatively large (Mabry, 1959).

If the term crowding is taken in the national or regional scale, it indicates the degree of population pressure upon natural or human resources. Pressure is caused by the imbalance between human numbers and needs and the physical and human resources of the area in question. As the demand exceeds the supply, resources become scarce and thus overcrowding is said to occur. Taken in this sense, on the urban scale the overcrowding index may indicate high rates of unemployment as well as a quantitative assessment of shortage of housing. Therefore, a planning goal "to decrease overcrowding" will cause many complicated economic implications on the national investment plans.

The concept of crowding and/or density can also be considered in the framework of population distribution within a spatial unit. Such a framework provides a broad range of spatial measures of population distribution from centralization-decentralization, concentration-deconcentration indexes (Gibbs, 1961), nearest-neighbour statistics (Garney, 1967), to the concepts of "social physics" such as the demographic potential, energy and force (Stewart and Warntz, 1958) and to the rank-size (Berry and Garrison, 1958) or the central place theory (Berry,1967). In the study of distributions, Wilson derived a family of gravity models by using èntropy maximizing techniques (Wilson,(1967; 1970).This framework in which a tremendous amount of work has been achieved would also provide us with quantitative tools for the analyses of population distribution or spatial organization.

Intensity, a similar concept to density, is technically used to give the overall structure-mass and open space relationships in a developed land or may be measured by relating annual inputs of labor, capital outlay on goods, etc., by annual outputs of crops, manufactured goods, retail sales (McLoughlin, 1969). The concept of intensity thus can be generalized to include the speed, flow, or tempo of life in an urban spatial unit. By this approach, the densities of the past cities cannot be directly compared with these of the present ones which had a higher technological level. Even at the present, an hectare of land in a small town has a different intensity of life than that of a large city, though they may both have the same density. R.Meier (1962) analyses the messages received by an urban man from different sources like television, radio, newspapers, books, and the present urban environment generally and suggests a new term, the "hubit", which is one bit of meaningful information. Meier's conclusion that an urban man today receives messages beyond his capacity to process may be taken to mean a general trend of increasing intensity upon man. Thus, a conceptual shift from intensity to stress leading to pathological behaviors as studied by social psychologists seems not difficult.

In the social sciences it appears that the present interdisciplinary concern with human crowding and density originated from several lines of thought. Traditional research on population crowding can be divided into two groups.

First, the positive view advocates that a high level of density enables a complex and specialized social organization which in turn facilitates innovation, industrialization and a high standard of living.

The Durkheimian point of view accepts high population density, along with high population size, as a prerequisite for the development of division of labor. The division of labor and high density are necessary conditions for the development of a great variety of aptitudes and also of needs; for bringing these together in reciprocal stimulation and for establishing adaptation between the techniques of the more and more specialized producers and the needs of the more and more diversified consumers. L.Wirth (1938), in his article "Urbanism As A Way Of Life", describes the city as "a relatively large, dense and permanent settlement of socially heterogeneous individuals". More recently, J.Jacobs asserts "the need for concentration whatever the purpose there may be there" and thinks that 100 dwellings per acre (i.e. 250 dwellings per hectare) will be found too low to produce city vitality and diversity (Jacobs, 1961).

In regional science, this point of view has also received theoretical and empirical elaboration in the central place theory and in interaction models based on gravity. The central place theory developed by W.Christaller (1933), Lösch (1944) and recently by Isard (1956) explains how an hierarchical structure of urban centers supplying central goods emerges. It argues that a certain number of consumers are necessary within a given radius of a center for the support of a specific good or service. Whether a specific good becomes a central one depends on the population density of the area, that is, on whether it will find a sufficient number of consumers within its range. The concept of the "range of a good" has an important role in the theory and has a minimum threshold and an upper limit beyond which the central place is no longer able to sell the good or service. Some recent developments in this theory suggest that it may apply to the hierarchical distribution of retail services within the city (Berry, 1961).

In the usual form of the gravity model, Interaction$=KP_1P_2/d^x$, the interaction is directly proportional to the population sizes P_1 and P_2 and inversely proportional to the distance separating them raised to some power X and thus, the aggregation of more population means increasing information or traffic flows between centers (Isard, 1960).

Secondly, the negative view argues that urbanization and industrialisation produce psychological strain generated in an overcrowded existence lacking both privacy and close personal ties. Pathologists have produced some quite astonishing information from studies of animal behavior. Calhoun (1962), in an attempt to investigate the relationship between animal behavior and population characteristics, finds that as the population begins to reach half its maximum, "pathological togetherness" occurs. I.McHarg (1962) accepts that the correspondence between the kind of diseases in rats and 20th century urban man is startling. H.Winsborough (1965) and O.R.Galle, et al. (1972) studied the implications of Calhoun's findings and tested with 1950 and 1960 data for Chicago respectively. These studies suggest that overcrowding may have a serious impact on human behavior and that social scientists should consider overcrowding when attempting to explain pathological behaviors. R. Schmitt's (1966) research of 42 census areas in Honolulu casts doubt on the validity of J.Jacobs' thesis regarding the relative importance of population densities and overcrowding to health and social disorganization by correlating five independent variables with nine dependent variables. In the study of relation between behavior and crowding, cultural, economic, climatic and technological factors are emphasized in many researches (Hall, 1962; Hall, 1966; Doles Report, 1965; Doles Report, 1966; Doles Report, 1967). However, most of the recent researchers note that specific knowledge about causal links is lacking since some data both support and refute the hypothesis that human density is directly related to social pathology. D.Carnahan, et al. (1974) and A.R. Gillis (1974) raise some questions on the validity of population density as measured by persons per sq. mile at least in the city centers.

For planning purposes, some authors do not consider the high-density and the low-density arguments as alternative to each other, but they prefer to give a range of optimum densities outside of which development is likely to be under-used or over-used. For H. Blumenfeld (1967) the acceptable range is between 50 and 250 persons per hectare (pph) (i.e., 12,000 and 60,000 persons per sq. mile), where below the lower limit of 50 pph, he finds the over-extension of the urban area, isolation of daily life and difficulty in finding labor for industry or commerce. K.Lynch (1962) thinks that acceptable net densities may range from 3 families per hectare to 300 families (i.e. 1 family to 120 families per acre) where there are a number of breakpoints defining smaller ranges in which a certain type of structure is suitable. Although J. Jacobs (1961) thinks 250 dwelling units per hectare (i.e. 100 units per acre) would be too low and rejects "in-between densities" from 50 to 250 units per hectare that fit neither suburban life nor city life, agrees that the truth differs in specific instances. If the problem is viewed in the national scale, H.Richardson (1972) points out that even from a theoretical standpoint the search for the optimum city size is unsound, and suggests to try to identify efficient ranges of city sizes between a minimum threshold and a maximum size, especially in a dynamic setting. He makes distinctions among the rank size rule, central place hierarchies, and the spacing of cities, and notes that there is no unique spatial distribution of population which can be said to achieve the given goals of a national urban policy.

In the search for optimum, a recent account is given by L.Martin (1972), and L.March (1971) and his colleagues in the University of Cambridge, challenging the conventional methods in finding densities. They propose to study built-forms which are not buildings in a series of constantly changing internal-external relationships by applying set theory, graph theory and other related mathematics.

As seen in the foregoing discussion, problems of overcrowding and density are rooted in the dynamic relationship between the demands of population and the supply of natural and economic resources of the area in question. The earth has 510 million sq.kms, out of which 140 million sq. kms represent land surfaces of all kinds. If we substract areas covered by ice, deserts, steep mountains, jungles and other types of inhospitable land, we are left today with about 40 million as habitable, out of which human settlements occupy approximately 400.000 sq.kms, or, only 1% of the habitable surface. The figure of 40 million sq.kms of habitable land represents a very concrete ceiling for available space which can only be exceeded by means of adequate technological and economic development. By an optimistic estimate the present habitable space may probably be increased to somewhere between 60 to 80 million sq.kms. We can estimate the present densities over the area occupied by human settlements as 3.2 billion/40 million sq.kms.=80 inhabitants per hectare. This average overall settlement density has been decreasing since ancient times. Overall densities in both rural and urban settlements have for thousands of years been fairly constant: around 100 pph in rural settlements, with variations from 70 to 200 pph; and around 150-200 pph in the urban areas with variations from 100 to 300 pph. This decrease seems to be accelerated in our times when trends are clearly toward a marked reduction in densities as cars and other transportation technology enter the scene. Thus, two opposite trends are operative at the same time:

a) The earth is a closed system, thus the world population and density is increasing,

b) On the other hand, there is a trend toward decreasing densities at the urban-metropolitan level; and this trend has progressed to the point of a drop to about half of what previous densities were (Figure 1).

Areal growth rates of urban centers are higher than their rates of population growth. Thus, cities are dispersing at a rate faster than new people are coming into them by natural increase and migration as a world-wide phenomenon. As a result of this areal expansion agricultural land seems to lose its competition with urban and industrial land with rates of urban growth in Asia 400% higher than in the West. In India, the urban area is spreading to valuable agricultural lands; in Pakistan suburban subdivisions are occupying soil capable of producing four crops a year. The Çukurova and Bursa plains in Turkey are troubled by the pressure of housing and industry. Robin Best (1968) finds that the American urban area is extending 2.2 hectares and British urban area 0.32 hectares per 1000 population per year, which means 22 sq.m. per person and 3.2 sq.m. per person each year respectively. Should urban densities be increased to save the fertile agricultural land? Long debates took place and analyses showed that variations of residential density between possible and practicable extremes of high and low could make no significant difference either to the total amount of agricultural land or the total output. W.Zelinsky (1970) in his critisism of eleven assumptions indicates that the assumption that rapid population growth per se will damage man's habitat remains unproven. The architectural profession associated itself with the debate, too, and recommended high-rise buildings to save more land. An analysis of the relationship between the population density and the required area gives interesting results; area plotted as a function of density would yield a rectangular hyperbola in a convex shape. Such a curve shows that beyond a certain limit, the policy of saving land by increasing density cannot be maintained (Figure 2) (Esmer, 1970).

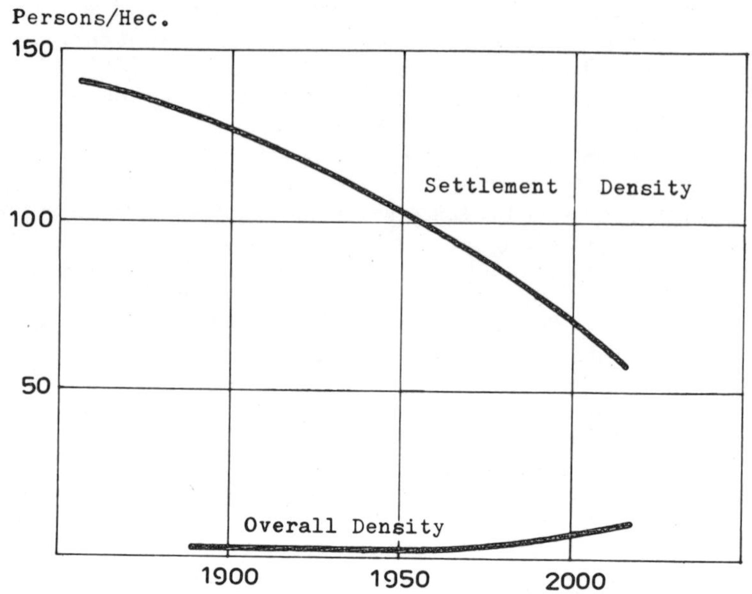

FIGURE 1. DENSITY AND WORLD POPULATION

The overall urban density of Britain's new towns is at an average rate of only 13 dwellings unit per hectare (i.e., 5.3 units per acre), or approximately 40 pph. J. James (1967) finds that modern planning standands produce a town density of somewhere near this figure. C.Clark (1967) suggests that planned densities of British new towns should be reduced or certainly not be raised any further. Figure 3 shows that the present population of Turkey (40 million) could be accomodated in a square of 100 kms x 100 kms at a given overall urban density of 40 pph. In the same manner, it can be estimated that the entire population of the world, 3.5 billions, could be housed at the density of 45 pph within the area of Turkey (780.576 sq-kms), assuming that the whole land were available.

From the arguments above, it can be recapitulated that:

. Overall urban densities are decreasing and the rapid urban expansion is a world-wide phenomenon,

. An urban policy of higher densities to control the spread cannot be successful because the residential density is responsible for a limited part of the situation, and what really causes urban spread is the existence of extensive non-residential uses and the vacant or agricultural land separating each little group of settlements in the Western metropolitan areas, along with the sporadic development of squatter housing around the large cities due to tides of migration from the rural areas in the case of the developing countries. However, the characteristics of urbanization differ in developing

FIGURE 2. RELATIONSHIP BETWEEN DENSITY AND LAND AREA

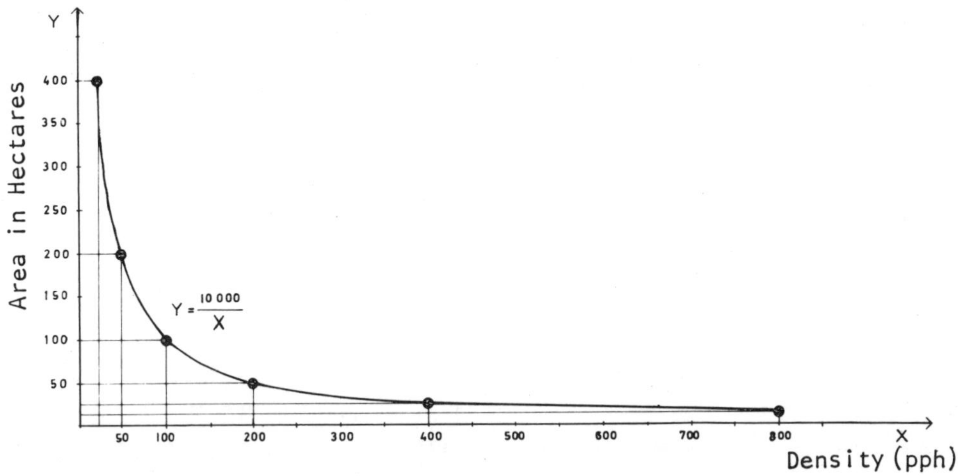

countries. E.Brush's study (1970) offers evidence for the marked decline of India's urban growth rate during the census decade ending in 1961 which was contrary to expectations. In Turkey, although the percentage of population in urban centers (i.e., all municipalities with more than 10,000 inhabitants) has been increasing since the 1950's and has reached approximately 42% in the 1975 census, both the proportion of the three largest centers (Istanbul, Ankara, Izmir) in the total urban population and the rateof growth of their totalshave decreased during the 1970-75 period (Table 1). Analysis has shown that people migrating from rural areas have changed their targets from those three centers to the subsequent four centers (i.e., Adana, Gaziantep, Bursa, Eskişehir) which have populations between 250,000-500-000 in the 1975 census (T.C.D.P.T., 1977).

Despite the overall drop in densities, social-economic problems relating to the human crowding have become even more complex. The policy of decreasing overcrowding, i.e.,creating employment opportunities, increasing the housing standards or generally, attempts to establish a balance between the demand and supply of human and natural resources, would have many economic implications and would mean additional millions of dollars of investment for the governments. Thus, setting the planning goals for crowding indexes seems to be more strategic than determining the crude population densities of urban areas.

DENSITY GRADIENT AS A TOOL FOR ANALYSIS

The previous section of the paper discussed some problems relating to the definitions of the terms and pointed oft two opposite trends in population distribu-

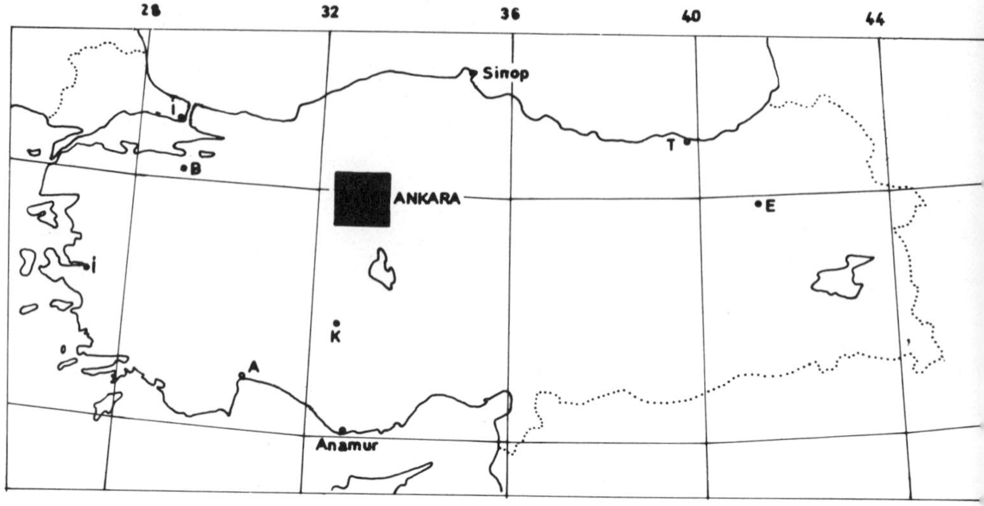

FIGURE 3. POPULATION OF TURKEY (1975) AT THE DENSITY OF 40pph

TABLE 1. GROWTH RATES OF TOTAL POPULATION OF 3 METROPOLITAN CENTERS, TURKEY, 1960-75.

Census Years	% of Total Urban Population (Turkey) (1)	Population of 3 Metropolitan Centers Istanbul, Ankara, Izmir,		
		Total Population	Annual Increase (%)	By % of Total Turkish Urban Population
1960	25.9%	2,531,975	–	35.2%
1965	29.8%	3,059,216	20.8%	32.7%
1970	35.8%	3,977,107	30.6%	21.2%
1975	40.0%(2)	4,869,459	22.1%	28.8%

Commision Report For The Fourth-Five Year Development Plan, State Planning Organization, Ankara, November 1976 (mimeo.)

1) Ratio of total population to the total population of settlements with 10,000 or more inhabitants.
2) Temporary results.

tion in the present metropolitan centers. Within metropolitan areas population dispersal has been a considerable concern. A city has a multitude of space-occupying activities seeking close contact. If these activities compete for limited areas of maximum accessibility, that competition results in a spatial gradient in a monocentric metropolitan area. We should expect their density, or intensity of space use, to be at a peak at the center, i.e., the point of optimum total access, and to fall off in all directions with increasing distance from the center. Such a tendency can be described by a density gradient, where density is a negative function of radial distance. Since the contribution of Colin Clark in 1951, extensive research has been made to test the hypotheses imbedded in his formulations on urban densities. Clark produced evidence in support of his argument that "regardless of time or place the spatial distribution of population densities within cities appears to conform to a single empirically derived expression,

$$D_x = D_o e^{-bx}$$

where D_x is population density at distance x from the city center, d_o is central density, and b is the density gradient. Clark provided 36 examples in which the equation expressed in a natural logarithm appeared to be a good fit to the sample data. B.Berry, et al. (1963) provide the theoretical rationale and assert its universal applicability by comparing the cross-sectional and temporal data for the Western and non-Western cities. The most intensive statistical analysis of the urban density gradients is that of Richard Muth (1961). After a series of tests of fit and linearity applied to density figures for forty-six U.S. cities in 1950, he concluded that a negative exponential function would seem to be the best simple approximation to the pattern of population decline. This concensus of opinion has been extendedinto a systemof axioms and equations by P.H. Rees (1970) as follows:

Axiom-2. The Density Gradient Declines with Time.

Equation derived is

$$b_t = b_o e^{-ct}$$

where b_t - distance - density gradient at time t.
b_o - distance - density gradient at time t_o
e - exponent

From the two axioms above, B.Newling (1966) goes on to deduce a number of necessary consequences about the density of urban populations. In his third theorem he asserts that there is a critical density above which growth is negative and below which growth is positive. By analysing the results in Kinston, Jamaica, he argues for a link between such a critical density (i.e., 32,000 persons per sq.mile, or 123 pph) and social conditions for 1950-60 data.

The two parameters, D_o, indicating concentration or crowding at the center, and (b) indicating compactness, vary from city to city. In any temporal cross-section, central density appears to be determined by the growth history of the city up to that time; and the density gradient appears to be a function of city size. B.Berry, et al. (1963) make cross-sectional and temporal comparisons among selected Western and non-Western (particularly Indian) cities. This analysis shows how Western and non-Western cities differ in the ways in which D_0 and the gradient change through time. More recent research by John E.Brush (1970) support Berry's generalization that non-Western cities follow a pattern of concentrated growth and increasing residential congestion in contrast to the Western patterns of growth, which are accompanied by residential dispersion or suburbanization.

It has been suggested by Newling (1969) that a quadratic regression of the logarithm of density on distance is a better generalization, because it accounts for the density crater at the center of the city. Thus, Clark's equation would become,

$$D_x = D_0 e^{bx - cx^2}$$

In a recent study, the second-degree negative exponential model was found to have better descriptive capabilities of density patterns of the Tel-Aviv metropolitan area than the first-degree Clark function (Schackar, 1975) or E.Casetti's (1969) higher degree of functions.

The next part of this paper applies the first-degree exponential function to 1965 and 1970 data for Ankara.

DENSITY GRADIENTS FOR ANKARA, 1965-1970

The population of Ankara has been rapidly increasing in the last 50 years (Table 2). In 1927 the predicted maximum population for 1977 was 300,000, a figure which was thought to be too large but possible. By 1975, the actual population was six times as high as the predicted maximum, and the state planning organization now predicts the present level, a total of 1,750,000

The total urban area is 17,622 hectars (i.e., 68 sq. miles) of which total 3,703 hectares are vacant and 13,919 hectares are developed land; excluding Atatürk Park which covers 4000 hectares, the residential area covers almost half of the total settlement land (45%). The next most intensive use is by the governmental services (10.2%), if we omit the vacant (21.3%) and open spaces (11.25%),(acc. to 1970 land-use figures).

The residential land is not homogeneous and shows variations depending on the location and the income groups occupying the area. Approximately 65% of the net residential area consists of squatter housing which is called the "Gecekondu" and has a peripheral location. Extensive land use patterns by governmental buildings, education, military areas and railway create discontinuities in the formation of homogeneous land uses.

Ankara consists of 195 "mahalle" (administrative meighbourhood) units within the municipal boundaries. These units are grouped into 33 zones by the Ankara Metropolitan Planning Bureau (AMPB), considering their socio-economic characteristics.

The overall urban density is 1,236,152:17,622=70 pph, whereas the gross residential density is 155 pph. The net residential density is higher, i.e. 183 pph. (AMPB gives the gross and net residential areas as 8,000 and 6,745 hectares respectively as of 1970).

Ankara has two central business areas - Ulus and Kızılay. Physical factors like the railway, warehouses, topography, some other extensive uses and the historical development pattern justify not only the division but the kind of specialization between the centers. Ulus is the old commercial center serving mainly people coming from low and middle-income groups, as well as consumers from the neighbouring settlements of Ankara. R.Keleş (1971) describes how the population of Old Ankara around the citadel and Ulus has remained almost the same during the last 40 years in spite of its inner high rates of mobility, i.e., changes in residences within the same area. On the other hand, Kızılay has not only more specialized shops but also the offices and financial organizations. The rate of employment and turn-over increase between the years 1969-1972 has been higher in Kızılay, though the Ulus Center has higher values in absolute terms. In this study, in determining distances between the center and the statistical units Ulus shall be taken as the center of the overall area.

TABLE 2: POPULATION INCREASES IN ANKARA, 1927-1995

Year	Population	Population Growth Index
1923	21,500 (estimate)	–
1927	74,800 (census enumerations)	–
1935	122,700	100
1940	157,200	128
1945	226,700	182
1950	288,500	235
1955	451,200	368
1960	650,000	532
1965	905,600	737
1970	1,236,152	1007
1975	1,788,617*	1457
1980	1,890,000 (projection)	2030
1985	2,890,000	2430
1990	3,340,000	2720
1995	4,750,000	3850

(*) Projection for 1975 was 1,890,000 which was higher than this temporary enumeration figure. (The difference between the temporary and the final results is not very great. For 1970, the difference was 1,236,152-1,209,000=27,152.) Similarly, the growth index decreased from an estimated 1540 to 1457 for 1975. This indicates a decreasing rate of growth during the five-year period. (See also Table 1 for an analysis of the same tendency in 3 metropolitan centers).

Canberra, which was selected as the capital of Australia about the same dates (i.e. 1920's) reached a population of 56,000 in 1961.

As the sizes of the 33 zones determined by the AMPB are quite dissimilar and large, using the air distances from Ulus Center to the mean center of each zone, the density profile would not produce a good fit, especially around the center. Therefore, in the analysis the whole area was divided into four sectoral parts using 195 mahalle units. This procedure was carried out for 1965 and 1970 data. Table 3 gives the density gradients and the parameters of the first degree negative exponential functions of the respective sectors.

Table 3 shows four important results:

a) In all cases, densities decline with distance from Ulus Center, as in Clark's formulation of the density pattern.

b) Central densities increased in three sectors, excluding the North-East (The Ulus-Altındağ-Citadel area) sector where "gecekondu" housing is dominant. This indicates a trend of deconcentration in the parts near Ulus. The Old Ankara and Altındağ area should be losing its population and requires a closer analysis of its social conditions. Critical density, in Newling's sense, may be operative around this area. (Our preliminary calculation gave a critical density of 185 pph which is 1800 meters far from the center).

Furthermore, the central densities of both eastern (N-E and S-E) sectors are higher than the central densities of other two (N-W and S-W) sectors.

c) Density gradients are all negative as expected from the negative exponential function. Moreover, there is a decline of the gradient in the 1965-1970 period except the South-East (Kızılay-Cebeci) sector. This indicates more concentration of population around the Kızılay Center during this period. Gradients of the Eastern sectors

TABLE 3: THE DECLINE OF POPULATION DENSITY AND GRADIENT-ANKARA, 1965-1970

Sectors / Parameters	North-West (N-31)		North-east (N-100)		South-West (N-27)		South-east (N-37)	
	1965	1970	1965	1970	1965	1970	1965	1970
a	4.669	4.827	6.988	9.3936	4.927	5.163	5.333	5.847
b	−0.36	−0.26	−1.1	−0.8	−0.44	−0.32	−0.51	−0.67
D_o	106	125	811	598	138	174	207	346

N= Number of "mahalle" units

a= Intercept of regression line with density axis measured in $\log_e$ units.

D_o= Central density in persons per hectare (pph).

b= Density gradient measured in units of natural log/kilometer (i.e. the fall of natural log of density per km. of distance).

x= Distance in kms. from Ulus Center. Thus the first-degree density function for the N-W sector in 1965 becomes:

$D_x = 106e^{-0.36x}$

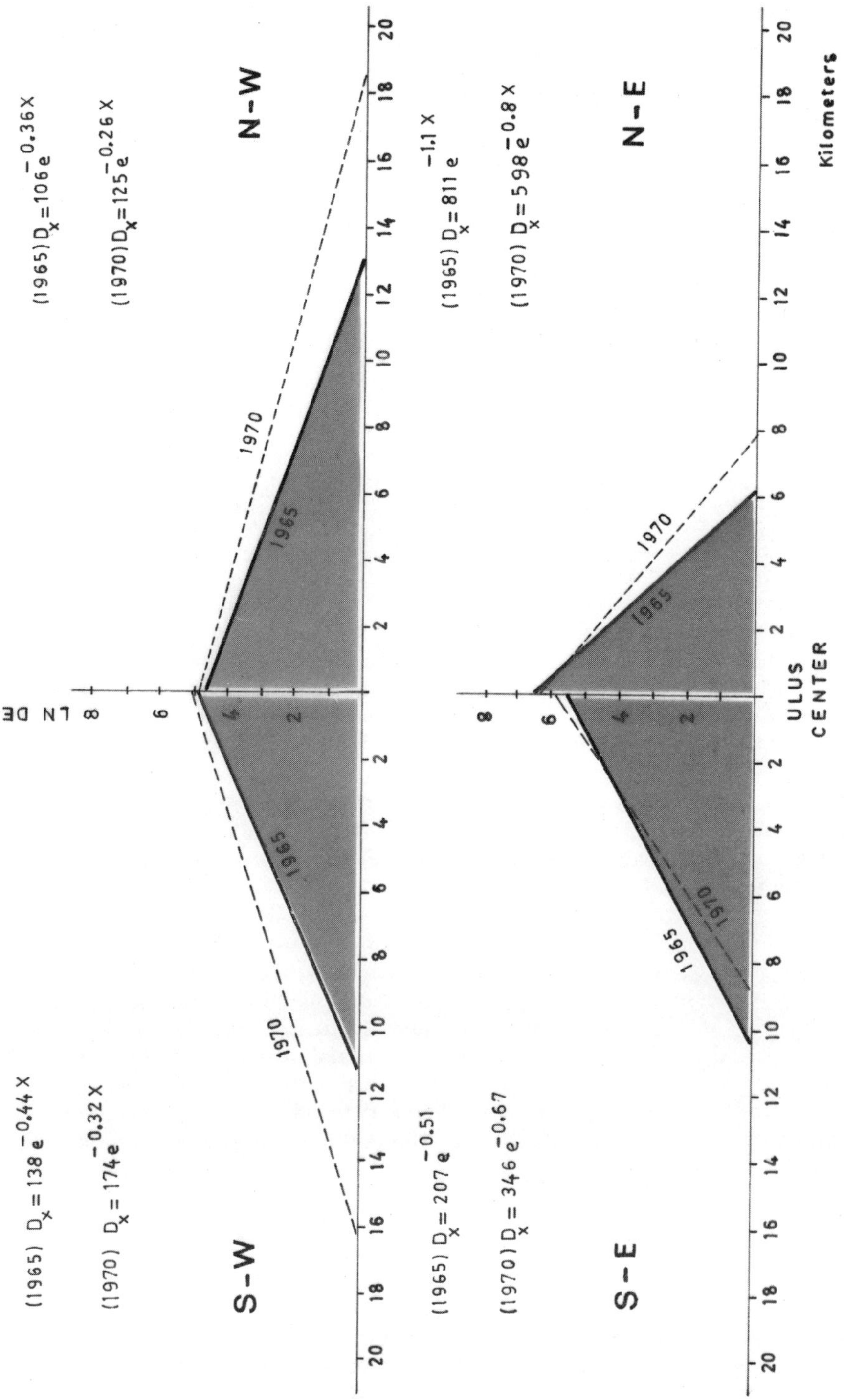

Figure 4: POPULATION DENSITY - DISTANCE GRAPHS FOR ANKARA

are steeper than those of the Western sectors.

d) Expansion of Ankara along the North-West (Ulus-Yenimahalle) and South-West (Kızılay-Bahçelievler) directions can be seen from their decreasing gradients with time.

(See also Figure 4).

However, a 5-year period is short for such generalizations. Yet, the analysis for Ankara has the characteristics of the Berry-type Non-Western city with increasing central densities indicating crowding or concentration, and sharp density gradients diminishing in time.

Further research on the second or higher degree of negative exponential functions may reveal how the two types of centers, Ulus and Kızılay, set in the urban setting of the Ankara Metropolitan Area.

CONCLUSIONS

Urban and regional planners frequently face questions relating to the optimum level of population density for which to plan. Yet, as this paper first tried to indicate, the functions of crowding, density or other related concepts, due to the lack in technical and spatial contents, cannot precisely be determined for spatial planning affairs. To assist planning in making choices or developing alternatives, the research should include appropriate spatial dimensions. At the micro-spatial level, authors emphasize more empirical data to obtain causal links between the quality of living and other variables. On the other hand, a simple process of aggregation of the results does not account for the behaviors at the higher level of spatial units. Therefore, this paper proposes that these terms should be taken within the context of population distribution, or more importantly, of spatial organization where distance can also be introduced as a function.

At the metropolitan level of observation a negative exponential function of the first or higher degrees can be accepted as an interpretable quantitative measure of population concentration. Moreover, cross-sectional and temporal comparisons of the parameters of the density functions such as the gradients can provide meaningful results about the sfecific variables on the spatial distribution of population and developing urban growth strategies. As seen from the density profiles for Ankara, 1965-70, cities will have different gradients determined by socio-economic factors. The presence of critical densities, as in the Ankara case, indicates areas for further research.

Most of the urban land-use models treat density as an exogeneously determined variable. In urban land-use and transportation models recent developed and applied, a maximum density level is introduced as one of the constraints in the set of equations (Lowry, 1964; Echenique, 1968; Schlager, 1966; Chadwick, 1971). For urban planners, therefore, optimum densities are necessary planning criteria; their definitions should provide the necessary tools for planning.

PART II

EXPERIMENTAL STUDIES OF CROWDING CONSEQUENCES

This set of papers opens with a discussion by Chalsa Loo on the consequences of crowding in children. Working with groups of 5- and 10-year-old subjects, Loo finds negative effects of high density conditions to be more pronounced in the younger age group. One theme in Loo's work that is also reflected in the next paper, by Bharucha-Reid and Kiyak, is the systematic investigation of individual difference variables within the same design with different environmental states. In this way, personal characteristics can be studied in their interaction with crowding. An interesting wrinkle on this design is provided by the next author, Rikard Küller, when he shows that one consequence of crowding is to influence the subject's perceptions, not only of the physical environment itself, but also of colleagues who are experienced within that environment. Thus, the interrelation of individual difference variables with environmental density is a two-way street; personal psychological characteristics both determine how we respond to a crowded environment, and are themselves determined by it.

One factor that may provide a theoretical bridge to the understanding of person perception within the context of high population density is provided by attribution theory. Nerenz, et al. report on a series of experiments investigating when college students attribute a violation in personal space to the intentions of the person, rather than to the environment. This "attribution" theme continues in spirit, if not in name in the next two papers. Rall, et al.s' distinction between "neutral" and "personal" crowding is highly related to the same distinction and Schultz-Gambard's experiment is certainly consistent with the notion that, if the source of the crowding is seen as an impersonal, objective factor, experiences of high density may have few consequences in terms of such psychological characteristics as anxiety, depression or insecurity.

THE CONSEOUENCES OF CROWDING ON CHILDREN

Chalsa M.Loo

University of California, Santa Cruz

California, U.S.A.

ABSTRACT: A series of studies was conducted to investigate the effects of density on childen and to determine whether there were differential effects of density on various person characteristics (age, sex, personal space preferences, and level of behavioral normalcy). The statistical procedures used, which incorporated factor analyses and multivariate analysis of variance, provided a more comprehensive analysis of the effects of density. Negative effects of density were significant for 5-year-old children and greater than effects for 10-year-old bovs, suggesting that negative consequences of crowding are greatest for younger children. Among 5-year-olds, there was more Negative Affect-Aggression, more Desire-to-Leave-a-Crowded-Room, and less Activity-Toy Play in the small room than in the large room (spatial-density comparisons). There was more Activity-Aggression-Anger, more Distress-and-Fear, more Negative Feelings, and less Positive Interaction among the 8-person groups of children than among the 4-person groups of children (social-density comparisons). The activity level of "hyperactive' children was intensified in the high-density condition whereas no effect of density on activity level was found for normal children. High-anxiety and impulsive children were emotionally distressed by high density; low-anxiety and high-motorinhibitors, in contrast, reacted to high density with adaptive, coping behaviors rather than with emotional displyas of distress. Differential effects of density on sex and on personal space were found for some factors.

The Consequences of Crowding on Children

The search for answers to questions concerning the consequences of crowding on children was the focus of a three-year research grant funded by the National Institute of Mental Health. The primary purpose of the project was to determine whether or not crowding had consequences for children.

A secondary purpose of the project was to investigate differential effects of density as a function of person characteristics. It is assumed that the environment does not have the same effect on all persons; in keeping with this assumption, the interaction between the person and the environment was an important area of exploration in this project. Several characteristics of the person were explored: age, sex, personal space preferences, and level of behavioral normalcy. This research project sought to answer the question "Are there differential effects of density on boys versus girls, 5-year-old children versus 10-year-old children, far- versus close-personal-space children, and normal versus disturbed children?" Considering the characteristic of normalcy, investigation was made into the types of behavioral disturbance that would be affected by crowding. Main effects for sex and personal space were also analyzed, although their role in this project was secondary to our concern with density.

In addition to the applied relevance of studying individual differences this investigation had a theoretical purpose which was to test Freedman's (1975) density-intensity hypothesis. This hypothesis maintains that already existing predispositions are intensified in a high density or crowded condition. Thus, by comparing behaviors and perceptions evidenced in the low-density condition with those evidenced in the high-density condition, one could determine whether predispositions that existed in the low-density condition were intensified, minimized, or remained unchanged in the high-density condition. This could be determined for sex, personal space, and types of behavioral disturbances of children.

In regard to sex differences, differential effects of density on boys versus girls had been investigated (Loo, 1972); more extensive research on differential effects of density on adult men and women had also been conducted. Among children, effects of density on aggression were found for boys but not girls (Loo, 1972). Among adults, two studies found that men felt more crowded than women in high-density conditions (Baum and Koman, 1976; Freedman, Levy, Buchanan and Price, 1972), and one study found that women reacted more negatively in a high-density condition while men displayed more aggression (Schettino and Borden, 1976).

The paradigmatic model for investigating differential effects of density on personal space is provided by Dooley's (in press) research on effects of social density on far- and close-personal-space men. Personal space refers to an area surrounding a person's body into which intruders may not come (Sommer, 1969) or the area around a person within which anxiety is produced if another enters (Horowitz, Duff, and Stratton, 1964). While Dooley found significant differences for affect, perceptions, and aftereffects on a performance task between far- and close-personal-space men, no interaction effect for density and personal space was found. In other words, high density did not affect far-personal-space men any differently than it affected close-personal-space men. No research on the differential efects of density on close- and far-personal-space children had been conducted previous to this project.

Considering the effects of density among children who vary on the dimension of behavioral deviancy, only one study has been published that investigated this issue. Hutt and Vaizey (1966) compared the effects of three social densities on three groups of children: normal, brain-damaged, and autistic. While this study speaks to the needs of more severely disturbed children, the behavior problems most commonly found in elementary schools include those of hyperactivity, anxiety, impulsivity, or hostile-aggressive behavior. Investigation into the effects of density on children who evidence such behavior problems had not been conducted prior to the present research project.

The statistical approach used for the research studies conducted in this project differed from that of previous research. In past experimental-laboratory research on crowding effects, single dependent variables had been analyzed; such an approach is limited in scope and does not provide a comprehensive view of the complexity of human behavior. In the search for comprehensive approaches to the study of crowding consequences and in response to appeals for the use of multivariate experimental designs (Baldassare and Fischer, 1977; Patterson, 1977), the statistical procedures used in this project were intended to better reflect the broad integration and organization of human behavior. For these research studies, a multi-level and multivariate approach was used that organized various dimensions into their related units. The following dimensions were investigated: social and non-social behavior, activity level, body positions, perceptions, emotional reactions, coping strategies, quailty of play, and quality of interactions. In addition, children's perceptions of each density condition were obtained through a structured post-experimental interview. A factor analysis of the rated behaviors and reported perceptions determined the factors that provided maximum interpretation. A multivariate analysis of variance was then performed on the selected factors as a whole, followed by a univariate analysis of variance on each individual factor.

A literature review of the research on density effects on children will not be provided here since it can be found elsewhere (Loo and Smetana, in press). Because the statistical design used in these studies provides for factors rather than for single dependent variables, the relatedness of previous research to this project is not exactly parallel. However, previous findings are relevant to the present research and those findings that are appropriate will be cited in the discussion of the results of all studies in this project.

Common Method for All Studies

Participants: Participants were children from schools in Santa Cruz, California, whose parents had given written permission for their child's participation. The children were generally from a middle-class background.

Experimental Situation and Design: A non-repeated measures design was used for all studies. The children came for one session of free play in an adult-free situation lasting 54 minutes.

Apparatus: The unfurnished rooms were equipped with a one-way mirror and microphones (hung from the ceiling) permitting the children to be seen and heard in the adjoining room. A variety of new and attractive toys was provided. The floor was marked off into 63 floor squares; each square was labeled in order to identify the location of each child at timed intervals.

Procedure: 1. Personal space. Personal space was determined by Horowitz, Duff and Stratton's (1964) measure of body-buffer zone for use with younger children. Each child was instructed to stand on a fixed spot and told that the female experimenter would talk towards him/her and that he/she should say "stop" as soon as the adult was getting "too close." Distance from the child's toes to those of the adult was recorded.* A total personal space score was obtained by summing the scores for eight angles (front, back, right, left, and four diagonals). Within each classroom, the highest-scoring children were put into far-personal-space groups, and the lowest-scoring children were put into close-personal-space groups. There were no significant sex differences for personal space; this findings is consistent with Bass and Weinstein's (1971) finding for children ages five to nine. A test-retest correlation using Pearson r for 10 children who were immediately retested for personal space was r=.88. Thirty children were retested after two and a half weeks; their test-retest correlation was .77.

2. Behavior ratings; experimental session. The children were told that they would be in the playroom for about an hour and that they could play with anything in the room. In the adjoining room research assistants rated the children's behavior. Each child was observed by each research assistant for equal portions of time. Research assistants rotated their assigned child at equal intervals in order to equally distribute any rater biases among all children.

3. Post-experimental interview. Immediately after each experimental session, each research assistant individually interviewed one child using a structured questionnaire format.

4. Interrater reliability. The intraclass correlation formula (Ebel, 1951) was used to calculate interrater reliability for the observed variables. Considering all studies conducted in this project, the interrater reliability ranged from .60 to 1.00.

Specific Method and Results for Study IA

The first study investigated the effects of spatial density (larger versus smaller rooms) on 5-year-old boys and girls in mixed-sex groups of six children each. With a total sample pool of 72 children, six groups of six children were observed in the high-density condition and six groups of six children were observed in the low-density condition. Each group was composed of three girls and three boys. Half of the groups in each density condition were far-personal-space groups ($\bar{X}$ approach distance = 24.1 inches) while the other half were close-personal-space groups ($\bar{X}$ approach distance= 8.4 inches). The low-density condition allowed 43.4 square feet per person and the high-density condition allowed 21.8 square feet per person. Since only 14% of the children havin 21.8 square feet person reported feeling crowded, high density and the perception of crowding could not be equated in this study.

A factor analysis of the behaviors and perceptions was performed which yielded five factors: Verbally-Abusive Interaction, Activity-Toy Play, Avoidance, Negative Affect-Aggression, and Desire-to-Leave-a-Crowded-Room. These factors

----

*A pilot study was conducted on 21 girls and 22 boys, 5 and 6 years of age, to measure comparability of an adult and a peer approacher. Adult approach scores correlated significantly with same-sex peer approach scores (r=.70, p<.001) and with opposite-sex peer approach scores (r=.56, p<.001).

accounted for 47% of the variance. Verbally-Abusive Interaction was composed of social interaction, non-solitary behavior, verbal abuse, and facing in. Activity-Toy Play was composed of toy changes, activity, non-onlooking, toy play, and walking. Avoidance was composed of standing, non-helpful behavior, escape attempts, location on the fringes of the room, and non-interruptions. Negative Affect-Aggression was composed of non-happy affect, anger, distress, physical aggression, and recipient of verbal abuse. Desire-to-Leave-a-Crowded-Room was composed of two items from the post-experimental interview, wanting to leave and feeling crowded.

A step-wise discriminant analysis was performed to predict low or high density as a dichotomous variable. The greatest predictive weight was held by Negative Affect-Aggression, Activity-Toy Play, and Desire-to-Leave-a-Crowded-Room ($D = .52X_2 - .67X_4 - .45X_5$; Wilks' $\lambda = .68$, $\chi^2 = 26.41$, $df = 3/68$, $p < .001$). More simply stated, scores that were high on Negative Affect-Aggression, low on Activity-Toy Play, and high on Desire-to-Leave-a-Crowded-Room were significantly predicted to occur in the high-density condition and the reverse predicted the low-density condition. Verbally-Abusive Interaction and Avoidance did not add much more in terms of predictive power in discriminating between density conditions.

A multivariate analysis of variance, analyzed by group scores, revealed significant main effects for density ($F = 24.77$, $df = 5/4$, $p < .01$) and for personal space $F = 6.64$, $df = 5/4$, $p < .05$). There was no significant effect for sex or for any of the interactions.

Since the results of the multivariate analysis of variance were significant for density, a univariate analysis of variance for each factor was performed. Spatial density had a significant effect on 5-year-olds; this effect was found for some factors but not for others. There was significantly more Negative Affect-Aggression in the high-density condition than in the low-density condition, that is, the intercorrelated variables of aggression, anger, distress, and verbal abuse were found more frequently in the high-density condition than in the low-density condition.

The discrepancy between the present finding and the previous finding (Loo, 1972) of more aggression in the low-density condition suggests that effects of density on aggression may be due to those behaviors that are concomitant with aggression. Aggression that is concomitant with anger may be found more frequently in a high-density condition than in a low-density condition, whereas playful aggression concomitant with motoric activity may be found more frequently in a low-density condition than in the high-density condition. Had a factor analysis been conducted on the 1972 study, such a discrepancy could be resolved. Presently we can only speculate as to the difference in the type of aggression that was demonstrated in the 1972 study.

Through the factor analysis, it was determined that physical activity and toy involvement were correlated. There was more Activity-Toy Play in the low-density condition than in the high-density condition. Specifically, there were more toy changes, more activity across the room, more toy play, more walking, and less on-looking in the low-density condition than in the high-density condition. This result supports prior findings (McGrew, 1970; Smith and Connelly, 1972) of greater physical activity in the low-density condition than in the high-density condition. The factor analysis allows us to understand what other behaviors accompanied the greater physical activity, which is not indicated in previous studies. A Sex X Density interaction was significant; a high-density condition substantially reduced the amount of Activity-Toy Play of girls whereas boys were affected to a much smaller extent.

There was also a significantly greater Desire-to-Leave-a-Crowded-Room in the high-density condition than in the low-density condition, and girls showed a greater effect of density on this factor than did boys.

The significant Sex X Density interaction effects for Desire-to-Leave-a Crowded-Room and Activity-Toy Play indicate that high-spatial-density resulted in a reduction of physical activity and toy involvement and an increase in negative feelings about the situation for girls; boys were similarly affected but were less sensitive to or less affected by density changes for these behaviors.

There was more Verbally-Abusive Interaction in the high-density condition than in the low-density condition. Density had the strongest effect on far-personal-space boys; Verbally-Abusive Interaction was greatest among the far-personal-space boys in the high-density condition and significantly greater than in the low-density condition.

Differences between far- and close-personal-space children were found for Negative Affect-Aggression, Activity-Toy Play, Avoidance and Verbally Abusive Interaction and were not found for Desire-to-Leave-a-Crowded-Room. Far-personal-space boys showed more Negative Affect-Aggression than any other Personal Space X Sex group. In addition, far-personal-space children showed significantly less Activity-Toy-Play, more Avoidance and more Verbally-Abusive Interaction than close-personal-space children.

Density differentially affected far-personal-space children from close-personal-space children for Avoidance and Verbally-Abusive Interaction. For the other factors personal space seemed to be a personality-related preference that was not situationally dependent.

Method and Results for Study IB

A substudy of Study I was conducted to investigate the differential effects of density on children who were normal compared to those having some behavior disturbance. A fuller description of this substudy can be found elsewhere (Loo, in press-a). The same children who participated in Study IA were tested for motor impulsivity using the Draw-a-Line-Slowly test (Maccoby, Dowley, Hagen, and Dagerman, 1965) and were rated on dimensions of hyperactivity, anxiety, hostility-aggressiveness, and behavior disturbance based on teacher ratings on the Preschool Behavior Questionnaire (Behar and Stringfield, 1974).

A median split was performed on the scores for each of the behavior problem measures: hyperactivity-distractibility, hostility-aggressiveness, anxiety, behavior disturbance, and motor inhibition. For each measure or scale, the 36 children who scored below the median were designated as "lows" and the 36 children who scored above the median were designated as "highs". The distributions for these variables were skewed in the direction of normalcy, thus our range of participants did not include severely disturbed children.

The average "high" -scoring hyperactive child was rated as being restless and inattentive. The average "high" -scoring anxious child was rated sometimes worried, miserable, fearful, fussy, and having a speech difficulty. The average "high" -scoring hostile-aggressive child was rated as one who is destructive of belongings, not liked by others, fights with others, hits others, and blames others. Behavior disturbance scores combined the ratings for hyperactivity, anxiety and hostile-aggressiveness. All scales on the teacher ratings were significantly correlated. Motor inhibition was not correlated with any of the teacher ratings.

For the analysis of this substudy, the factors derived from the factor analysis were not used. The original single variables were used for the analysis to avoid small cell sizes.

High-anxiety children were found to express significantly more negative affect (distress and anger) and to like others less in the high-density condition, compared to the low-density condition, while no density effect was found for the low-anxiety children for negative affect or for interpersonal attraction. High-anxiety children tended to respond to a high-density condition with emotional helplessness whereas low-anxiety children responded with motoric coping behaviors of reduced walking and increased facing out.

Impulsive children showed more negative affect than high-motor-inhibitors in the high-density condition. In the high-density condition, high-motor-inhibitors adaptively altered their motoric behaviors in response to a high-density condition; they stood more, attempted to escape more and were located on the fringes of the room.

High-hyperactive children, especially boys, were significantly more active (sat less and walked more) in the high-density condition than in the low-density condition. In contrast, low-hyperactive children showed no effect for density on sitting or walking.

Density did not differentially affect low- from high-hostile-aggressive children. In the high-density condition, high-behaviorally-disturbed children engaged in significantly more walking and less toy play than the low-behaviorally-disturbed children. Differential effects of density were not found for social interactions nor for instability of activity for the individual differences investigated.

In general, normal children motorically adjusted to a high-density condition to a greater degree than children with behavior problems. Anxious and impulsive children were especially distressed by a high-density condition compared to non-anxious and motorically-controlled children.

The differential approach to the study of density effects proved to be a fruitful approach for selecting "high risk" groups of children who are most stressed by high-density conditions. This approach also revealed differing modes of coping with high density that vary depending upon the interaction of person characteristics and density.

In pursuit of a further understanding of the effects of density on children with behavior problems a study on the effects of spatial density on austistic children was begun. However, allowing teachers to observe the experimental sessions as "participant advocates" led to a termination of the study. The teachers felt that the stress of the condition was too harmful to the children. This experience demonstrated the difficulties of investigating the effects of crowding on powerless and vulnerable participants in a non-hospital, experimental setting when ethical considerations for participant welfare must be taken into account (Loo, in press-b).

Specific Methods and Results for Study II

The first study was an investigation of the effects of spatial density (larger versus smaller room area) on groups of six children. The second study, a collaborative effort with Denise Kennelly, was an investigation of the effects of social density(four versus eight people groups) on a different sample of 5-year-old boys and girls in mixed-sex groups. The low-density condition provided 33 square feet per child; the high-density condition provided 16 square feet per child. As in Study I, each group was composed of equal numbers of boys and girls; half of the groups in each density condition were far-personal-space groups and half were close-personal-space groups. Seventy-five percent of the children in the high-density condition reported that it was crowded, therefore high-density and perceived crowdedness were considered comparable in this study. The effects of social density were analyzed for five factors that were derived from the factor analysis of rated behaviors and self-reported perceptions. These factors included Activity-Aggression-Anger, Negative Feelings, Avoidance, Positive Interactions, and Distress-and-Fear. They accounted for 47% of the variance. Activity-Aggression-Anger was composed of physical aggression, non-sitting, running, playful aggression, activity, and angry expressions. Negative Feelings was composed of five items from the interview: (a) being the target of aggression and feeling bad as a result, (b) feeling sad, (c) feeling angry, (d) being the target of someone else's anger, and (e) feeling crowded. Avoidance was composed of escape attempts, standing, facing out, and bored/tired. Positive Interaction was composed of positive-social-overtures, social interaction in a group, and non-solitary behavior. Distress-and-Fear was composed of distress, non-toy changes, non-toy play and feeling scared.

The step-wise discriminant analysis demonstrated that Activity-Aggression-Anger, Positive Interaction, and Distress-and-Fear were best able to discriminate the density conditions ($D=.71X_1 - .70X_4 + .42X_5$; Wilks' $\lambda = .68$ $\chi^2 = 26.27$, df=3/68, $p<.001$). Behavior which was high on Activity-Aggression-Anger, low on Positive Interaction, and high on Distress-and-Fear was significantly predicted to occur in the high-density condition and the opposite predicted in the low-density condition. Negative Feelings and Avoidance did not add much in terms of predictive power in discriminating between density conditions.

The multivariate analysis of variance demonstrated a significant effect for density ($F=8.79$, $df=5/4$, $p<.05$) and for sex ($F=32.13$, $df=5/4$, $p<.01$), but there was no significant effect for personal space or for any of the interactions. A univariate analysis of variance was performed for each factor. There was significantly more Activity-Agression-Anger in the crowded condition than in the uncrowded condition; thus there was more physical and playful aggression, more activity across floor squares, more anger, and less sitting in the crowded condition than in the uncrowded condition.

There was a significant Density X Sex interaction for Activity-Aggression-Anger. In the uncrowded condition, no sex difference for this factor was found. However, in the crowded condition, boys showed significantly more Activity Aggression-Anger than girls. Thus boys showed greater effects of density on Activity-Aggression-Anger than girls, and sex differences were greater in the crowded than in the uncrowded condition.

There was significantly more Positive Interaction in the uncrowded condition than in the crowded condition. Children in the uncrowded condition engaged in more positive-social-overtures and group interaction and less solitary behavior than did those in the crowded condition.

Under crowded conditions emotional reactions of distress and fear occur along with the failure to engage in normal task behavior for 5-year-olds, which includes toy play and social interaction. There was significantly more Distress-and-Fear in the crowded contidion than in the uncrowded condition. The crowded condition resulted in more distress, less toy involvement, and more self-reports of feeling scared than were found in the uncrowded condition.

As already noted, Negative Feelings and Avoidance were not found to contribute much to density prediction. The analysis of variance showed no effects for density on Avoidance. However, for Negative Feelings the crowded condition was associated with significantly more reports on feeling bad, sad, angry, crowded, or being the target of someone else's anger than were associated with the uncrowded condition. There were also greater sex differences in the uncrowded than in the crowded condition for Negative feelings. Boys reported significantly more Negative Feelings than girls in the uncrowded condition but no sex differences were found in the crowded condition. Comparing these findings to those on adults suggests that 5-year-olds show smaller sex differences in their reported reaction to crowding than do adult men and women.

There were significant sex differences found for Activity-Aggression-Anger and Negative Feelings; boys displayed significantly more such behaviors and feelings than girls. No sex differences were found for Positive Interaction, Distress-and-Fear, or Avoidance. No significant differences for personal space were found in this study.

Specific Methods and Results for Study III

The third study (Loo and Smetana, in press) investigated the effects of spatial density on 80 10-year-old boys in groups of five boys each. The differential effect of density on strangers versus acquaintances was also analyzed. For each density condition, half of the groups were strangers, half were acquaintances. The low-density condition allowed 52.1 square feet per person; the high-density condition allowed 13.6 square feet per person. Eighty-eight percent of the 10-year-olds who had 13.6 square feet per child reported feeling crowded. From a factor analysis of variables from the observational session and post-experimental interview, five factors emerged: Discomfort-and-Dislike-of-Room, Active-Play, Avoidance, Positive-Group-Interactions and Anger-and-Aggression. These factors accounted for 50% of the variance.

Discomfort-and-Dislike included variables from the post-experimental interview: discomfort, dislike of room, unhappiness, perceiving the room as "too small," and desire to leave the room. Active Play included non-onlooking, running, creative play, playful aggression, toy play, toy changes, and not feeling crowded. Avoidance included the variables of escape, standing, facing out, and non-sitting. Positive-Group-Interaction included non-solitary behavior, social interaction in a group, and positive-social-overtures. The last factor, Anger-and Aggression, included non-happiness, anger, and aggression.

A step-wise discriminant analysis was performed and two factors met the minimum criteria, Active-Play and Avoidance ($D = .88X_2 - .33X_3$; Wilks' λ .82, $\chi^2 = 14.99$, $df = 5/66$, $p < .01$). This demonstrated that the greatest weight was held by Active Play, followed in a less discriminating fashion by Avoidance. The other factors (Discomfort-and-Dislike-of-Room, Positive-Group-Interactions, and Anger-and-Aggression) did not contribute much to further density discrimination.

The multivariate analysis of variance for group means yielded no significant effects for density, personal space, or degree of acquaintance. An univariate analysis of variance for individual scores was performed for each factor. Crowding resulted in greater Discomfort-and-Dislike-of-the-Room, greater Avoidance, and less Active Play for 10-year-old boys but had no effect on Positive-Group-Interactions nor on Anger-and-Aggression. Thus, 10-year-old boys tended to feel greater discomfort, dislike, unhappiness, and desire to leave and perceived the room as smaller in the crowded condition than in the uncrowded condition. They also used more stress-reducing strategies in the crowded condition; they attempted to leave the room, visually blocked out stimuli by looking out the window or into the mirror, and faced out while standing or walking. Ten-year-old boys also engaged in more running and rough-and-tumble play in the uncrowded condition; they played with more toys, changed toys more frequently, and were more creative in their play in the uncrowded condition. In the crowded condition they engaged in more onlooking and felt more crowded.

While there were no Density X Personal Space interactions for any of the factors, a *t*-test that was run on the scores for each density condition revealed significant differential effects for personal space for one of the density conditions. First, in the crowded condition, far-personal-space boys felt significantly more Discomfort-and-Dislike-of-the-Room than close-personal-space boys, demonstrating that crowding is more stressful to far-personal-space boys. Secondly, while no personal space differences were found for Positive-Group-Interaction in the crowded condition, close-personal-space boys engaged in significantly more Positive-Group-Interactions than far-personal-space boys in the uncrowded condition. In this case, crowding minimized predispositional differences. A significant Density X Personal Space X Acquaintance Interaction for this factor demonstrated that these effects were identical and significant for close- and far-personal-space strangers as well as for close- and far-personal-space acquaintances. Thirdly, in the crowded condition, no differences between strangers and acquaintances were found for Avoidance. Yet, in the uncrowded condition, acquaintances engaged in more Avoidance than strangers.

No differential effect of density on personal space was found for Active-Play, Avoidance, or Anger-and-Aggression.

A significant Density X Acquaintance effect was found for Active-Play but for none of the other factors. The Active-Play of acquaintances was more strongly affected by density than that of strangers. In the uncrowded condition acquaintances engaged in more Active-Play than strangers but in the crowded condition differences between strangers and acquaintances disappeared; the crowded condition restrained the Active-Play of both groups alike.

As for main effects for personal space, close-personal-space boys engaged in more Positive-Group-Interactions than far-personal-space boys. In addition, far-personal-space boys felt greater Discomfort-and-Dislike-of-the-Room than close-personal-space boys.

As for main effects for acquaintance, acquaintances engaged in Avoidance and more Positive-Group-Interactions than strangers. A Personal Space X Acquaintance effect was found for Anger-and-Aggression; close-personal-space strangers showed the smallest amount of Anger-and-Aggression of all Personal Space X Acquaintance groups.

Discussion of the Results of All Studies

Effects of Density: The primary purpose of this research project was to determine whether or not crowding had consequences for children. In general, our findings indicate that high density had significant consequences for children's behaviors and perceptions. Greater specificity, however, is required to accurately address this question; crowding consequences are dependent upon the types of behaviors and perceptions that are investigated and the age of the participants. Besides age, the person characteristics of level of behavioral normalcy also affects whether or not crowding has consequences. Differential effects of density for girls versus boys and for close- versus far-personal-space were found for some of the factors. The effects of density as a function of these person characteristics will be separately dealt with following some general comments about overall effects of density.

For 5-year-olds, significant effects for density were found in both the spatial- and social-density studies when a conservative type of a multivariate analysis of variance was performed. Results for the spatial- and social-density studies for 5-year-olds were essentially similar. That is, crowding was associated with more aggression, negative feelings and expressions of anger and distress. Avoidance was not found to be density related. Positive-verbal-interactions occured in the low-density condition while negative-verbal-interactions occured in the high-density condition.

Differences that existed between the spatial- and social-density studies of 5-year-olds may be attributed to differential effects of spatial versus social density. When room size varied while size of the group remained constant, personal-space-differences were found for four of the five factors. But when group size varied while room size remained constant, no personal-space-differences were found. Perhaps crowding which is attributed to numbers of people rather than available space reduces any impact of personal-space-preferences.

Another interesting difference between the spatial- and social-density studies is that sex differences were more frequently found in the social-density study. Sex differences were greater in the 2:2 and 4:4 sex ratio groups than in the 3:3 and 3:3 ratio groups. Differences may be due to the even number of each sex represented in the groups in the social-density study; when there is equal companionship for members of each sex, sex differences will more frequently be found.

Another difference between the spatial- and social-density studies involved activity level. When room size varied, activity level correlated with quality of toy play. But when size of the group varied, activity level correlated with aggression and anger. Room size apparently affects quality of toy play more than group size.

Effects of Density as a Function of Age

Besides age, the main difference in group composition between the studies of 5-year-olds and the study of 10-year-olds was that the groups in the former case were of mixed-sex composition while the groups in the latter case were all-boy groups. This difference confounds age with sex composition. However, age differences can be roughly assessed by comparing results of Studies IA and II with results of study III because the presence of girls has been found to be an inhibitor of aggressive behavior of boys (Maccoby and Jacklin, 1974) and girls show less anger, aggression, and activity than boys (Loo, 1972; Maccoby and Jacklin, 1974). Therefore if the 5-year-old studies had been composed of all-boy groups, differences between the 5- and 10-year-old studies would have been greater than differences found in our research. Thus our comparative interpretations are on the conservative side because of the mixed-sex composition of the 5-year-old groups.

Comparing Studies II and III, we found no difference in how crowded the children felt; therefore differences in behaviors were due to age and not to differences in the perception of the situation. Since the more conservative multivariate analysis of variance revealed a significant effect for density on 5-year-olds in two studies and no density effect for 10-year-olds, the evidence strongly suggests that the negative effects of crowding are greater for younger persons. Since younger organisms show greater suceptibility to distraction and have smaller channel capacities (Evans and Eichelman, 1976), younger children would be less capable of coping adaptively with crowded conditions. Adaptive strategies develop with age and serve to reduce negative consequences of crowding.

At this point, a summary of the findings of the discriminant analyses and the analyses of variance for all three studies is provided. The discriminant analysis revealed that for the studies on 5-year-olds, factors composed of aggression and anger variables held the greatest weight of all factors in predicting the density condition. In contrast, the discriminant analysis performed on the data of 10-year-olds revealed that the factor composed of aggression and anger variables did not contribute to density discrimination and the analysis of variance showed no significant effect for density as well.

With the exception of one factor, Avoidance, results of the univariate analysis of variance demonstrated that effects that were significant for 5-year-olds were not significant for 10-year-olds. For 5-year-olds, there was more Negative Affect-and Aggression in the high-spatial-density condition than in the low-spatial-density condition than in the low-social-density condition. In contrast, for 10-year-olds there was no effect of density on the Anger-and-Aggression factor. Since the 10-year-olds felt more Discomfort-and-Dislike in the crowded condition, the fact that no density effect for Anger-and-Aggression was found for them does not imply that they did not feel stressed. Apparently, 10-year-olds who feel crowded do not act out as much as younger children, instead they use avoidant strategies in crowded conditions. Thus we found that while density affected 5-year-olds more than 10-year-olds in terms of physical aggression and negative affect, 10-year-olds used avoidant strategies in response to density more frequently than 5-year-olds. There was significantly more Avoidance displayed by 10-year-olds in the high-density condition than in the low-density condition; in contrast, no effects of density on Avoidance were found for 5 year-olds in either the social- or spatial-density studies.

Other significant effects of density were found for 5-year-olds. In the high-sptatial-density condition there was less Activity-Toy Play, more Desire-to-Leave-a Crowded-Room and more Verbally-Abusive Interaction; in the high-social-density condition there was less Positive Interaction, more Distress-and-Fear, and more Negative Feelings than in the low-density condition.

Our results argue strongly against generalizations from findings on adults to other age groups. Our findings demonstrated that what might hold true for adults in terms of crowding consequences is not applicable to children. In this study, consistent evidence for negative effects of density were found for 5-year-olds which has never been found for adults.

Effects of Density as a Function of Level of Behavioral Disturbance

Besides suggesting that crowding has greater negative consequences for younger children than for older children, our findings suggest that crowding has greater negative consequences for more emotionally- and behaviorally-disturbed children than for children without such disturbances. In particular, effects of high density were detrimental to "high"-anxiety children and impulsive children but were not so for children without these problems. For both types of children, crowding resulted in emotionally-helpless responses, contrary to the more adaptive and less emotionally-negative responses of the healthy children. Freedman's (1975) density-intensity hypothesis was not confirmed for the hostile-aggressive dimension but was confirmed for the hyperactivity and anxiety dimensions. That is, differences that existed between "high" and "low" scorers on hyperactivity and anxiety in the low-density condition were intensified in the high-density condition.

An attempt to test for effects of crowding on autistic children provided further evidence that negative crowding effects are greater for more emotionally-vulnerable persons. This attempt also suggested that ethical considerations for participant welfare make it difficult, if not impossible, to investigate the effects of crowding on vulnerable participants in a laboratory setting. Consequently, such investigation is probably best conducted in a natural setting instead.

Effects of Density as a Function of Personal Space Preferences

For children, personal space was a significant differentiating variable in relationship to density for very few factors. Far-personal-space boys proved to be the most verbally-abusive, felt the most Discomfort-and-Dislike-of-the-Room and showed more Avoidance in the high-denisty conditions than any other personal space group. Although Evans and Eichelman (1976) suggest a connection between the size of an individual's personal space and his/her susceptibility to being crowded, our results show that this is the case for only the aforementioned behaviors and feelings.

Freedman's (1975) density-intensity hypothesis suggests that differences in personal space that represent normal predispositions are intensified in a crowded condition. This hypothesis was not confirmed for all factors since effects were in a mixed direction.

Effects of Density as a Function of Sex

Since sex differences in response to density conditions were only studied for 5-year-old children, no developmental implications can be made about the interaction of density and sex. However, for 5-year-olds, high-spatial-density reduced the activity level and toy involvement (Activity-Toy Play) of girls significantly more than it did for boys. High-spatial-density also increased negative feelings about the room (Desire-to-Leave-a-Crowded-Room) for girls significantly more than it did for boys. No Sex X Density interaction was found for Negative Affect-Aggression, Verbally-Abusive Interaction, or Avoidance for the 5-year-olds who participated in the spatial-density study.

In the social-density study, high-density increased the level of Activity-Aggression-Anger for boys significantly more than it did for girls. There were no Sex X Density interactions for Positive Interaction, Distress-and-Fear, Negative-Feelings, or Avoidance.

Comparing the two 5-year-old studies, the small-room condition increased the Negative Affect-Aggression of boys and girls to the same degree relative to the large-room condition. In contrast, the large-group condition increased the Activity-Aggression-Anger of boys substantially more than girls, relative to the small-group condition. Apparently it is the presence of more children rather than restricted space that caused a sex difference in response to crowding.

Freedman's (1975) density-intensity hypothesis held true for a few but not for most of the factors in these studies. For some variables, significant sex differences that existed in the low-density condition disappeared in the high-density condition.

For other variables no sex differences were found in the low-density condition while significant sex differences were found in the high-density condition. Finally, for a few variables, sex differences that existed in the low-density condition were intensified in the high-density condition.

Effects of Density as a Function of Degree of Acquaintance

Although degree of acquaintance is an interpersonal rather than a person characteristic, the research findings provided some useful data in this area of investigation. From one study of 10-year-olds, it appears as though crowding does not affect groups of strangers much differently from groups of acquaintances. Where effects were found, crowding reduced differences that existed between strangers and acquaintances in the uncrowded condition. This runs contrary to Freedman's density-intensity hypothesis.

Our findings appear discrepant with previous findings. Previous research using projective figure placement tasks given to children have consistently demonstrated that children depict smaller interpersonal distances between acquaintances than between strangers (Bass and Weinstein, 1971; Guardo, 1969; Little, 1965; and Meisels and Guardo, 1969). Previous research studies on adults that have used figures in model rooms (Cohen, Sladen and Bennett, 1975) or questionnaires (Iwata, 1974) have found significant effects for degree of acquaintance; strangers felt more crowded than acquaintances in a high-density condition.

Discrepancies between our findings and those of previous research may be due to a non-relationship between projective figure-placement or imagined responses and actual behavioral responses of children. It is also possible that 10-year-old children may be less affected by interpersonal variables than adults. The interdependence and meaningfulness of an acquaintance may be less for 10-year-olds than for adults; therefore environmental effects may be less influenced by the interpersonal relationship of 10-year-olds. Further research on the behavioral effects of degree of acquaintance on 5-year-olds and adults is needed to provide a developmental perspective to the interplay of crowding and interpersonal variables.

This research project was supported by research grant MH 25522 from the National Institute of Mental Health and by a summer Faculty Fellowship and Faculty Research Funds granted by the University of California, Santa Cruz.

Correspondence should be directed to Chalsa Loo, Ph.D., 1728 Laguna Street, San Francisco, California 94115.

HUMAN AND NON-HUMAN COMPONENTS OF CROWDING

Rodabe Bharucha-Reid, Department of Psychology and Program in Environmental Studies, Wayne State University

H. Asuman Kiyak, Department of Psychology, Wayne State University

ABSTRACT: A laboratory study of task performance and affect under conditions of environmental stress varied social, spatial density and noise, as well as locus of control. Interaction effects of environmental stressors and the personality variable suggested the need to examine multiple environmental and individual factors in crowding research. Background characteristics of subjects –including number of persons they presently live with, largest number lived with, and preference for privacy– were found to interact with environmental variables. Results are discussed in terms of adaptation level theory; the concept of "dissonant space" is introduced.

\-------

Many investigators have looked at the psycho-social impacts of environmental density conditions. A review of either the sociological or psychological literature leaves one with the impression that density effects may be neutral or detrimental. It is the thesis of this paper that density emerges as "crowding" (Stokols, 1972) or "non-crowding", depending upon the individual and other environmental variables that interact with density. As the following research indicates, density has been dealt with in a vacuum, without the above-mentioned interactions that are found to be significant, and which must be considered if models of crowding are to be reality-based (Bharucha-Reid, 1975; Bharucha-Reid and Kiyak, 1976; Bharucha-Reid and Kiyak, 1977).

Research on the effects of high density may be traced to the pioneering work of Faris and Dunham (1939), who assessed the relationship between living in high density environments and various forms of mental disorder. They found density to be a contributing factor, but not the primary cause of mental disorders. More recent urban sociological and epidemiological research by Schmitt (1966) in Honolulu and Mitchell (1971) in Hong Kong has continued the tradition of studying individuals in the macro-environment, and human reactions to high density living conditions. Mitchell concluded the overcrowding produces superficial signs of strain, but does not cause deeper levels of strain and hostility. Schmitt's study suggests a need to differentiate between dwelling unit density (people per dwelling unit) and a real density (people per acre or square mile); the former had no effects, while the latter correlated with death rate, TB, VD, delinquency, illegitimate births, mental hospital and prison admissions. A conflicting set of findings by Galle, Grove and McPherson (1972) in their study of Chicago suggests the need to select appropriate outcome variables and to examine competing variables. The latter study found dwelling unit density related to social pathology indicators (only two of which were the same as those in Schmitt's study), but areal density did not relate when income and ethnicity were controlled. More conflicting findings are reported by Levy and Herzog (1974), who studied the effects of high density under relatively affluent conditions in the Netherlands. They found mortality and morbidity rates to increase with a real density, but to decrease with dwelling unit density.

The difficulty with large-scale urban density research lies in the low potential for precisely controlling density levels and social conditions. Criteria for high density vary across nations and across studies; ethnic and cross-cultural differences, family composition and styles of social interaction influence the relationship between density and pathology. The findings of Winsborough (1965) and Altman (1974) are relevant in this regard. The former study found high density to be nonsignificant when income, jobs, education and racial segregation were controlled. Altman (1975) suggests that interaction of persons within the dense setting –its nature, quality and rate, are important contributors to the effects of density upon the individual.

These findings suggest that it is both the social background of individuals and their physical environment which determine density effects. The interaction of these effects may be more profitably examined under controlled laboratory conditions. The major shortcoming of laboratory studies of crowding is their short-term nature and the relatively mild form of stress which may be produced under such artificial conditions. Nevertheless, the potential for controlling a number of environmental and individual characteristics, while examining a variety of outcome measures, makes this an appealing alternative to large scale field studies. The laboratory setting also offers one the opportunity to examine process (Altman, 1974), as well outcomes of high density conditions. In the field, one may only assess outcomes. However, laboratory investigators should utilize the information gained in field studies, particularly in the examination of social and demographic factors which contribute to density effects. The impact of previous experience under density, social and personality variables are often not considered by laboratory researchers. The present study was an attempt to examine various personality and demographic variables which interact with environmental factors to produce behavioral and affective responses.

Laboratory studies of density received their impetus from the early work of Calhoun (1962) and Christian (1960). Calhoun's striking finding that crowding produces a "behavioral sink," aggressive and anti-social behavior, as well as physiological changes such as enlarged adrenal glands and inability to bear offspring, and Christian's findings of suicide under high density led psychologists to consider high density as a major cause of psychological and physiological pathology. The fact that Calhoun's sample consisted of Norway mice and Christian's of sika deer did not deter researchers, who argued that, if crowding can produce such devastating effects on animals, it must surely be more detrimental to humans in even more diverse ways.

During the past 15 years, we have seen the growth of laboratory research in crowding. This trend may be observed in the steady increase of articles on crowding in professional journals. In fact, the tremendous output of research in this area has spawned the development of a new area within psychology and the emergence of new journals in environmental psychology and sociology. Despite the shortcomings of laboratory research which attempts to simulate real-life crowding, the advantages of this approach which were described above make this a useful strategy in investigating crowding effects.

The effects of high density have been examined with respect to psychological and physiological stress reactions (Stokols, 1972; Esser, 1971, 1972; D'Atri, 1975), on task performance (Epstein and Karlin, 1975; Freedman et al., 1971, 1972), as well as on interpersonal behavior (Hutt and Vaizey, 1966; Stokols, Rall, Pinner and Schopler, 1973; Saegert, 1975) and on social affect (Saegert, et al., 1975; Freedman, et al., 1972). Unequivocal statements regarding the effects of high density upon all of these variables cannot be made. However, findings of these investigators suggest that decrements in task performance are minimal; aggressive behavior intensifies among men but not among women; social affect becomes more positive for women but declines for men under high density conditions.

In defining crowding, Stokols (1972) has emphasized the role of two different forms of physical density: social density (or number of people per given space), and spatial density (absolute size of physical space). Both forms of density represent violations of personal space, but the individual's perception depends on the extent of both spatial and social density. It should be noted in this regard that Stokols distinguishes between the physical state of high density and the individual's perception, which he labels "crowding." The present researchers have also attempted to reserve the term "crowding" to the psychological state, whereas density will be used to decribe the manipulable, physical environmental variable.

Unfortunately, while the crowding literature has grown extensively, little integration with other major environmental and individual state variables has been undertaken. The interaction of types of density with noise and personality factors, while recognized implicitly as being important, have not been investigated. It is suggested by the present investigators that the mixed and often equivocal findings of researchers in this area result from the lack of recognition that psychological "crowding" is a function of various environmental variables interacting with diverse personality and demographic factors.

As indicated earlier, laboratory studies of crowding phenomena often are limited in their ability to simulate real-life crowding. Researchers often manipulate only one environmental variable and examine the interaction of this variable with sex, age, emotional state and other individual variables. However, in order to replicate the effects of environmental stressors on the individual, one must manipulate a number of variables.

Thus, one must covary characteristics of the individual. As decribed above, researchers have found crowding to affect different types of individuals differently. Further research on personality and demographic variables which determine reactions to environmental stressors is necessary. Similarly, density must be varied in different ways (e.g. social and spatial density) for the real world includes both forms of density. Noise, temperature, illumination are other environmental variables which must be examined in conjunction with density in order to assess their combined effects, as they exist in the "real" world. Research by Glass and Singer (1971) on short-term and long-term effects of noise has examined stimulus qualities such as continuous vs. intermittent, and controllability of the noise source.

The combination of density, noise and personality variables (such as ability to control a situation) are variables that must be considered together. Thus, controllability is an important factor in determining the reaction of individuals to other urban stressors as noise suggests that a systematic study of the effects of locus of control on environmental response would be worthwhile. Locus of control has been defined by Rotter (1966), one of the first researchers in this area, as the extent to which an individual perceives events in his life as being a consequence of his own actions or as being contingent on forces outside himself (i.e. fate, luck, other persons). The former belief defines an "internally-controlled" individual, the latter is "externally-controlled." Stokols (1972) has found externally-controlled individuals to manifest greater distance needs than internals, while Walton (1972) has found that density affects externally-controlled individuals more negatively. The work of Duke and Nowicki (1972) also supports these results. The latter have attempted to explain the externals' response by hypothesizing that externals are afraid of being controlled and therefore have greater distance needs and negative affect toward crowding. However, previous research by one of the present authors (Bharucha-Reid, 1972) suggests the internal-like individuals will be more negatively affected by density and noise than will external-like individuals. This follows from the findings of Glass and Singer that controllability is an important variable in post-adaptative effects of noise. Lack of control over the source of noise should affect externals less because such a person never expects to have control, and powerlessness is "normal" for him. Similarly, we would speculate the internals would be more affected by a situation over which they have no control, since they perceive themselves as determining outcomes.

It was reported earlier that adaptation to noise occurs, and that such adaptation may be expected to occur to high density as well. Evidence for this prediction is found in the research by Schmitt (1963, 1966) and Mitchell (1971). These investigators reported few detrimental consequences of living in high density urban centers. Apparently, individuals who have lived all their lives under conditions which seem unbearably crowded to the newcomer, have adapted to such objectively high levels of density and demonstrate few declines in functioning which would be analogous to Calhoun's findings with rats.

Another study of residential density suggests that "adaptation" may occur but at serious costs to the individual. In a study of children living in high-rise apartments, Cohen, Glass and Singer (1973) found that cognitive skills (especially reading ability) deteriorated the longer the child lived under high noise conditions. Thus, it appears that Calhoun's findings may have parallels in research with humans, but the effects of high density may be less overt and less immediate than with animals.

Thus, we have mixed evidence that adaptation does occur, whether the stimulus in physical pain, noise or density. It will be the aim of the present investigators to understand which variables among a number of personality, demographic and environmental factors play a role in effecting reactions and adaptation to environmental stressor. Specifically, we have evidence that locus of control should be examined, along with demographic factors such as family composition and previous experiences in high density environments. Dependent variables examined in this series of studies were task performance, self-reported stress, social and physical affect. The following hypotheses are suggested by the review of research evidence in this field:

1. Individuals who score in the internal range of a test locus of control will be more negatively affected by density and noise. Thus, one would expect greater decrements in task performance, greater stress and more negative affect among internals than among externals.

2. The combined effects of high social density, spatial density, and high noise levels will be more negative than each of these variables taken separately.

3. Previous experience with high density environments will mitigate the negative effects of high density. Thus, individuals who have lived with large numbers of people, and who are now living with many others, will experience less stress, fewer decrements in performance and affect under high density conditions than will individuals who have had little previous experience with crowding.

4. Similarly, individuals who prefer minimal privacy and have high noise tolerance will be less affected by environmental stressors than those with a high need for privacy and preference for quiet environments.

The present investigators have attempted to approximate real-world conditions to a greater extent than previous researchers by covarying three environmental variables: noise, social and spatial density, and a number of demographic and personality variables in a series of studies. These studies (Bharucha-Reid and Kiyak, 1976; Bharucha-Reid and Kiyak, 1977) involve looking at different outcome variables related to the emotional and cognitive behaviors of the individual.

Method

Subjects: Eighty-six male undergraduates volunteered to take part in the experiment for psychology course credit. Previous research in this field provides considerable evidence for sex differences in reaction to high density. For this reason, it was decided to include only males in the sample. A pre-test of locus of control was used to assign subjects to conditions. Subjects were assigned randomly to conditions, with the only criterion being that one-half of each group should consist of internals, the other half externals.

Materials: The pre-test consisted a nine-item Internality-Externality Scale developed by Mirels (1970). These are actually statements from Rotter's (1967) I-E Scale which were found in a factor analysis by Mirels to load highest on the dimension of "Belief Concerning Control over One's Own Life." It was felt that removal of those Rotter items having to do with "Belief Concerning Control Over

Political Institutions" would provide a unidimensional I-E scale, increasing the variance attributable to individual differences (Mirels, 1970), and a more valid measure of locus of control as it correlates with self-efficacy. Similar selectivity of research onblacksand females (Gurin et al., 1969).

Noise tolerance and privacy preference were measured by a questionnaire revised from Marshall's (1971) Privacy Preference Scale. A sixteen-item semantic differential scale of "Attitudes toward others in the Room" (Social Affect) and a six-item scale of "Perceptions of the Room" (Overall Room Affect) were also administered. Three of the adjective pairs in the latter scale measured psychological affect (Good, Pleasant, Comfortable), while the other three measured physical room affect (Spacious, Adequate, Well-arranged).

Spatial density was varied by using two sizes of rooms. The smaller rooms measured 80 square feet each, while the two large rooms measured 240 square feet each. Equipment for producing noise consisted of a tape of intermittent white noise, a reel-to-reel tape recorder, two 12-inch speakers and a SPL meter. The latter instrument was used to maintain the level of white noise at 105 decibels, A scale.

Procedure: A 2x2x2x2 factorial design was used, with two levels each of social density (6 vs. 16 subjects), spatial density (80 vs. 240 square feet), noise (105 decibels white noise vs. no noise), and locus of control (internal score vs. external score on Rotters' Dimension I). This produced eight experimental conditions, with internals making up one-half of the subjects in each condition, and externals comprising the other half.

The same procedure was used for each group. Thus, when subjects arrived for the experiment, they were assigned a number and told that this number should be indicated on all test forms. No other identifying information was used. After all subjects arrived for each condition, they were taken into the first experimental room as a group and told to sit wherever they wished. Each room had 16 desk-chairs, regardless of social density. After subjects were seated, the experimenter told them that the experiment was a study of people's behavior in groups, and that further information would be provided at the completion of the study. They were also told that the experiment would last approximately 60 minutes, but they would receive two hours credit for research participation. Subjects were informed of their right to withdraw from the experiment at any time, but that they would receive only partial research credit if they did so.

A brief sentence completion task and measures of personality characteristics as well as a demographic questionnaire were administered during the first 15 minutes of the experiment. Because the noise apparatus had been set up in an adjacent room and because movement under stress was one of the dependent variables of interest, subjects were asked to move to the second room after the first tasks were completed. The cognitive tasks were described to subjects before thy proceeded to the second room. They were told to complete the 10 problems in any order they wished, and told that they would have 30 minutes to complete the task. In the no-noise condition, subjects were told that the first room would be used by another experimenter; in the noise condition, the noise had already been turned on, indicating a change in environmental condition.

Subjects performed the cognitive task during the next 30 minutes in the second room, under noise or no-noise conditions. After this time, the recorder was turned off. Subjects completed a questionnaire regarding their feelings about the experiment, the environment and about other people in the room, as well as noise tolerance and privacy preference questionnaires (the latter revised from Marshall's 1972 Privacy Preference Scale) during the last 10 minutes of the experiment. They were then debriefed regarding the purpose of the study, and questions were answered, after which time they were thanked for their participation in the experiment and told they could leave the lab.

Results

A series of four-way factorial analyses of variance (Winer, 1972) were conducted to test the effects of social and spatial density, noise and locus of control (LOC) on task performance, social and room affect, as well as on self-perceived stress.

Table 1 illustrates the results of these analyses. No main effects were found for task performance, but two interaction effects appeared to be significant. An interaction of LOC x Noise was found, with $F(1,64)=8.06$, $p<.005$. The results were in the predicted direction, such that noise was more conducive for externals than no-noise, whereas internals performed better under no-noise than under noise conditions. No differences were found between internals and externals in task performance under social and spatial density. A spatial x social density interaction was also significant: $F(1,64)=6.96$, $p<.01$; the direction of this effect suggests that task performance was better in the small room when social density was low, and better in the large room when social density was high.

Self-perceived stress was also predicted by a social x spatial density interaction, with $F(1,64)=5.44$, $p<.05$. The effect was in the same direction as that for task performance. Noise was the only significant main effect for stress, with $(F1,64)=16,75$, $p<.001$.

Self-perceived crowding was predicted by a four-way interaction: $F(1,64)=4.82$, $p<.03$, such that internals experienced greatest crowding under high spatial density, high social density and high noise, least under low social and spatial density and no noise. Externals reported greatest crowding under high spatial density and high noise, regardless of social density. Affect toward others in the room (social affect) was not influenced by locus of control, neither as a main effect, nor as an interaction term. However, a main effect for LOC was found in assessing affect toward the room ($F=4.06$, $p<.05$), such that internals reported lower room affect across all conditions than did externals.

Turning now to background variables which affect reactions to environmental stimuli, a series of stepwise multiple regression analyses were conducted. Analysis of variance was not used to answer these research questions because of the unequal sized cells produced by many of the state variables of interest.

In an analysis to predict task performance, an interaction of Noise x Number of Persons Subject now lives with emerged as the best predictor, with Beta=-.19, $R^2=0.4$

Table 1

Analysis of Variance Results for Performance, Stress, Affect

Source	df	F(pref.)	F(stress)	F(Soc.Aff.)	F(Room Aff.)	F(crowding)
A. Spatial Density	1	.43	.03	.10	10.6***	6.23**
B. Social Density	1	.31	.13	4.84*	.01	31.50****
C. Noise	1	1.08	16.75****	.62	3.43	1.19
D. Mirels	1	.33	.03	.11	4.06*	.42
AB	1	6.96**	5.44*	2.37	3.26	1.05
AC	1	.46	1.76	.09	.97	1.06
AD	1	.18	2.78	.43	.01	.39
BC	1	.08	.02	1.01	2.73	1.99
BD	1	1.67	.16	.75	.00	.62
CD	1	8.06**	.01	.89	.49	.25
ABCD	1	1.57	1.09	.93	.15	4.82*
S/ABCD	64					

*p<.05 **p<.01 ***p<.005 ****p<.001

The beta-weight suggests that under noise conditions, the fewer persons with whom an individual lives, the more difficulty he will have with cognitive tasks. Noise and a Noise x Locus of Control interaction were the second and third best predictors of task performance respectively. The standardized regression coefficient (Beta) for noise was -.27, and for Noise x LOC was -.09.

A second regression analysis to predict "Affect Toward Other Persons in the Room" resulted in significant state variables in interaction with environmental variables. An interaction effect of Privacy Preference x Social Density was the primary predictor, with Beta= -.29, $\underline{R}^2$=.08, and $\underline{F}(1,78)$=7.17, $\underline{p}$<.01. The direction of this effect suggests that, under density conditions, these with high preference for privacy express lower affect toward people around them than do those with low preference for privacy. The largest number of persons with whom a Subject has lived in the past interacts with Social Density to predict Social Affect, although the predictive power of this effect is weaker than the first (Beta=.06, $\underline{R}^2$=.003). Thus, under high social density conditions, social affect is lower for people who have lived with small numbers of people in the past.

The individuals' subjective evaluation of their physical environment (Room Affect) was best predicted by Privacy Preference in the small room. This interaction term accounted for 11% of the variance, with Beta=-.33, $\underline{F}(1,78)$=9.22, $\underline{p}$<.005. Locus of control and spatial density also emerged as significant predictors, explaining 4% and 2% of the variance in room affect respectively. These three predictors in combination produce a significant equation, with $\underline{F}(3.76)$=4.85, $\underline{p}$<.01.

A self-assessment of stress was measured at the completion of the experiment. It was found that main effect of noise (R^2=.17, Beta=.42, $F(1,78)$=16.53) and number of persons subject is presently living with (R^2=.07, Beta=.27), best predicted self-reported stress. Interaction effects of Social x Spatial Density (R^2=.05, Beta=-.14) and largest number S lived with x spatial density (R^2=.02, Beta=.12). These four predictors together accounted for 27% of the variance, with $F(4,75)$= 7.08, $p<.001$.

Discussion

On of the major conclusions to be drawn from the findings presented above is the role of multiple environmental stressors. Thus, task performance was worst social and room affect most negative, and self-reported crowding and stress were greatest when social density, spatial density and noise were highest. It was also found that a reduction in one form of density may alleviate the impact of the other stressor. Thus, task performance deteriorated when both social and spatial density were high, but performance improved if only one of the variables was aversive but not the other. The finding of a similar pattern of effects for self-reported stress suggests that it may be this variable which is most critical, i.e., the individual is less stressed and therefore performs better under social density when spatial density is low, and vice versa. However, the lack of a similar relationship for noise and either form of density suggests that reducing social or spatial density may not be effective in alleviating the detrimental impact of noise.

The finding of a significant interaction between social and spatial density for stress and task performance suggests the need to consider both forms of density simultaneously. An examination of Bechtel's (1974) concept "optimum size" may be useful at this point. This term refers to behavioral settings which are neither undermanned nor overmanned, in which the appropriate number of people is available for the task at hand. While behavior setting has physical characteristics (e.g. big or small school or church), space by itself is not a major characteristic. In this paper the interaction of people and space assumes considerable importance. The concept of dissonant space (Bharucha-Reid and Kiyak, 1977) helps to clarify this situation. Thus for example, undermanning theory discusses ratios of setting size to organization size, while concept of dissonant space addresses itself to the relationships between setting size and the individual's perception of the appropriateness of the space for the activities taking place within it. Dissonant space may be defined as the disjunction in the individual's perception about the appropriateness of the ratio of space to the utilization of the space. We should note that adequate space is not so much the issue as appropriate space. The interaction of space and individuals is the focus relative to the task to be accomplished. Hence if a large space is more appropriate with a larger number of people and a small space is more appropriate with fewer people, then it seems logical that large numbers of people in a small space would result in poor task performance, as would few people in a very large space. That this in fact is the case is shown by the results of this study.

The importance of considering characteristics of the individual in assessing the human consequences of high density and noise is emphasized by the findings of the present investigators. Locus of control was found to interact with noise, social and spatial density in task performance, self-perceived crowding and room affect. The number of persons with which the individual presently lives interacts with environmental variables in predicting task performance and stress; while the largest number of people one has lived with has an impact on one's evaluation of oneself and of others in the room under high social and with spatial density. The individual's preference for privacy must also be considered in predicting their reactions to high density, particularly in their affect toward their social and physical environment.

These results are in line with our hypothesis that environmental effects are mediated by personality variables. One can immediately point to possible implications of these findings for applied industrial studies or for highway or housing studies. It is not enough to know that stress is produced by density and noise; differential effects may be produced on individuals. Thus for example, in an industrial setting, personality and previous experience in high density and noisy environments may well become an important variable to consider in hiring a worker.

The results are also consistent with adaptation level theory proposed by Helson (1964). According to this theory, performance is best and stress least with a repeated stimulus, i.e. when the individual has adapted to that stimulus. In the research reported here, it appears that humans may adapt to environmental conditions as well as they do to any other stimulus. Thus, the individual who now lives or has lived in the past in a dwelling unit or a city with a large population is considerably less affected by short-term crowding effects than is the individual who is not "adapted" to high density.

The individual's tolerance for high density and noise is determined not only by his previous experience but also by his personality. Thus, one is less likely to feel stressed or crowded, and more tolerant of others and better able to perform in objectively stressful environments if one is an external rather than an internal, and if one prefers low levels of privacy.

In summary then, the major findings of the present study are that environmental variables must be considered in combination, and that individual characteristics play an important role in predicting human consequences of stressful environments. Further research is needed on these environmental and individual variables, as well as on additional stressors in the environment and on other personality characteristics which interact with environmental characteristics. Finally, research in this area has reached a stage where it is now critical to conduct longitudinal studies of individual reactions to environmental stressors in industrial, residential and institutional settings.

REACTIONS TO PERSONAL SPACE INVASION: EXPERIMENTAL TESTS OF AN ATTRIBUTIONAL ANALYSIS

David R. Nerenz and William Ickes, University of Wisconsin-Madison

Norbert L. Kerr and Deborah F. Oliver, University of California, San Diego , U.S.A.

ABSTRACT: A series of experiments was conducted to test the hypothesis that reactions to an invasion of personal space would depend on the meaning attributed to the invader's action. In all three studies, a confederate invaded the personal space of subjects under conditions such that the invasion could either be attributed to dispositions of the invader or to a situational constraint. In two of the three studies, subjects indicated greater discomfort, either by leaving the situation or by self-report ratings of their own feelings, in the conditions where the invasion could be attributed to dispositions of the invader.

Personal space has been defined by Sommer as "an area with invisible boundaries surrounding a person's body into which intruders may not come (1969, p. 26)." Although personal space zones can vary widely across individuals and situations, most authors agree that the reaction to an invasion of this zone is a negative one, usually characterized by discomfort and stress (Altman, 1975; Linder, 1976; McBride, King and James, 1965).

Unfortunately, theoretical statements have been somewhat equivocal with respect to the question of what psychological mechanisms (if any) may underlie a person's responses to the invasion of his personal space. Indeed, the major statements to date (Hall, 1966; Sommer, 1969) suggest that once a norm of personal space has been established for a given class of interactions, an intrusion on this space will elicit a negative reaction on the part of the invadee, often without his conscious awareness of the true cause of his discomfort. There are some problems with this "invisible bubble" approach in that it has little to say about discomfort due to abnormally large interaction distances and does not easily account for positive reactions to close proximity between people. We suggest that a more adequate model of the processes involved in a person's response to invasion of his personal space may have to take into account the person's cognitive interperetations of the meaning of the invasion.

A greater emphasis on the meaning ascribed to interpersonal distance can be found in the work of Mehrabian and others (e.g. Argyle and Dean, 1965; Mehrabian, 1968) who suggest that interpersonal distance is one of many nonverbal behaviors that can indicate intimacy, friendship, or some other relation between the participants in an interaction. Assuming that people do assign meaning to interaction distances and that these meanings generally derive from attributions about the intentions and sentiment of the invader, it seems reasonable to propose that reactions to such events as personal space invasions may be determined more by the meaning attributed to the other's distance than to the distance per se. In other words, an invasion of personal space should not be viewed as an event which always produces stress or negative reactions directly, but rather as an event whose meaning is mediated by the intentions and sentiments which are attributed to the invader. Thus, while these attributed intentions and sentiments may often lead the person whose space is invaded to experience stress or other negative reactions, they will not always do so, and may frequently have the opposite effects. We are not proposing that the attributed meaning of an invasion is the only datum on which behavioral responses to the invasion will be based, since the close proximity of others may involve other factors such as behavioral constraints, shortage of resources, the presence of unpleasant odors, etc. (Stokols, 1976), which can also contribute to one's reaction to the situation. Rather, we are suggesting that it is one very important and heretofore overlooked source of data that should be included in any adequate account of human spatial behavior.

The importance of attributed meaning in response to personal space invasion may be tested quite simply under the following rationale. If reactions to personal space invasion are determined by attributions about the invader's affective sentiments, there should be little or no reaction to an invasion if such attributions are not made. However, if the act of invasion is itself the eliciting stimulus for the reaction, one would expect a reaction to the invasion regardless of whether it conveyed any information about the invader's psychological dispositions. In terms of attribution theory, an important factor may be whether the invadee sees the invasion as caused by the invader personally or by some factor(s) in the environment. If causality for the invasion is located in the invader, then the act should have interpersonal meaning, and the invadee should make attributions about this meaning, and react accordingly. However, if causality is located in the environment (e.g., the invader is perceived as having little or no choice about where he stands or sits), the invasion will not be interpreted as having interpersonal meaning, and there should be little emotional reaction on the part of the invadee.

EXPERIMENT 1

As a first test of this attributional model of interpersonal distancing, a field experiment was conducted in which a confederate invaded the personal space of subjects sitting in a lounge area under one of two conditions: (a) the room was nearly empty so that the confederate supposedly could sit in any one of several seats (low social density condition); or (b) the room was nearly full so that the seat next to the subjects was one of very few available seats (high social density condition). Because of our belief that interpretations of one male's invasion of another male's personal space would nearly always tend to be negative, we decided to confine our initial investigation to the male-male situation in order to provide the least equivocal test of our hypothesis. Our prediction was that, following an invasion of their personal space, subjects would leave the area more quickly in the low social density condition (where the invasion can be attributed dispositionally) than in the high social density condition (where the invasion can be attributed to the situation).

Method

Setting. The experiment was conducted in a lounge area of a large student union building. This area, a room 30' x 30', was separated by an open hallway from a club-type dining area which afforded a clear view of the lounge. The lounge was small enough to be wholly observable at all times, but it was not so small as to be constantly crowded. It contained three long, upholstered couches that were each flush against one of the three walls. The front of each couch was aligned with a row of 5-8 small tables with facing chairs. In the interior of the room, several larger tables were arranged in four parallel rows of two tables each, with chairs spaced evenly along the sides. The physical arrangements were such that about 60 people could be seated comfortably in the room at any one time.

Subjects and Design. The subjects were 30 male undergraduates who were strangers to the male experimenter. They were randomly chosen from the population of males who were currently occupying the lounge area, within the constraints established by the social density of the room and the availability of a seat next to the subject. The between-subjects design contrasted two conditions of social density (low vs. high), with 15 subjects in each condition.

Procedure. When seated in the dining area across the hallway from the lounge, the experimenter was able to unobtrusively view the entire lounge area without being noticed himself. From this vantage point, he waited until a condition of low or high social density was established in the lounge. Operationally, the lounge was considered to have low social density when it was occupied by no more than eight people who were not grouped together but were spaced throughout the room. The lounge was considered to have high social density when it was at least three-quarters full. In the high density condition, it was virtually impossible for the experimenter to sit anywhere without violating someone's zone of intimacy (Hall, 1966). When conditions in the lounge area were such that one of these two levels of social density had been established, the experimenter would leave his seat in the dining area and take a seat immediately beside a preselected male subject who was not occupied with studying or talking with friends. There were never more than five inches of distance separating the experimenter from the subject.

After sitting down, the experimenter attempted to keep his behavior as similar as possible for each subject. Avoiding eye contact with the subject, he would first glance at his watch and then begin drinking a soda. At no time did he attempt to converse with the subject or make any sign of recognition. If spoken to, the experimenter acknowledged the subject's comment as briefly as possible and then re-estabished his non-involved posture (noting those cases in which the subject spoke to him).

When the subject got up and left the room, the experimenter again glanced at his watch and recorded the amount of time that had elapsed since his violation of the other's space. This procedure was varied only when a previously specified limit of five minutes had been reached. On these occasions, the experimenter simply recorded the time as 300 seconds and then moved on. He then returned to his seat in the dining area and waited until a complete changeover in the population of the lounge had taken place. When conditions in the lounge permitted a new trial to be run, the procedure was repeated.

Results

Consistent with the attribution hypothesis, subjects whose personal space had been invaded left the area significantly sooner in the low social density condition ($\underline{M}=54.9$ sec.) than in the high social density condition ($\underline{M}=266.3$ sec.; $\underline{t}(28)=10.8$, $\underline{p}<.001$). A chi-square analysis corroborated this finding indicating that 12 out of 15 subjects remained for the entire five minute test period in the high social density condition, whereas none of the 15 subjects in the low density condition remained that long (χ^2 (1) = 20.0, $\underline{p}<.001$).

In addition to leaving the area more quickly, three of the subjects in the low social density condition attempted to engage the confederate in conversation, whereas none of the subjects in the high social density condition did so. This finding is at least suggestive in its implication that subjects in the low social density condition may have been motivated to talk to the confederate in order to ascertain if there were any dispositional reasons for his invasion of their space.

Discussion

The result of the first study provided some support for the attributional model of personal space behavior, since subjects in the low density condition were apparently considerably more uncomfortable than subjects in the high density condition. This result would not be predicted by a theory which considers personal space violations to be automatic elicitors of negative reactions, or by a theory which relates discomfort directly to the social density in a room. It seems most reasonable to assume that subjects in this experiment attributed different meanings to the act of spatial invasion in the two conditions, and that their degree of discomfort was in large measure dependent on the nature of these attributions.

Support for the model must be qualified, however, because of design problems in Experiment 1 which make interpretation somewhat ambiguous. These include: (a) a possible bias due to non-random subject selection such that subjects who felt comfortable with other people close by were more likely to be in the high social density condition, whereas subjects who would be most bothered by a spatial invasion were more likely to be found only in the low social density condition, (b) no control data to indicate that rate of leaving was equal in the two density conditions, and (c) no direct evidence to indicate that subjects were in fact making attributions about the intent and meaning of the invader's actions. Although such attributions could be inferred on the basis of the subjects' behavior, more direct evidence such as subjects' self reports would be desirable.

In order to address these problems, two further experiments were conducted: a large-scale field experiment which included no-invader control groups and extended the design to include invaders of both sexes, and a laboratory experiment which addressed all three problems mentioned above.

EXPERIMENT 2

In addition to expanding the design of Experiment 1 to include both male and female invaders, we attempted to assess the relative impact to one vs. two invaders, under conditions in which conversation with the subject was either facilitated or inhibited. Although no specific predictions were made concerning the simple or interactive effects of these variables, pilot observations indicated that number of invaders and possible conversation were likely to have an effect on subjects' attributions about the nature of the invasion and their subsequent behavior.

Method

Subjects. Subjects in this experiment were 270 male patrons of the Del Mar Race Track, Del Mar, California. They ranged in age from late teens to approximately 70 years. Only those patrons were used who met the following criteria: (a) seated alone with an empty space adjacent, (b) not obviously waiting for the return of a companion, and (c) sitting in a position where an observer could record their behavior in response to an invasion by a confederate. An attempt was made to include female subjects in the study, but there were so few unaccompanied females at the track that systematic data collection for female subjects was impossible.

Setting. Observations were made while subjects were seated at one of two areas: an open area between the track and the grandstand which held five sections of 30 ft. benches, approximately 20 rows of benches per section, and an enclosed area under the grandstand near the betting windows with several 10 ft. benches placed against the walls. Pilot observations indicated that people in these areas left their seats fairly often in order to place bets, collect winnings, or get refreshments. Trials were always run so that no armrest or other object was between the subject and the invader, and trials within a given condition were equally divided between the two areas. Trials were run during late afternoon hours, but never within five minutes of the completion of a race or five minutes of the next post time.

Design. A four-way factorial design was employed which included number of invaders (one vs. two), sex of the invader (s), population density in the area (0-20% of seats occupied, 20-50%, and 50% or more), and the possibility of conversation between the subject and the invader (high or low). In the one-invader conditions, the possibility of conversation was varied by having the invader either look at a newspaper or not; in the two-invader conditions, the invaders were either conversing or not. During the course of the experiment, confederate availability made it impossible to have two male invaders, so the designs used for statistical analysis reflected the presence of these empty cells. A no-invader control group was also run which consisted of 10 male subjects in each of the three density conditions.

Procedure. Confederates in this study were white college students ranging in age from 20 to 26 who were casually dressed and without unusual appearance features such as large beards or unusual haircuts. Confederates worked in pairs or triplets, alternating the roles of invader and observer. All confederates were trained in the use of the coding categories for subjects' responses until satisfactory agreement was obtained (usually .90 or above).

After the invader and observer had selected a subject according to the criteria mentioned earlier, the observer positioned himself* so that he could unobstrusively record the subject's reactions to a spatial invasion. This often required that the observer sit to the side of or behind the subject so that recording of facial expressions was impossible. After the observer was seated, the invader approached and sat next to the subject at a distance of approximately 10". The invader did not speak to the subject unless spoken to, and then kept any response as brief as possible. In the two-invader conditions, the invaders approached the subject together and both sat on the same side of the subject with the nearest invader approximately 10" from the subject. The invader(s) remained seated until the subject left the area or until five minutes had passed, at which point they left and went to a different area of the track to run a new trial.

Observers recorded the subjects' behavior by coding responses into several categories. Categories were constructed on the basis of pilot observations so that they were as precisely defined as possible, included the behaviors in which subjects commonly engaged, and were possibly indicative of discomfort or stress. These categories included:

Turn toward or Turn away. Any rotation of the subject's body more than 10" toward or away from the invader.

Touches self. A grooming or scratching movement directed to any part of the body with the hands.

Leaves. Subject gets up and leaves the area.

Scoring for the first two measures consisted of simple frequencies, while time to leave was measured in seconds from the time the invader sat down. For control subjects, the same coding scheme was used, with a position to one side of the subject denoted as the invader's position for behaviors which involved "toward" or "away" distinctions.

Results

Because the first two nonverbal measures of discomfort occurred with very low frequency, and because they did not show significant effects for treatment conditions, further discussion in this section will focus primarily on the time-to-leave variable.

----

* For simplicity, the pronouns he, him, and his will refer to both male and female confederates.

Sex of invader(s)			
Male	Female	Female-Female	Male-Female
224.4	261.6	261.6	250.4

Table 1. Average time to leave (in seconds) for the four sex conditions

For purposes of analysis of variance, the sex of invader(s) factors were combined to yield a single factor with four levels; male, female, female-female, and male-female. Thus the design became a 4 x 3 x 2 factorial with sex of invader, density, and possible conversation as factors. Contrary to the results of Experiment 1, there was no effect of density on time to leave in either the experimental or control conditions ($Fs<1.25$, n.s.).

It did appear, however, that subjects were more likely to leave when invaded by a male, as indicated in Table 1. The sex of invader main effect approached significance, $F(3,216)=2.26$, $p<.07$, and separate analyses for the one- and two-invader conditions indicated that the difference between the male alone and female alone conditions was significant, $F(1,108)=4.20$, $p<.05$, whereas the female-female and male-female conditions did not differ from each other ($F<1$).

Because the distributions for the time to leave measure were not normal due to a high proportion of subjects staying the full five minutes, data for this measure were also analyzed using the method for multidimensional contingency tables suggested by Fienberg (1970). Since this procedure is able to accomodate empty cells, the analysis was able to include density, the possibility of conversation, sex of the invader closest to the subject, and sex of the second invader (none, male, or female) as factors. When the time to leave data were dichotomized into subjects staying the full five minutes and those leaving earlier, the contingency table analysis confirmed the ANOVA results in suggesting that subjects were more likely to leave when their space was invaded by a male ($\chi^2(1)=3.96$, $p<.05$), particularly when the male was alone ($\chi^2(1)$ for sex of first invader x set of second invader = 3.58, $p<.07$).

An interaction between density, sex of nearest invader, and the possibility of conversation was also obtained ($\chi^2(2)=6.46$, $p<.025$), which suggests that the effects of sex of invader and the possibility of conversation were more salient in the low density condition than in the moderate or high density conditions. This interaction, plotted in Figure 1, suggests that subjects in the moderate and high density conditions, who had a situational explanation for the invader's behavior, were not affected by other factors, while subjects in the low density condition who may have been motivated to explain the invader's actions, used the cues of sex of invader and possible conversation to determine their reaction. However, since no direct evidence is available on this point, interpretation of the interaction must remain speculative.

Discussion

Although reliable sex effects were obtained in this study, the results provided very little support for an attributional model of spatial behavior. Although the interaction in Figure 1 suggested an attributional interpretation, the straightforward

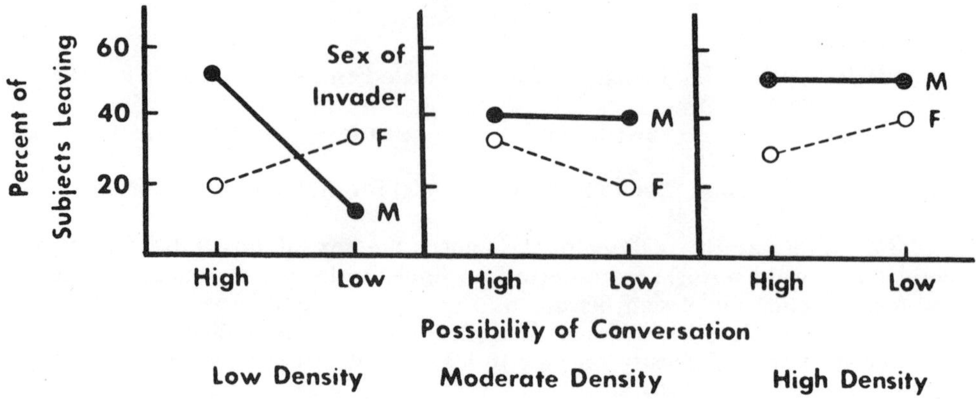

Figure 1. Three-way interaction from Experiment 2.

effects found in Experiment 1 were not replicated. On the whole, the vast number of differences in setting, subject population, etc., between Experiment 1 and Experiment 2 make discussion of possible reasons for the different findings a speculative venture. Because of the nature of the situations in the first two experiments, it was not feasible to collect the self-report data needed to assess subjects' attributions, so we were forced to infer attributions based on behaviors which did not necessariliy reflect the attribution process accurately. It was apparent that there was a need to control as many extraneous factors as possible and get more direct data concerning subjects' attributions about the intent and meaning underlying the invader's actions. These needs suggested that a laboratory experiment would be desirable to provide a more incisive look at the issues raised in the first two experiments.

EXPERIMENT 3

In addition to the sex of invader factor, which seemed to be the most important determinant of subjects' responses in Experiment 2, sex of the subject was also varied in Experiment 3. Factors such as number of invaders and possible conversation which did not produce reliable effects in Experiment 2 were omitted. Since it was not possible to manipulate social density in the situation of Experiment 3, a different procedure for varying perceived control was instituted in which an experimenter either directed a confederate to sit next to the subject (causality located in the environment) or apparently allowed the confederate to choose any seat out of several which were available (causality located in the invader).

Method

Subjects. Ninety-three undergraduate stutents at the University of Wisconsin served as subjects in this experiment. The data for four subjects were not used because of obvious suspicion about the experimental procedures, and the data for four additional subjects were not used because they sat in a position in the room which made it impossible to continue the experiment.

Design. A 2 x 2 x 2 between-subjects design was employed which included sex of subject, sex of invader, and perceived control of the invader (choice vs. no choice) as factors. A no-invader control condition was included in order to assess baseline rates for those measures which did not refer directly to the presence of another person.

Procedure. To ensure that subjects would clearly feel that their space had been invaded, it was necessary to seat the subject first, and then bring in a confederate to sit in an adjacent seat. Pilot testing indicated that subjects were often suspicious when the confederate was introduced as a late-arriving subject, so the following procedure was developed to reduce suspicion to acceptable levels.

Subjects arrived individually at a waiting room. Two minutes before the start of a session, a confederate (one of three male and three female psychology students) entered the waiting room and explained that he was also to be in the experiment but would be delayed because he had to make a phone call. The confederate then left and did not sit down during this phase of the procedure in order to avoid establishing a norm for seated distance between himself and the subject.

When the confederate had left the waiting room, the experimenter entered the room and asked for both the subject and confederate by name. Subjects uniformly told the experimenter that the other person would return shortly. After a moment's deliberation, the experimenter told the subject that he could begin the session immediately and that the other subject could begin when he returned.

The subject was taken to a 14' x 8' room with 10 armless chairs arranged in a semicircle along three of the four walls. On a table in one corner, concealed by some cardboard boxes, was a videotape camera used to record subjects' nonverbal behaviors. As the experimenter entered the room, he walked to a chair along one of the long walls which had two questionnaires on the seat, picked up the set in the adjacent seat, and motioned for the subject to sit next to him so that the experimenter could explain the questionnaire. As a result of this seating procedure (used to avoid suspicion about a camera), 12 subjects did not sit in the proper seat and were out of camera range. The session continued for these subjects, but no videotape data was available. The subjects were fairly evenly distributed across experimental conditions and no significant differences were observed on any measures between subjects in the proper seat and those seated elsewhere. When the subject was seated, the experimenter directed him to read a cover sheet which described the study as an experiment in aesthetic judgment, and told the subject not to proceed with the attached questionnaire until further instructions were given.

The experimenter then left the room, met the confederate in a control room, and waited until the subject had finished reading the cover sheet. The experimenter and confederate then entered the experimental room. As they entered, the experimenter gave the confederate instructions to read the cover sheet and not to proceed further. At this point, the experimenter gave one of two instructions: the confederate was told either to "take a seat anywhere" (accompanied by a sweeping gesture to indicate the whole room) or to "take a seat right there so that the next part of the experiment will be easier" (accompanied by a gesture pointing to the seat directly next to the subject, which was approximately 6" from the subject's chair). In both conditions, the confederate took the seat next to the subject.

After the confederate had read the cover sheet, the experimenter took a measure of perspiration from the middle fingertip of both the subject and confederate according to the method proposed by Strahan, Todd and Inglis (1974). The experimenter then told the confederate and subject that they could begin the questionnaire and that he would return in about 10 minutes. He then left the room, and both subject and confederate completed a questionnaire about the arrangements in their apartment or dormitory room which was supposedly related to the aesthetic judgments they would be making later. As in the previous experiments, the confederate did not speak to the subject unless spoken to, and kept any responses as brief as possible. The confederate timed his progress so as to finish at approximately the same time as the subject. At the end of the questionnaire was a series of bipolar adjective items in which the subject rated his current feelings on six dimensions, the experimental room on four dimensions, and the confederate on three dimensions.

When the questionnaires had been completed, the experimenter returned, took another sweat measure using the corresponding finger on the subject's opposite hand, then led the confederate from the room. He returned to give the subject a post-experimental questionnaire which consisted of several open-ended questions about the subject's thoughts during the experiment and a manipulation check item. When the subject had finished this questionnaire, the experimenter returned and debriefed the subject thoroughly as to the true purpose of the experiment.

Procedures for the control subjects were identical to those for experimental subjects with the following exceptions: (a) no confederate appeared in the waiting room, (b) the experimenter returned to the experimental room after ostensibly looking for the other subject and told the subject that the experiment could continue with only one participant, (c) no videotape data was collected, and (d) the subject ignored items on the questionnaires which referred to the other person in the room.

Results

Manipulation check. As a check on the choice-no choice manipulation, subjects were asked on the post-experimental questionnaire: "To what extent did you think the person sitting next to you had control over where he/she chose to sit?" Subjects placed a mark along a 90 mm line whose ends were labeled "had complete freedom to choose" and "had no choice at all". Responses were scored as the distance from the left end of the scale such that a low score indicated high perceived choice and a high score indicated little perceived choice. As expected, subjects in the choice condition saw the invader as having more control over his actions ($M=22.6$) than did subjects in the no-choice condition ($M=58.5$, $F(1,63)=35.27$, $p<.001$).

Self-report measures. Subjects rated their own feelings by marking a line between six sets of bipolar adjectives: comfortable-uncomfortable, happy-unhappy, nervous-calm, crowded-uncrowded, friendly-unfriendly, and confident-uncertain. The significant correlations between the six measures suggested that a multivariate analysis

	Condition		
	Choice	No Choice	Mean
Male Subjects	37.22	29.94	33.58
Female Subjects	27.98	29.69	28.83

Table 2. Self-report ratings averaged across the six measures.
(Low scores indicate more positive feelings)

of variance would be appropriate. As shown in Table 2, males reported greater discomfort than females (Multivariate $\underline{F}(6,55)=2.56$, $\underline{p}<.03$). As indicated by a nearly significant sex of subject by condition interaction (Multivariate $\underline{F}(6,55)=2.56$, $\underline{p}<.08$) males felt particularly uncomfortable in the choice condition, whereas females were not similarly affected. In fact, examination of the individual measure means showed that the male-male-choice condition was the most negative on five of the six measures and was the second most negative on the other. If one considers the male-male pairs alone as a replication of Experiment 1, it was noted that the subjects in the choice condition reported more discomfort on all six measures than did subjects in the no-choice condition. These differences were statistically significant for three of the six measures: calm-nervous, ($\underline{F}$ $(1,16)=4.33$, $\underline{p}<.054$; friendly-unfriendly, $\underline{F}(1,16)=4.64$, $\underline{p}<.04$; and confident-uncertain, $\underline{F}(1,16)=11.34$, $\underline{p}<.004$.

Male subjects rated invaders of both sexes as more unfriendly in the choice than in the no-choice condition, while this trend was reversed for female subjects (Interaction $\underline{F}(1,60)=3.27$, $\underline{p}<.07$). Similarly, male subjects viewed the invader as less attractive in the choice than in the no-choice condition, whereas female subjects viewed the invaders as less attractive in the no-choice condition (Interaction $\underline{F}$ $(1,60)=6.72$, $\underline{p}<.025$). Means for both these measures are presented in Table 3.

<u>Open-ended questions</u>. On the post-experimental questionnaire, subjects were asked: "How did you feel when the person sat next to you?", "Did that action seem unusual?", and "Did you wonderwhyhe/she chose to sit there?". Because a large number of naive coders were unavailable, simple yes or no responses to the second and third questions were analyzed as two levels of a dichotomous variable, while the

	Measure	
	Attractive	Friendly
Male Subjects		
Choice	52.29	40.47
No Choice	38.12	36.83
Female Subjects		
Choice	37.87	33.63
No Choice	41.53	44.59

Table 3. Subjects' ratings of invaders.
(Low scores indicate more positive impressions.)

	Sex of Subject			
	Male		Female	
Tone of Response	Choice	No Choice	Choice	No Choice
Positive	2	2	8	3
Neutral	5	12	7	8
Negative	11	4	3	7

Table 4. Responses to first open-ended question, coded or positive, neutral, or negative on affective tone.

very small number of ambiguous responses (N=2) were discarded from analysis. Responses to the first question were coded by a rater (who was blind to experimental conditions and unaware of any hypotheses) as being positive, neutral, or negative in affective tone. Since most answers to the first question were obviously neutral (i.e., "I felt nothing at all"), coding was not difficult and ratings were in substantial agreement with those of the senior author.

Data from the first question are presented in Table 4. Although the most common reaction was a neutral one, several female subjects reported positive reactions to personal space invasion in the choice condition. The greatest number of negative reactions were reported by males in the choice condition and by females in the no-choice condition.

As indicated in the top half of Table 5, most subjects did not report the invader's action as unusual. The exceptions to this occured for male subjects in the choice condition, who often viewed an invasion by a member of either sex as inappropriate. The data in the bottom half of Table 5 suggest that subjects in the choice condition were often motivated to seek an explanation for the invader's behavior, whereas subjects in the no-choice condition, who had been given a situational reason, did not seek further explanation.

	Sex of Subject			
	Male		Female	
Subject's Response	Choice	No Choice	Choice	No Choice
	Question 2			
Yes	10	3	2	2
No	7	15	15	16
	Question 3			
Yes	12	4	8	2
No	6	14	10	16

Table 5. Subjects' responses to Question 2 ("Did the action seem unusual?") and Question 3 ("Did you wonder why he/she chose to sit there?").

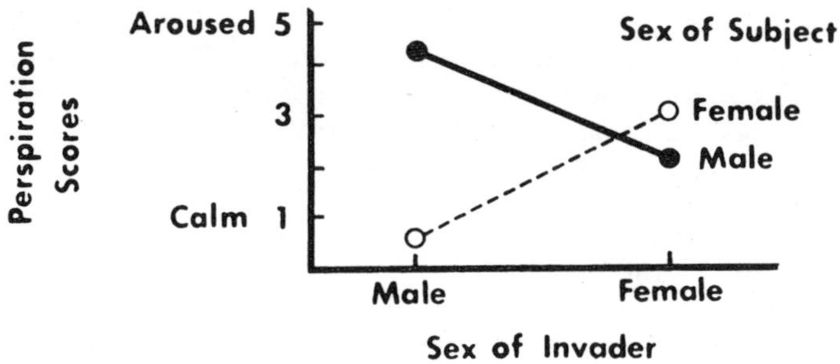

Figure 2. Digital multimeter readings of fingertip perspiration scores (arbitrary scale).

Sweat data. Analysis of variance on the perspiration scores indicated no significant main effects for experimental conditions, and no significant change between the first and second readings. However, a significant interaction was obtained between sex of subject and sex of invader, as indicated in Figure 2, $F(1,64)=4,93$, $p<.05$. Male subjects showed the highest rates of fingertip sweating when a male invaded their space, whereas female subjects perspired least in the presence of a male invader. There were no main effects or interactions attributable to the choice manipulation.

Nonverbal measures. Subjects' nonverbal behavior was coded by two trained observers using the same categories as in Experiment 2. Both observers were blind to the choice-no choice manipulation. As in Experiment 2, nonverbal measures were largely uncorrelated with each other and most behaviors were relatively infrequent. Furthermore, they were uncorrelated with either the perspiration measures, the manipulation check, or the self-report measures beyond chance levels. It was noted that although seven subjects in the choice condition turned away from the invader, no subjects in the no-choice condition did so ($\chi^2(1)=5.06$, $p<.025$).

Discussion

There was more support for the attributional model in Experiment 3, particularly in the male-male pairs. Several of the self-report measures indicated that the strong effects in Experiment 1 were not merely experimental artifacts, but rather reflected a real effect of attributed meaning on reaction to the close approach of another male. Results for the other sex combinitions were not as clear, however. The data from the open-ended questions suggests that subjects were making attributions in the mixed-sex and female-female conditions as well as in the male-male conditions, but these attributions did not lead as directly to a negative emotional state. In fact, some of the female-female data suggests a positive reaction.

It was disappointing, although not necessarily a serious problem, that the different types of measures (self-report, physiological, nonverbal) did not correlate with each other. If the measures are viewed as simple indicators of an underlying emotional state like discomfort, they should be correlated, but on the other hand, if the person is viewed as a homeostatic system (Altman, 1975), discomfort might be alleviated by engaging in any one of several behaviors. Engaging in one behavior may reduce or preclude engaging in another, so that correlations within a set of measures designed to tap the same emotional state would be low. Thus, in the situation of Experiment 3, subjects who felt discomfort about the presence of an invader may have either turned away from the invader, leaned forward, or changed position to avoid interaction in order to reduce their discomfort. Subjects who did none of these things may have appeared more fidgety and/or shown greater physiological and self-report signs of arousal. Under such a process, each measure may be affected in the same way by experimental conditions and yet be uncorrelated with the others.

Although results from the three experiments are not in complete agreement, a measure of support for the attributional model is found. In male-male pairs, where attributed intent appears to produce predominantly negative reaction, the behavioral and affective consequences of attributed meaning are fairly clear. However, in mixed-sex and female pairs it appears that attributions of intent to the invader can have positive or neutral as well as negative consequences. An important goal of future research using this approach will be to clarify situational and personal factors which control the type of meaning ascribed to an invader's action and consequent reaction by the invadee.

SOCIAL CROWDING AND THE COMPLEXITY OF THE BUILT ENVIRONMENT: A THEORETICAL AND EXPERIMENTAL FRAMEWORK

Rikard Küller

School of Architecture,Lund Institute of Technology, Fack 725,

S-220 07 Lund 7, Sweden

ABSTRACT: Based on neuropsychological considerations a model is suggested where man's capacity and activity are related to his built and social environments. Using methods from psychology and physiology the model has thus far been tested on several points. The paper presents a specific study where the visual complexity and coherence of the built environment are related to social intensity. Using four rooms of differing visual character, subjects, singly or in groups of two, each spent one hour in one of the rooms. Measurements included ratings of the rooms and of the social settings. By means of multi-dimensional analysis, it was possible to study the effects of the interaction of the social and built environments.

- - - - - - - - - - -

Crowding represents a large research field with efforts toward a theory formation and with areas of application which continually increase and become more significant. Several attempts have been made to incorporate crowding within presently existing ecological, sociological, psychological or physiological models. (See for example Stokols, 1972,; Saegert, 1972, Saegert, 1975; also refer to criticisms by Fischer, Baldassare, and Ofshe, 1975).

I will try to contribute to a somewhat broader view by relating crowding to a model of environmental psychology, that is based on neuropsychological considerations (Küller, 1976). I also intend to illuminate experimentally some of the properties of the model, notably the description of the built and social environment and their interdependence. It is appropriate to first define what is here meant by crowding.

A DEFINITION OF CROWDING

Stokols (1972) differentiates between crowding and density with density representing physical conditions while crowding implies the experiencing of these conditions. Rapoport (1975), not completely satisfied by this approach, differentiates between physical density, perceived density, and affective density, Perceived density according to Rapoport is the estimate of the number of people present in a given area, the space available, and its organization. Affective density is the evaluation of that perceived density against certain standards, norms, and desired levels of interaction and information. Crowding as well as isolation are the extreme values of affective density. Thus according to Rapoport's analysis, crowding is neither physical nor perceptual but affective.

This seems to make sense especially when one considers all the studies where crowding has been shown to be related to physiological reactions. D'Atri (1975) found for example that the enforced crowding prevailing in prison dormitories, often paired with an atmosphere of threat and conflict, resulted in higher blood pressure levels than for those inmates who stayed in single or double occupancy cells. Results like this and others obtained along similar lines in animal experiments are often explained in rather vague terms of overstimulation, physiological tension, or overloaded channel capacity. At other times very specific reactions such as increased blood pressure or adrenal glandular mechanisms are mentioned. However, the crowding concept might benefit by being specified in rigorous yet general neuropsychological terms. In doing this, it is advantageous to establish a connection between the concept of crowding and the emotional activation theory as it has been developed in recent years (e.g. Luria, 1973).

In short the theory of activation implies that every stimulation, irrespective of its coming from within or without the subject, causes a short temporary reaction (phasic arousal). Depending upon the nature of the stimulation, it can also give rise to an immediate pleasant or unpleasant feeling. As a result of repeated stimulation, the general arousal level (tonic arousal) may rise while on the other hand a lack of stimulation may lead to a decrease of the level. At the same time there are changes in the emotional state of the subject and in preparedness to react to the surroundings.

I will now propose that in a neuropsychological sense crowding be considered as an increase in the tonic arousal level caused by the presence of others in the perceptual field. Although the increase in tonic arousal level is used as an indicator of crowding, we must not confuse the one with the other. Crowding is an aspect of the environment giving rise to the effect and should not be confused with the effect itself. Two important inferences will follow from this proposition.

- If one is temporarily exposed to a high density situation but the tonic response is lacking because somehow one manages to control the situation, then there is no reason to speak of crowding.

- Although crowding implies an increase in tonic arousal level, this does not necessarily imply something negative. In cases where the tonic level initially is very low, an increase may instead mean that the individual achieves a more adequate level. Only when the crowding effect is combined with an already optimal or an excessively high level of arousal, may the effect of this combination become clearly negative. This simple facts explains a number of apparently contradictory results such as why crowding sometimes leads to improved performance.

CROWDING IN A BROADER CONTEXT

Utilizing arousal as a central concept, I will attempt to relate crowding to a model of environmental psychology. Consider the following illustration (figure 1).

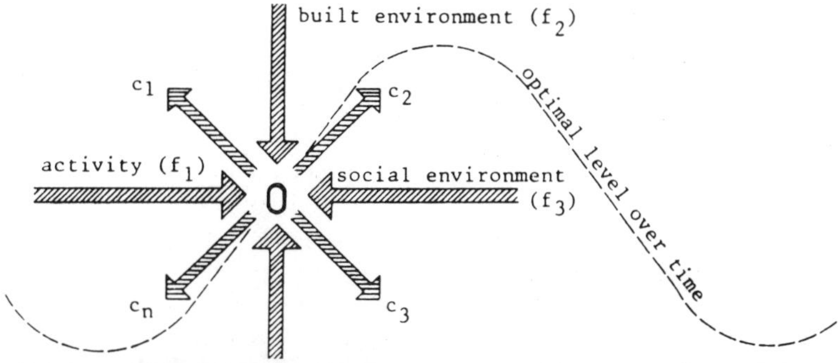

Figure 1. The model shows the interrelations between the individual and his environment. Factors f_1-f_n contribute to the arousal level of the system. Deviations from optimal arousal level (O) result in various adaptations or compensations c_1-c_n.

The optimal arousal level is related to the diurnal cycle and to the age of the subject, and is assumed to be a precondition for among other things, enjoyment and performance standards. In principle, the model implies that different individuals have varying arousal requirements, that the activity which the individual is performing itself contributes further arousal (this varies widely depending on the nature of the activity in question), and that the built and social environment both add to the arousal. Every combination of factors, if it is unsatisfactory, will engage mechanisms of either adaptive or compensating character. The direct effects of this will be manifested primarily as psychological and physiological reactions of emotional character. These in turn might lead to long-term medical and socio-psychological consequences. Thus, the demands of the individual and his activities can with the help of the model be related to the conditions of the built and social environment, and the consequences can be studied.

It is clearly apparent that crowding, according to the model, should be considered as one of the factors which contribute to the arousal level of the system. Crowding must actually be considered as part of f_3, social environment. It is true that Stokols (1972) differentiates between social crowding and non-social crowding with the latter implying "spatial factors including the amount and arrangement of space, stressor variables such as noise or glare which heighten the salience of physical constraints, and personal characteristics including idiosyncratic skills and traits". However, according to the present model, these aspects refer to f_2, built environment, and f_n, reaction tendency. Even if these factors do interact with crowding, they should not in themselves be considered as such.

In order to illustrate some of the properties of the model we shall now turn to some experimental findings.

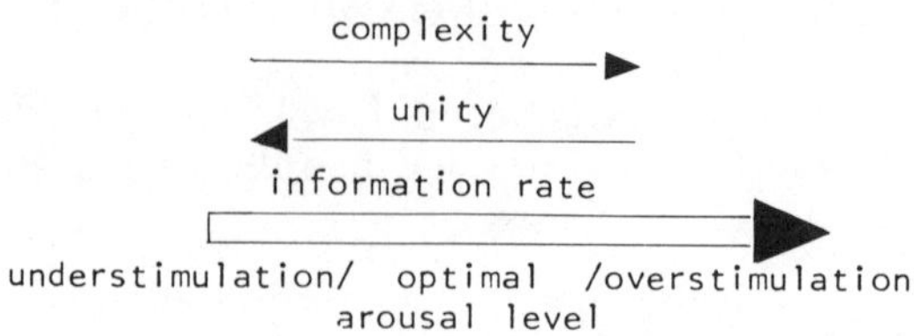

Figure 2. Relationship between complexity, unity, information rate and arousal.

THE EXPERIMENT

Four square rooms with floor spaces of 5.5 square meters and ceiling height of 2.4 meters were constructed in the laboratory. By working with floor and wall materials, colouration, decorative patterns and lighting, it was possible to obtain one subdued room of well arranged appearance (A), one lively and disorderly room (B), one subdued but disorderly room (C), and finally one lively and well arranged room (D).

Students (n=48) spent one hour sitting in one of the rooms, either alone or in pairs. They were told to complete a number of evaluations and tasks independently according to the following schedule. Instructions were given using a tape recorder.

Time in minutes:
00 experiment begins
05 semantic room description
25 first self report
39 work period (vigilance task for approx. 10 minutes)
51 second self report
55 ratings of social situation
65 end of experiment

The results obtained from the semantic room descriptions and the social situation ratings are briefly discussed in the following sections. Prior to describing the results, comments are made concerning the development of the measures.

THE SEMANTIC ROOM DESCRIPTION

Architectonic space –in the model corresponding to factor f_2– has in recent years been the subject of systematic perception research. In Lund we have studied how it is experienced and described from a number of different aspects. This has resulted in a terminology where space is described in eight orthogonal dimensions: pleasantness, unity (also named coherence), complexity, enclosedness, potency, social status, affection (familiarity) and originality (novelty) (Küller, 1972, 1973).

The choice of room design was governed by an intention to influence arousal. Of the various dimensions, complexity (the presence of numerous and intense stimulations) is considered to increase tonic arousal level. Unity, which implies the designing of the space into a coherent and functional unit, can be assumed to reduce tonic arousal level. Together, complexity and the degree of unity constitute a dimension of higher order which we call information rate (figure 2). It is further assumed that the dimension of affection reduces the occurrence of phasic arousal reactions. Thus if the individual is familiar with his surroundings he does not let himself be disturbed or upset. Originality is assumed to lead to phasic arousal reactions through unusual and unexpected stimulation (Berlyne, 1971; Küller, 1977).

Earlier experimental results point to the soundness of this description. In one study a room with high complexity and low unity was compared to a room with low complexity and high unity (Küller, 1976). Subjects spent three hours in each of the two rooms. Measurements of performance, EEG, EKG and self ratings of emotions, indicated higher arousal in the room with high information rate.

The test of semantic room description (Küller, 1972) administered in the beginning of the experiment gave the profiles shown in figure 3 for the four rooms. Looking at the profiles we see that room A which has low complexity and originality and high unity and affection ought to contribute very little to arousal. Room B, on the other hand, with a very high complexity and originality and very low unity, should contribute much more to arousal.

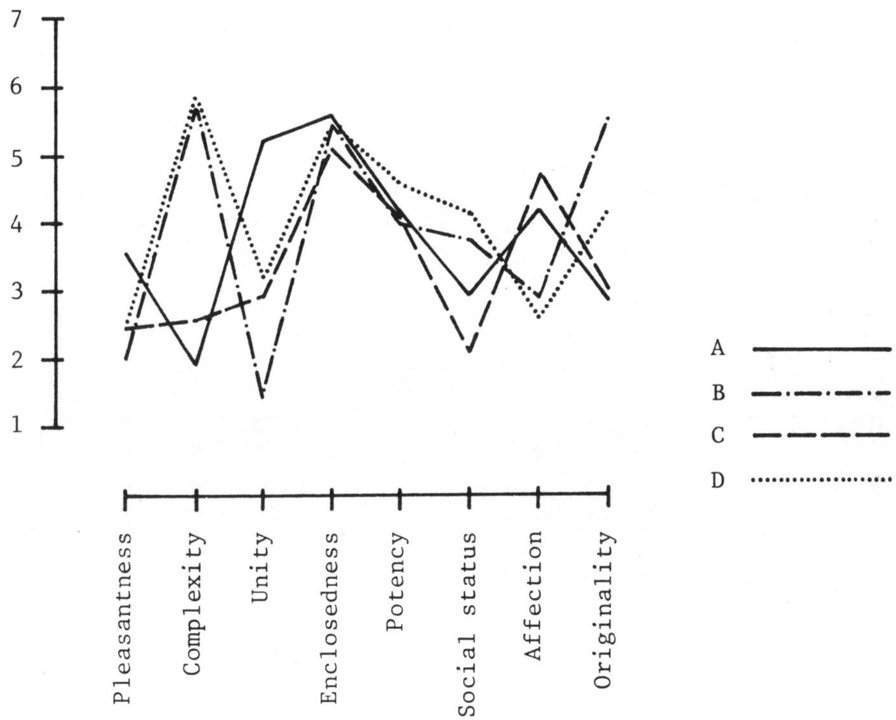

Figure 3. Profiles for four rooms according to the semantic milieu description test.

THE SOCIAL SITUATION

Descriptions of social situations, in the model corresponding to factor f3, were subjected to factor analysis in a study which included both large and small groups in formal as well as informal contexts (Küller, forthcoming). The following five dimensions emerged: social intensity, interpersonal stability, familiarity of the situation, coherence, and finally friendliness.

Of these dimensions, social intensity may be considered to be closely related to perceived density and accordingly is of interest in a discussion of crowding. There exists experimental evidence showing that the presence of other persons in the immediate vicinity tends to increase arousal. Eye contact for instance, has been shown beyond doubt to be arousing. In one study where EEG was used to assess arousal, different gaze-conditions were compared. When the experimenter looked into the eyes of the subject there was a distinct increase in arousal. If at the same time the experimenter smiled, arousal became even more accentuated (Gale et al., 1972).

In the present study we would expect the difference in physical density to have an effect on social intensity. In physical terms sitting alone in the room implied a density of .18, while sitting in pairs corresponded to .36 individuals per square meter. Even more important must be the fact of being alone or in pairs in a small room, as it is known to be the probability of interaction rather than the availability of space that is of most relevance (Freedman 1975). On the other hand the subjects were not expected to cooperate or compete, but to perform their tasks individually, which may have worked to suppress feelings of social intensity. With this in mind let us look at the students' ratings of the social situation for singles and pairs separately (figure 4).

Not on one single point does the difference between the curves reach significance although for friendliness there is what may be regarded a tendency, and for intensity the difference, although far from significant, is in the expected direction. Thus, being in the room alone or in pairs seems to have a negligible influence on the

Figure 4. Profiles for two social situations, singles and pairs (values are relative factor scores).

experiencing of the social situation. Altogether we will have to conclude that so far the results of this study have given us little reason to speak of two different social situations in reference to singles and pairs. However, let us consider the interaction aspects of the model before we take any definite position.

INTERACTIONS

In an unpublished study by Sorte, it was shown that the number of people present in an urban space actually influence the impression of the space itself. There was generally a decline in pleasantness and a rise in complexity as the number of people increased from very few to many, although in some environments this trend was weak or absent.

Some years ago Maslow and Mintz (1956), in their now classic study, showed that the character of the environment had an effect on how people are perceived. Subjects who spent 10-15 minutes rating a series of facial-photographs in a beautiful room rated the faces as having significantly more energy and well-being than subjects tested in either an average or an ugly room.

Taken together, these two studies indicate that a two-way effect exists. Not only does the presence of people have an influence on how the space is experienced, but the character of the space also influences how people in that space are being perceived. Do we find similar interactions in the present study?

Actually, room A was judged as much more pleasant by subjects who were alone in it ($m=4.53$) than by subjects in pairs ($m=2.71$). For the remaining rooms, the difference was less than one third of a unit. Thus, there is some agreement with Sorte's results, for at least in one of the rooms pleasantness declined as density increased.

But the influence also worked the other way. The social situation was judged to be more friendly and stable in room A, which was the most pleasant of the four rooms. This result seems to be similar to the one obtained by Maslow and Mintz. Further analysis hints at the possibility that the low complexity and high unity of room A must somehow be involved in an eventual explanation.

One final example of interaction should be mentioned. In the present study ratings of social intensity increased with the complexity of the room, but degree of unity as well as social density also played a role. Thus, social intensity had its very lowest value when single subjects rated room A, where complexity was low and unity high. Social intensity had its highest value when the subjects were present in pairs in room B, where the complexity was very high and the unity very low. This result certainly reveals the previously hidden impact of social density and supports an interaction hypothesis, and also leads us to expect the second situation to contribute much more to arousal than the first.

DISCUSSION

If these conclusions are correct then this naturally means that the so-called independent variables, e.g. space characteristics and social density, cannot in fact be considered as independent. Instead these, as well as the other components of the model, must be regarded as abstracted aspects of an integrated process. There is cause also to expect similar interactions to occur in the case of activity and reaction tendency. Stokols et al. (1973), for instance, found that subjects' perceptions of crowdedness were greater when the ongoing activity was competitive rather than cooperative. (For a detailed discussion of these and other aspects of the model refer to Küller, 1978. The model is presently being employed in a field study of ship interior environments.)

Attempts, like the one presented here, to relate crowding to existing theoretical models in the long run will most likely be more profitable than to continue treating crowding as a separate subject area, because crowding is only one of the many factors constituting the human environment. Furthermore, crowding interacts with other environmental aspects, and ought therefore to be studied in as wide a context as possible. Otherwise the risk is that the knowledge obtained will be contradictory and thus difficult to interpret. For example, negative effects of crowding have been reported by some authors, positive effects by others, while again others have found no effects at all. These are findings which if not taken in their proper context, will not make us at all the more wise.

EXPERIMENTAL INVESTIGATIONS OF THE DETERMINANTS AND PERSONAL CONSEOUENCES OF CROWDING

Marilyn Rall, Brooklyn College; Daniel Stokols, University of California; Robert Russo, Wayne State University; Alan Steinberg, Brooklyn College; Anthony Norbut, Brooklyn College, U.S.A.

ABSTRACT: Four laboratory experiments were conducted to test the contention that behavioral responses to crowding can be predicted by distinguishing between neutral and personal sources of the crowding experience. Manipulations of neutral crowding (relatively small rooms without windows) provoked a physical withdrawal from a target person and no negative emotional labeling besides "presently crowded". Manipulations of personal crowding (such as the threat of an intrusive interview) provoked complex spatial adaptations and many negative emotional labels in addition to the rating of "crowded".

— — — — — —

Recent research in the area of crowding has increasingly demonstrated that a distinction must be drawn between the experience of crowding and the high-density conditions which may contribute to it (Stokols, 1972). Other important dimensions which have an impact on the perception of crowding are: social intrusive , physical constraint, and individual differences (Stokols, Rall, Pinner, and Schopler, 1973).

A number of theoretical models have evolved which have attempted to specify the critical determinants of the crowding experience. Milgram (1970) has suggested that crowding is primarily mediated by the excessive arousal of stimulus overload. Proshansky, Ittelson, and Rivlin (1970) have focused upon the reduced behavioral freedom brought about by the presence of others. Wicker (1973) has proposed that the scarcity of resources in a situation is central to the experience of crowding.

Basic to all crowding models is the assumption that crowding is a form of psychological stress which motivates a person to augment his supply of space. The general prediction of a person's resulting behavior is that he will distance himself from others. Depending upon the circumstance, the individual is expected to escape, withdraw, isolate himself, put up barriers, force others to withdraw, or to psychologically redefine the situation as being less crowded.

In a partial attempt to provide a more specific basis for predicting behavior in a crowded situation, Stokols (1975) has proposed two classes of antecedent variables which would limit the adaptive options the individual would have open to him. The first class has to do with whether the individual's sense of crowding has been activated by a neutral or personal source. If the source is neutral, the individual has the option of finding a basis to cooperate with others in the situation or to simply ignore others. If, on the other hand, crowding stress is activated by some threatening action of another individual, cooperation is probably not possible and ignoring is dangerous. Under conditions of personal crowding, the individual must either defend himself or leave the situation.

The second class of variables proposed by Stokols has to do with the environment in which the crowding occurs. He makes the distinction between primary environments in which an individual spends most of his time engaging in important activities with people he knows well, and secondary environments which are relatively transitory, anonymous, and inconsequential. Since people have high investments in primary environments, crowding in this type of setting is particularly stressful and the option of escaping is usually costly. As a result, the predicted behavior in primary environments would be passive isolation, behavioral withdrawal, aggression, or, if possible, improved coordination with others.

The purpose of the present study is to empirically test some of the predictions of the Stokols typology. Since a laboratory study by definition is a transitory, anonymous situation, no attempt has been made at present to test the relative effects of primary vs. secondary environments. However, within the context of a secondary environment, an attempt has been made to measure the relative behavioral adjustment brought about by neutral and personal crowding.

Experimental Design

The basic design is an extremely simple one: The subject is asked to enter a room where one chair was stacked on top of another. He is told to take the top chair and to sit on it while waiting for an interviewer who will sit in the other chair. It is assumed that the distance the subject chooses to place himself from the interviewer's empty chair represents a spatial adaptation to any crowding he anticipates.

It is felt that keeping the interviewer's chair empty, as opposed to having an actual person sitting in it, removes extraneous reactions that the subject might have to any particular stimulus person. Baum and Greenberg (1975) have demonstrated that subjects respond to the anticipation of crowding. In fact, one might argue that using an anticipation manipulation is more informative than presenting the individual with an actual crowding experience, because he will draw upon his real-life memories of crowding as he prepares for the encounter.

Experiment I

Method

Subjects. The subjects were 40 male and 40 female introductory psychology students from the University of North Carolina at Chapel Hill. Most had lived in

North Carolina all their lives, and they generally came from small towns.

Independent Variables. Three factors were varied systematically:
1. A neutral, physical factor which consisted of a small room (5ft., 5 in. x 7 ft., 4 in.) vs. a large room (7 ft., 4 in., x 11 ft.).
2. A personal factor which consisted of an anticipated threatening interviewer vs. a non-threatening interviewer.
3. The sex of the subject.

Procedure. Subjects were recruited for an interview on the topic, "Should psychology be taught in high schools?" When the subject arrived he was asked to complete a self-rating form which consisted of nine semantic differentials: unfriendly-friendly, unpleasant-pleasant, submissive-dominant, lazy-ambitious, insincere-sincere, unattractive-attractive, introvert-extrovert, unsuccessful-successful, and tense-relaxed. The experimenter then explained that he was the laboratory assistant of the interviewer who was an advancedgraduate student and who happened to be the same sex as the subject.

In the non-threatening conditions, the subject was told that the interviewer was simply conducting a poll to see what college students think about teaching psychology at the high-school level. In the threatening conditions the interviewer was described as a clinical psychologist who had developed a new technique for detecting how sincere people are about their expressed beliefs. The interview was described as a vehicle for trying out the new technique, and it was explained that the interviewer would attempt to discover if the subject's rating on the "sincere-insincere" item of the preliminary questionnaire had been filled out honestly.

The experimenter then opened a side door and called out, "Are you ready to begin yet?" He returned to the subject and said, "He/she will be ready in a few minutes. Why don't you come on in to the interview room?" Another door was opened which lead into either the small or large room where two chairs were stacked against a wall. The interviewer then said, "Oh, the chairs are still stacked. Well, just go ahead and arrange them for the interview. While you're waiting you can fill this form out. It's a check to make sure you've understood my instructions and to see how you perceive the situation." After handing the subject a clipboard with a questionnaire, the experimenter closed the door behind him.

Five minutes later the experimenter returned, explained the true purpose of the experiment, measured the distance between the two chairs and recorded their relative angles.

Manipulation checks. The first item on the questionnaire was an open-ended question asking the purpose of the interview. A series of seven-point semantic differentials were also included: "The size of the discussion room is, large - small", and 'At this moment I feel: uncomfortable-comfortable, anxious-calm, tense-relaxed, insecure-secure."

Dependent variables. The second part of the questionnaire consisted of semantic differentials which were analyzed as dependent variables: "In this room I feel, not crowded -crowded; During the interview session while responding to the

Table 1
Cell Means for Distance (in inches)

	Threat		Non-Threat	
	Females	Males	Females	Males
Small Room	32.30	38.10	30.57	38.30
Large Room	24.85	30.55	22.45	33.83

interviewer's questions, I'll probably feel, not crowded-crowded; And I 'll probably be polite-impolite, intimate-detached, not hostile-hostile, sociable-unsociable, friendly-unfriendly, sincere-insincere, open-closed" The distance between the two chairs (taken from the two closest points) was also a dependent variable.

Results

A 2 x 2 x 2 multivariate analysis of variance was computed for the subjects' responses.

Manipulation checks. Responses to the open-ended question yielded no suggestion that subjects suspected that chair placement had anything to do with the study. It was found that subjects in the large room rated the room as significantly larger than subjects in the small room ($F=17.25$, $df=1/72$, $p<.001$). Unfortunately, there were no significant threat vs. non-threat main effects for the subjects' ratings of their current emotional state.

Dependent Variables. The distance between the two chairs yielded two main effects. Subjects sat further away in the small room than they did in the large room ($F=10.54$, $df=1/72$, $p<.01$), and males established a greater distance than females ($F=12.96$, $df=1/72$, $p<.001$). The cell means for the distance measure are presented in Table 1.

Threat vs. non-threat main effects on the questionnaire responses revealed that subjects expected to feel more hostile ($F=8.45$, $df=1/72$, $p<.01$) and less sincere ($F=9.11$, $df=1/72$, $p<.01$) in the threatening situation.

A surprising finding was that subjects in the large room tended to sit at right angles with the interviewer's chair, while subjects in the small room tended to sit straight across from the other chair. Unfortunately, the door in the small room was right next to the interviewer's chair, where as in the large room, the door was along a side wall relative to the interviewer. Thus, all subjects actually tended to sit facing the door, and, as a result, the spacial measure was seriously confounded.

Experiment II

Method

The second experiment was basically a replication of the first experiment with the following changes:

Table 2
Cell means for Distance in Experiment II

	Threat		Non-Threat	
	Females	Males	Females	Males
Small Room	27.1	28.4	29.6	34.7
Large Room	15.1	24.9	20.5	31.5

Subjects. Forty male and 41 female introductory psychology students from Brooklyn College participated.All but 15 said they had lived in New York all their lives. Ten lived in New York for over 10 years, and the remaining 5 claimed a residency ranging from 3 to 8 years.

Procedure. The experimental rooms were proportional rectangles with doors in the same relative position. The small room was 6' x 8' and the large room was 9' x 12'. The stacked chairs were placed at the center of the shorter wall, and the door was on the longer wall in the corner near the chairs.

In the non-threatening interview the subject was told he would hear a series of jokes, and his job would be to rate each joke in order to help the interviewer pick the funniest ones for a future experiment. In the threatening condition the interviewer was described as a clinician who was going to use a new, intensive technique to assess the subject's ability to develop mature, longlasting, intimate relationships.

In addition to measuring the distance between the two chairs, the experiment recorded the "revolution" or size of the angle between the natural line of sight on the interviewer's chair and the amount the interviewer would have to turn in order to see the subject. The "pivot" or angle between the subject's natural line of sight and the amount the subject would have to turn in order to see the interviewer was also recorded.

Results

Manipulation Checks. The large room was rated as significantly larger than the small room ($\bar{X}$=2.98 vs. $\bar{x}$=2.07; F=17.16, df=1/73, p<.001). The threat vs. non-threat manipulation was substantiated in that subjects who expected to hear jokes rated themselves as significantly more comfortable, ($\bar{X}$=5.12 vs. $\bar{x}$=4.02; F= 14.56, p< .001), calm ($\bar{X}$=4.80 vs. $\bar{x}$=3.83; F=8.19, p<.01), relaxed ($\bar{X}$=5.00 vs. $\bar{X}$=3.90; F= 12.24, p<.001), and secure ($\bar{X}$=5.30 vs. $\bar{x}$=4.44; F=8.76, p<.01) than subjects who expected the clinical interview.

Dependent Variables. Subjects sat further from the interviewer's chair in the small room than in the large room (F=9.40, p<.01); men placed themselves at a greater distance than did women (F=8.86, p<.01); and subjects sat closer when threatened (F=5.36, p<.05). The cell means for distance are presented in Table 2.

Table 3
Cell Means for Revolution in Experiment II

	Threat		Non-Threat	
	Females	Males	Females	Males
Small Room	10.5	15.1	16.2	8.9
Large Room	44.8	23.5	31.8	20.0

The extent to which subjects shifted their chairs away from the interviewer's line of sight (revolved) varied in a somewhat inverse relationship to the distance measure. Subjects revolved more in the large room than in the small room ($F=22.70$, $p<.001$), and women revolved more than men ($F=5.75$, $p<.05$). There was also a sex by room size interaction due to the fact that women revolved as much as men in the small room, but much more than man in the large room, ($F=4.40$, $p<.05$). The cell means for revolution are presented in Table 3.

The third spacial measure, the degree to which subjects pivoted their chairs to cause their own line of sight to swing away from the interviewer's chair, yielded a strong trend for a threat vs. non-threat main effect. Subjects pivoted further in the threat condition. ($\bar{X}=41.25$ vs. $\bar{x}=25.65$, $F=3.57$, $p<.06$).

The extent to which subjects said they felt crowded as they waited for the interviewer yielded a main effect for room size ($F=10.13$, $p<.01$) and an interaction of sex by room size ($F=4.14$, $p<.05$). Subjects felt more crowded in the small room ($\bar{X}=3.18$ vs. $\bar{X}=2.05$), but women felt more crowded than men in the small room ($\bar{X}=3.35$ vs. $\bar{X}=3.00$) and less crowded than men in the large room ($\bar{X}=1.5$ vs. $\bar{X}=2.6$). Subjects' ratings of their anticipated crowding upon the arrival of the interviewer yielded a threat vs. non-threat main effect. Anticipated crowding was higher in the threat condition ($\bar{X}=3.95$ vs. $\bar{X}=3.12$; $F=4.99$, $p<.05$).

Two additional measures from the post-experimental questionnaire yielded sex main effects: Women gave experimental rooms a larger rating than men ($\bar{X}=2.76$ vs. $\bar{X}=2.30$; $F=4.12$, $p<.05$); and women felt more comfortable than men ($\bar{X}=4.88$ vs. $\bar{X}=4.34$; $F=4.86$, $p<.05$). None of the questions which asked the subject to rate his anticipated emotions during the interview resulted in significant effects.

Discussion of Experiments I and II

The strongest finding borne out by the first two experiments is that people respond to a decrease in available space by increasing the physical distance between them and another person in the situation. In addition, there is a strong tendency in the situation with adequate space to revolve to a position which is more side-by-side with the target person.

Sommer (1965) has pointed out that most people prefer face-to-face positions in a competitive situation and reserve side-by-side positions for cooperative, affiliative situations. Thus, the physical response to "neutral" crowding in the 1st two experiments seems to be a position which signals a guarded withdrawal. The fact that women, who are traditionally viewed as more affiliative than men, consistently sit closer and revolve more than men is further evidence that the positions taken in large rooms symbolize friendlier and more intimate intentions than the placements in the smaller rooms. (The tendency for women to prefer side-by-side positions has also been documented by Byrne, Baskett, and Hodges, 1971.)

Another interesting finding is that subjects of both experiments perceived large and small rooms as significantly different in size, but only the New York subjects showed a room-size main effect on the rating of current crowding. One possible explanation for this discrepancy might be that New Yorkers, who confront inadequate space on a regular basis, have a more highly differentiated labeling process for crowding. If this is so, one might conclude that New York women are more sensitive to spatial differences than New York men, because the women rated the small room as more crowded and the large room as less crowded than did men.

Comparing the results of Experiments I and II on the social threat manipulation is somewhat problematic in that the manipulation was deliberately strengthened for Experiment II: Threatened North Carolinians anticipated feeling more hostile and insincere during the interview, while threatened New Yorkers reported heightened pre-interview feelings of discomfort, anxiety, tension, and insecurity, and expected to feel more crowded during the interview. Threatened New Yorkers also moved closer to the interviewer while pivoting almost 45 degrees away. One striking note, however, is that both samples of respondents reported negative affect when anticipating personal threat, while differences in affect were never associated with room size in either experiment. Stokols' suggestion that personal crowding is more stressful than neutral, spatial crowding seems to be supported by these results.

Another suggestion of Stokols', that personal and neutral crowding will evoke different physical coping responses, is also confirmed by the present experiments. Inadequate space consistently evoked a guarded withdrawal, while the personal threat of having one's ability intimately scrutinized evoked a spatial advance.

Experiment III

The third experiment was a partial replication of the second experiment with the following changes: Room size was eliminated as a manipulation, and a medium sized ($9^{1/2}$' x $10^{1/2}$') room was used in all conditions. In addition to conditions in which subjects expected an interviewer of the same sex, conditions where subjects expected interviewers of the opposite sex were included. A new form of personal threat was also introduced: Rather than presenting the subject with the expectation that intimacy would be viewed as a positive attribute, it was presented as a negative attribute.

Subjects. Thirty-one male and fifty-six female introductory psychology students from Brooklyn College participated.

Independent Variables. The experimental design generated three independent variables:

1. sex of subjects
2. sex of expected interviewer
3. expectation of a threatening interviewer vs. a non-threatening interviewer.

Procedure. Eleven male and eighteen female under-graduates in an experimental psychology laboratory class served as experimenters. Subjects were told that the interviewer would be the same sex as the experimenter. In the non-threat condition subjects were told they would be rating jokes. In the threat condition subjects were told that the interviewer was a clinician who was interested in discovering the extent to which college students participate in sexual perversions.

Results.

Manipulation checks. The threat vs. non-threat manipulation was substantiated in that subjects who expected to hear jokes rated themselves as significantly more comfortable ($X=4.88$ vs. $X=4.07$; $F=5.14$, $df=1/73$, $p<.05$), calm ($X=4.93$ vs $X=4.00$; $F=6.43$, $p<.01$), relaxed ($X=4.95$ vs. $X=4.01$; $F=6.56$, $p<.01$), and secure ($X=5.67$ vs. $X=4.51$; $F=15.88$, $p<.001$) than subjects who expected a clinical interview.

Dependent Variables. The measures of how subjects expected to feel during the interview itself also yielded significant threat vs. non-threat main effects. Subjects who were not threatened anticipated being less impolite ($X=1.73$ vs. $X=2.16$; $F=5.88$, $p<.05$), unsociable ($X=2.10$ vs. $X=2.78$; $F=7.69$, $p<.01$), insincere ($X=2.18$ vs. $X=2.81$; $F=5.33$, $p<.05$), closed ($X=2.27$ vs $X=3.37$; $F=17.05$, $p<.001$), and crowded ($X=3.78$ vs. $X=4.15$; $F=8.13$, $p<.01$) than subjects in the threat condition. Non-threatened subjects also pivoted more ($X=42.15$ vs. $X=30.15$; $F=8.86$, $p<.01$) and revolved more ($X=43.22$ vs. $X=32.45$; $F=3.78$, $p<.05$) than threatened subjects while establishing a marginally closer distance ($X=41.45$ vs. $X=44.45$; $F=2.47$, $p<$ 12).

Experiment III also yielded significant 3-way interactions for the measures of distance ($F=4.17$, $df=1/79$, $p<.05$) and pivot ($F=7.13$, $df=1/79$, $p<.01$). In the non-threat conditions subjects sat closer to same sex than to the opposite sex and pivoted more with the same sex. In the threat condition subjects sat closer to the opposite sex and pivoted more with the opposite sex. The cell means for the distance and pivot measures are presented in Table 4:

Table 4
Cell Means for Distance and Pivot in Experiment III

		Threat		Non-Threat	
		Female S	Male S	Female S	Male S
Distance	Female Int.	45.5	42.3	36.0	52.9
	Male Int.	44.1	45.9	40.7	36.3
Pivot	Female Int.	28.8	30.0	56.8	18.6
	Male Int	34.5	27.3	37.7	55.5

Experiment IV

Since it was discovered that the threat manipulation in Experiment III brought about a marginal withdrawal and a decrease in pivot (the reverse pattern of the one obtained in Experiment II), Experiment IV was designed to include the essential components of both Experiments II and III. Both of the earlier forms of personal threat were manipulated, i.e., the interviewer who was studying the ability to form intimate relationships and the interviewer who was interested in detecting sexual perversions. Two male experimenters were used; one who employed a slow-paced, friendly, personal style with a great deal of eye contact (similar to the style of the experimenters in Experiments I and II) and another who employed a hurried, formal, impersonal style with little eye contact (similar to the style developed by the laboratory students in Experiment III). Finally, since large and small rooms produced fairly consistent results in the first two experiments, a new potential source of neutral crowding was introduced: the use of a testing room with a window vs. a room without a window.

Method

Subjects. One hundred and twenty-three male and 132 female introductory psychology students from Brooklyn College participated in the study.

Independent Variables. The experimental design generated 5 independent variables:

1. sex of subject
2. sex of expected interviewer
3. a room with a window (providing a view of the sky and a few roof-tops) vs. a room without a window
4. Threat I (an expected interviewer studying the ability to be intimate) vs. Threat II (an interviewer studying sexual perversions) vs. no-threat (joke ratings).
5. A friendly vs. an impersonal experimenter.

Results

Manipulation Checks. The ratings of current emotional states yielded several main effects and a 3-way interaction. Male subjects were more relaxed ($X=4.61$ vs. $X=4.24$; $F=3.84$, $df=1/207$, $p<.05$) and secure ($X=5.07$ vs $X=4.63$; $F=6.67$, $df=1/207$, $p<.01$) than female subjects. Subjects felt more calm ($X=4.70$ vs. $X=3.84$; $F=18.62$, $p<.001$), relaxed ($X=4.62$ vs. $X=4.8$; $F=5.00$, $p<.05$), and secure ($X=4.99$ vs $X=4.65$; $F=3.74$, $p<.05$) with the friendly experimenter than with the impersonal experimenter. Subjects rated themselves as most comfortable ($X=4.99$), calm ($X=4.79$), and relaxed ($X=4.93$) in the non-threat condition; intermediate ($X=4.45$; $X=4.32$; $X=4.37$) in the Threat I condition; and as least comfortable ($X=3.91$), calm ($X=3.78$), and relaxed ($X=3.93$) in the Threat II condition ($F=11.40$, $df=2/207$, $p<.001$; $F=9.12$, $df=2/207$, $p<.001$; $F=8.67$, $df=2/207$, $p<.001$). Interactions of sex of subject by sex of expected interviewer by threat showed that subjects were most calm and comfortable when expecting a same-sex interviewer in the no-threat conditions, while same-sex pairings became less comfortable and calm relative to opposite-sex pairings in the threat conditions ($F=3.15$, $df=2/207$, $p<.05$;

Table 5
Threatment Means For Sex of Subject by Sex of Interviewer by Threat for Ratings of Comfort and Calm in Experiment III

		Male Subject		Female Subject	
		Male Int.	Female Int.	Male Int.	Female Int.
	Threat I	4.48	4.65	4.17	4.55
Comfort	ThreatII	3.85	4.15	3.96	3.67
	No Threat	5.35	4.96	4.30	5.43
	Threat I	4.57	4.70	4.13	3.90
Calm	Threat II	3.30	4.10	3.86	3.81
	No Threat	4.95	4.32	4.44	5.52

$F=4.74$, $df=2/207$, $p<.01$. The treat ment means for the sex of subject by sex of expected interviewer by threat are presented in Table 5.

Dependent Variables. The measures of how subjects expected to react during the interview and the spatial measures yielded a variety of main effects. Male subjects expected to be more hostile than female subjects ($\bar{X}=2.34$ vs. $\bar{X}=1.86$, $F=6.90$, $p<.01$). Subjects anticipated being more detached ($\bar{X}=4.83$ vs. $\bar{X}=4.44$; $F=4.5$, $p<.05$), unfriendly ($\bar{X}=2.95$ vs $\bar{X}=2.54$; $F=5.7$, $p<.05$), and closed $\bar{X}=2.95$ vs. $\bar{X}=2.62$; $F=.4.56$, $p<.05$) with male interviewers than with female interviewers. Subjects sat closer to the interviewer's chair in a room with a window than in a room with no window ($\bar{X}=43.31$ vs. $\bar{X}=46.92$, $F=3.75$, $p<.05$). Subjects anticipated feelings most crowded ($\bar{X}=3.77$) in the Threat II condition, intermediate ($\bar{X}=3.60$) in the Threat I conditon, and least crowded ($\bar{X}=3.16$) in the Non-Threat condition. Finally, when subjects were given directions by an impersonal experimenter, they reported feeling more crowded in the present ($\bar{X}=3.10$ vs $\bar{X}=2.62$; $F=4.71$, $p<.05$) and crowded in the future ($\bar{X}=3.83$ vs $\bar{X}=3.22$; $F=7.80$, $p<.01$). These subjects also distanced themselves more ($\bar{X}=47.65$ vs $\bar{X}=43.37$; $F=5.89$, $p<.05$) and pivoted further away ($\bar{X}=38.76$ vs $\bar{X}=31.43$; $F=4.10$, $p<.05$) from the interviewer's chair.

The dependent variables also yielded interactive effects: Sex of subject by sex of interviewer interactions revealed that female subjects planned to be more sociable, friendly, sincere and open with female interviewers, while male subjects planned to be equally sociable, friendly, sincere and open with male and female interviewers. All subjects sat closer to the chair of an expected same-sex interviewer than to an opposite-sex interviewer.

Sex of subject by sex of expected interviewer by threat interactions revealed that subjects expected to be least crowded and most polite, friendly, and sincere with same-sex interviewers in the no-threat condition. In the threat conditions subjects rated themselves as more crowded and less polite, friendly, and sincere in same-sex pairings relative to opposite-sex pairings. The treatment means and tests of significance for the interactive effects are presented in Table 6.

Discussion of Experiments III and IV

The manipulation checks for Experiments III and IV substantiate the findings

Table 6
Treatment Means and Significance Tests for Interactive Effects in Experiment IV

Dependent Variables		Conditions			
		Male Subject		Female Subject	
		Male Int.	Female Int.	Male Int.	Female Int.
Unsociable					
2-way F=5.58, df=1/207, p<.05		2.26	2.31	2.90	2.13
Unfriendly					
2-way F=3.99, df=1/207, p<.05		2.25	2.19	2.73	2.05
Insincere					
2-way F=7.71, df=1/207 p<.01		2.15	2.45	2.53	2.00
Closed					
2-way F=4.12, df=1/207, p<.05		2.82	2.81	3.11	2.39
Distance					
2-way F=4.61, df=1/207, p<.05		44.56	49.13	45.49	42.36
Future Crowding	Threat I	4.14	3.50	3.22	3.55
3-way interaction	Threat II	3.55	4.15	3.96	3.38
F=4.61,df=2/207,p<.01	Non Threat	2.50	3.91	3.35	2.81
Impolite	Threat I	1.81	1.80	2.00	2.45
3-way interaction	Threat II	2.20	1.85	1.92	1.91
F=4.76,df=2/207,p<.01	Non-Threat	1.60	2.63	2.17	1.38
Unfriendly	Threat1	2.52	1.90	2.57	2.60
3-way interaction	Threat II	2.55	2.05	2.83	1.91
F=6.44,df=2/207,p<.01	Non-Threat	1.65	2.59	2.78	1.67
Insincere	Threat I	2.29	2.25	2.44	2.25
3-way interaction	Threat II	2.25	2.30	2.46	2.19
F=4.03,df=2/207,p<.05	Non-Threat	1.90	2.77	2.70	1.57

of Experiment II in that an anticipated personal interview which is described as a probe into an individual's private life significantly evokes feelings of discomfort, anxiety, tension and insecurity. These feelings of discomfort also seem to be heightened when privacy invasion is presented as a search for negative information about the individual as opposed to a search for positive information. In addition, it appears that the population of students at Brooklyn College who are more comfortable with their own sex in non-threatening conversations become equally uncomfortable with males and females when their privacy is invaded.

Consistent with the earlier experiments in this study is the findings that a neutral, physical manipulation (a window vs. no window) is not associated with any emotional reactions. A new findings from Experiment IV is that a friendly experimenter reduces feelings of discomfort, insecurity and tension.

The results associated with the label "crowding" in Experiments III and IV consistently parallel the pattern of results associated with the discomfort of personal threat. Subjects anticipate feeling most crowded in the privacy invasion where negative information will be sought and least crowded when no personal questions are anticipated. The decrease in comfort associated with the same sex in the threat conditions is accompanied with an increase in the rating of anticipated crowding. And the friendly experimenter reduces expressions of both current and anticipated crowding.

The significant effects involving the spatial measures in Experiments III and IV partially substantiate and partially qualify earlier findings. The increase in distance in a room with no window vs. a room with a window parallels the distancing associated with a neutral source of crowding (a smaller room) in Experiments I and II. However, spatial adjustments associated with personal crowding appear to be highly variable. In Experiment II personal crowding was associated with a decrease in distance and an increase in pivot. In Experiment III personal crowding was associated with a marginal increase in distancing and a decrease in pivot and revolution. Finally, in Experiment IV the personal crowding elicited by anticipated threatening interviews was not associated with any systematic spatial adaptations, while the crowding elicited by an impersonal experimenter was associated with an increase in distance and pivot.

It is interesting to note that pivoting was continually associated with personal crowding as opposed to neutral crowding. In Experiments II and III it varied inversely with distancing, while in Experiment IV it varied directly with distancing. Variations in revolution, on the other hand, were associated with both neutral crowding (Experiments I and II) and personal crowding (Experiment III) and always appeared in an inverse relationship with distancing.

Conclusions

Taken together, the four experiments in the present study lend considerable support to Stokols' contention that responses to crowding can be more accurately predicted by distinguishing between neutral and personal sources of the crowding experience. Table 7 summarizes the major findings for the four experiments.

The prediction that neutral crowding is less stressful than personal crowding was substantiated. The three manipulations of neutral crowding were only accompanied with the label "present crowding", while the four manipulations of personal crowding consistently evoked additional negative emotional labels.

The spatial adaptations to neutral crowding were extremely consistent. Relatively small rooms and rooms without windows brought about a withdrawal from the anticipated target person. The spatial adaptations to personal crowding, on the other hand, were much more complex. Factors such as the experiment's style, the specific nature of the threat, and the sex of the expected intruder seemed to interact in producing spatial advances and retreats with complementary or compensatory pivots away from the intruder.

Table 7
Summary of the Major Effects for all Experiments

Source of Crowding	Emotional Labeling	Physical Response
Personal		
Threat of Privacy Invasion	Future Crowding and other negative emotions (Exps. I, II, III, and IV)	none (Exps. I and IV) Advance and Pivot (Exp. II) Withdrawal with less Pivot and Revolution (Exp. III)
Impersonal Experimenter	Present and Future Crowding with other negative emotions (Exp. IV)	Withdrawal and Pivot (Exp. IV)
Neutral		
Small Room	Present Crowding (Exp. II)	Withdrawal and Revolution (Exps. I and II)
No Window	none (Exp. IV)	Withdrawal (Exp. IV)

SOCIAL DETERMINANTS OF CROWDING

Dr. Jürgen Schultz-Gambard

Universität Bielefeld - PH-Westf.Lippe

Bielefeld, W.Germany

ABSTRACT: A review of the crowding literature shows that crowding has effects on social behavior. There also is some evidence that social factors might be important mediators for crowding. Yet there is a lack of empirical support. The present study investigates the separate effects of spatial density, group formation and interference on the experience of crowding. The results show that although the high density condition was given a negative evaluation and although deleterious aftereffects that indicated high density to produce stress were observed, there was neither a corresponding increase in selfrated anxiety, lethargy, depression, insecurity and arousal nor was there an increase in aggressiveness due to a density x interference interaction. In the group formation condition more positive social effects could be observed in high than in low density. Consequences for urban design are discussed.

Recent public concern with "overpopulation" and the "urban crisis" has created a growing interest in the effects of crowding. A basic problem throughout the crowding research has been whether high density produces stress and in what way other situational or non-situational variables relate to the emergence of crowding stress. Much theoretical emphasis has been placed on the role of social factors as mediator variables but there still is a lack of empirical support. The early investigations of human crowding phenomena have been highly influenced by results of crowding experiments with animals. Especially a study by Calhoun (1962), who found dramatic effects of extreme density on the behavior of rat populations, has given rise to generalizing speculations about the effects of crowding on human behavior and has even served as a model for research with humans (e.g. Galle, Gove and McPherson 1972). In this study, a complete breakdown of social organisation and behavior patterns as well as different forms of physiological pathologies could be observed and were considered to be caused by extreme density. It should be noted, however, that —on closer observation— not even in animal populations can the cited negative effects be attributed to high density alone and perhaps not even primarily to high density. In those cages in Calhoun's experiment where dominant male rats could take command and social structure remained intact none of the deleterious effects showed. It could also be observed that animal populations with fixed social hierarchies can obtain a larger population size than populations with transient hierarchies (Dubos 1970). Also the aggressiveness of the population (Christian, 1963), the introduction of strange animals (Southwick 1964), or conflicts over scarce resources (Calhoun, 1962) seem to have an influence on the emergence of crowding stress in animal populations. Therefore it is stated by Dubos (1970, p. 207) that "in fact, crowding per se, that is, population density is probably far less important in the long run even in animals than is the intensity of social conflicts or the relative peace after social adjustments have been made".

Effects of high density in studies with human subjects are even far less unequivocal than in studies with animals. By correlating a number of density measures (e.g. number of people per dwelling unit) with indices of social pathology (e.g. crime rate) survey studies have tried to reveal that high density causes social pathologies. But aside from a study by Galle, Gove and McPherson (1972), the reported evidence is not convincing and most studies are criticized for methodological weaknesses (Freedman, Heska and Levy, 1975). There is, however some evidence that the reported density effects are influenced to a considerable extent by social and cultural factors (Schmitt, 1963; Mitchell, 1970 and 1971; Galle, Gove and McPherson, 1972).

Experimental research on human crowding has gone from univariate approaches of the early crowding experiments, that tried to relate different behavior measures directly to density and failed to display any consistent density effects (e.g. Freedman, Klevansky and Ehrlich 1971), to multidimensional approaches of crowding where density is considered a necessary but not sufficient condition for the emergence of crowding stress (Stokols, 1972). Crowding is viewed as a syndrome of stress generated through the interaction of social, personal, spatial and other situational factors. The results obtained in experimental crowding research show that apart from effects on stress, performance and subjective moods and feelings crowding has effects predominantly on social behavior. Crowded subjects, as compared to their non-crowded conterparts, tend to withdraw from social interaction and show more negative interpersonal affects (Baron, Mandel and Adams, 1976; Baum and Greenberg, 1975; Baum and Koman, 1976; Griffitt and Veitch, 1971; Loo, 1972) and more aggressive acts (Hutt and Vaizey, 1966).

Further a consistent pattern of sex differences could be observed, indicating that men respond more negatively to crowding than women (Baum and Greenberg, 1975; Epstein and Karlin, 1975; Freedman, Klevansky and Ehrlich, 1971, Freedman, Levy, Buchanan and Price, 1972; Ross, Layton, Erickson and Schopler, 1973; Stokols, Rall, Pinner and Schopler, 1973). Traditionally, high density has been thought to have only negative effects on social behavior that can perhaps be weakened through other variables. In discussing their observed sex differences however, Epstein and Karlin (1975) suggested that crowding can have positive as well as negative behavioral consequences depending on the underlying group processes. They thought men and women to be differently affected by high density environments because of differing coping reactions. Monitored through sexrole norms, women form cohesive, cooperative groups whereas men show a fragmented competitive orientation when crowded. The observed social affects were considered to have been mediated by these different group processes. The present research will extend this line of reasoning by controlling for comparable group processes independently of sex. Summarizing, it can be said that crowding has been shown to have effects on social behavior and that social variables are used in explaining found sex differences. Nevertheless, there are only surprisingly few studies that introduce social factors as independent variables (Baum and Koman, 1976; Seta, Paulus and Schkade, 1976 and Sundstrom, 1975). These studies show only limited and fragmented evidence that social factors really are important mediators of crowding. More persistent empirical support can be found in the studies of personal space – a concept closely related to the concept of crowding (Altman, 1975; Worchel and Teddlie, 1976). This research shows that preferred interaction distances vary according to underlying social processes (Aiello and Cooper, 1972; Albert and Dabbs, 1970; Guardo and Meisels, 1971; Little, 1965; Lott and Sommer, 1967; Mehrabian, 1968; Sensing, Reed and Miller, 1972).

In the present research, the independent effects of spatial density and different forms of social interaction will be investigated. It is hypothesized that a high density environment produces stress and will be evaluated negatively independent of social processes. But social affects are suggested to be influenced by interactions of density and social structure. By controlling for verbal communication different group processes will be induced. To the extent that the members of an experimental group are allowed to communicate they will share the experienced distress and will discover similarities: e.g. that they all have to cope with a stressful density situation. The perceived similarities and the experienced common fate will lead to positive social effects. But if the subjects are not given the possibility of communication, they will remain alone, will not be able to share their distress and no social structure will emerge. As a result, less positive social affects will be observed. Further the absence of communication in the high density/no-communication condition is assumed to be a source of additional stress. Very close interaction distances signal intimate and personal interactions (Hall, 1966) and the absence of verbal communications in such situations will violate existing expectancies about appropiate interaction distances.

Another contributor to crowding stress – frequently cited in theoretical discussions (Zlutnik and Altman 1972)– is the amount of interference in a situation of high density. Depending on the ongoing activities space has a certain instrumental value. The perception of a situation as a crowding situation and the resulting responses depend largely upon the amount of restrictions in using the surrounding space to successfully achieve personal goals and satisfy personal needs. Whenever space is limited through the presence and the activities of too many others in such a way that goal oriented behavior patterns are inhibited, conflict arises and causes feelings of anger and aggression. Thus severe stressreactions and interpersonal conflict in a high density situation are to be expected for activities that require a large amount of body movement, active use of space and behavioral coordination. In spite of this, the tasks frequently used in crowding experiments are of a paper and pencil or group discussion type with fixed seats, no movement and no physical confrontation. The present research will use a task that is more density relevant in that it requires movement and active use of space. The task will also allow to control for interference. Interference is assumed to be the crucial variable for the emergence of anger or aggressiveness in a high density condition.

Method

The experiment employed a 2 x 2 x 2 factorial design manipulating spatial density (room sizes were 4.06 m^2 or .69 m^2/person vs. 16.24 m^2 or 2.71 m^2/person), group formation (communication vs. no communication) and interference (task with interference, task without interference and no task). Communication was manipulated through instructions by telling the subjects either to feel absolutely free about talking or not to have any verbal communication at all. All groups were asked not to engage in any self-occupation with notebooks or the like during the experiment.

Interference was manipulated through two different yet basically equivalent tasks. Each subject was assigned a geometric form as a personal sign. There were numbered black boxes with numbered push buttons and opaque glass screens in the experimental room and during the experiment the geometric signs appeared on the screens in fixed sequences but electronically randomized patterns. In the interference condition, the subjects had to observe where their sign appeared, move to that box as fast as possible, push the button on that box and sit down again. In the no-interference condition, the subjects had to watch their sign appear on a box with a certain number, turn around and push the corresponding button on their own box. There also was a condition in which no task was given and which served as a control condition. Each sequence was introduced by a buzzer sound and there were 90 sec. breaks between sequences to let the communication condition have some effect.

In order to assess the different dimensions of reaction to crowding stress different types of measurement were used. A measure of frustration tolerance that was adapted from Feather (1961) was used as a stress measure. Secondly there was a four scale questionnaire which pertained to the participants' perceptions of the aesthetic qualities of the experimental room: negative evaluation of room, positive evaluation of room, perception of size (two scales). Thirdly there was a mood questionnaire containing the following 10 scales: arousal, activity, lethargy, happiness, confinement, depression, aggressiveness, social adjustment, insecurity and anxiety; further several impression scales assessing the perceived similarity with others, the evaluation of their presence, difficulties of self-disclosure and attribution of subjective feelings (others vs. situation). The last measure was interpersonal judgement questionnaire assessing sympathy, aversion, crowding and competitiveness.

288 male subjects in groups of six were placen in hexagonal shaped rooms(*). They were given instructions and worked at the ascribed tasks for 37 minutes. The subjects were then led to separate small booths, where they worked on the stress measure and completed the described questionnaires.

Data were analyzed with separate Anova analyses for the different dependent measures followed by Duncan post-tests when more than two cells were involved.

(*) The rooms were of hexagonal shape in order to obtain the same amount of social stimulation for every subject.

Results

The manipulation checks indicated that the density manipulation was successful. The high density condition was rated as being more confined ($F=82.88$, $1/276$ df, $p<.001$) and the room being smaller ($F=52.46$, $1/276df$, $p<.001$) than the low density condition. The communication manipulation can also be considered to have been successful. Subjects in the communication condition as compared to those in the no - communication condition reported more feelings of happiness ($F=32.81$, $1/276df$, $p<.001$) and described themselves as being more socially adjusted ($F=13.59$, $1/276df$, $p<.003$), described the others to be more sympathetic ($F=16.35$, $1/276df$, $p<.001$), as having enjoyed their presence in the experiment more ($F=10.07$, $1/276df$, $p<.002$), thought that the others facilitated the display of subjective feelings more ($F=9.25$, $1/276df$, $p<.003$) and thought their own feelings during the experiment to have been influenced more by the other group members than by the situation ($F=12.99$, $1/276df$, $p<.001$). The results indicate that there was more group formation and emergence of social structure in the communication than in the no-communication condition. Subjects in the interference condition seemed to rate the others as more crowding and interfering than subjects in the no-interference condition ($F=3.94$, $2/276df$, $p<.10$) and subjects in the no-task condition ($F=3.94$, $2/276df$, $p<.05$). Interfering subjects also felt more crowded themselves than subjects in the no-task condition ($F= 3.57$, $2/276df$, $p<.05$), yet there is no corresponding difference between interference and no-interference. The results give support to the experimental manipulation of interference but it must be noted that this manipulation does not appear to be as powerful as intended.

The results of the frustration tolerance test show negative aftereffects in the high as compared to the low density condition. Subjects who were placed in the smaller experimental room showed less frustration tolerance than subjects in the larger room ($F=51.95$, $1/276df$, $p<.001$), indicating that high density produces stress.

Also, the no-communication condition proved –as predicted– to be more stressful than the communication condition ($F=6.79$, $1/276df$, $p<.009$), and the interference condition caused more stress than the no-interference condition ($F=3.45$, $2/276df$, $p<.05$).

Subjects also described the experimental room to be less pleasant, less comfortable and less cozy in the high than in the low density condition ($F=8.86$, $1/276df$, $p<.003$). In the high density condition, the other group members were perceived as being more crowding ($F=22.67$, $1/276df$, $p<.001$) and as more unpleasant ($F=4.46$, $1/276df$, $p<.034$). It should be noted, however, that the obtained values are rather low. Means on 6 point scales from 0 to 5 range from 1.23 to 1.94 for crowding and from 1.03 to 1.23 for unpleasantness.

Subjects in high density also feel more crowded themselves than subjects in low density ($F=9.28$, $1/276df$, $p<.003$). It should be noted however, that although subjects in the high density condition felt more crowded, considered the others to be more crowding and liked the rooms less, there were no self reports of higher arousal, more depression, more lethargy, more anxiety or more aggressiveness, nor could the predicted interactive effects, indicating more interpersonal conflict in the interference/high density condition, be observed.

It was predicted that, given group formation and emergence of social structure, high density might even have positive effects on social orientations. It has already been proven that by controlling for communication group formation processes were successfully manipulated. A density x communication interaction could now be observed ($F=6.31$, 1/276df, $p<.012$), showing that in the high density condition subjects in the communication condition showed higher sympathy ratings of the other group members than subjects in the no-communication condition ($p<.01$). There is no significant difference, however, in the sympathy ratings in the low density condition.

Conclusions

The present research clearly demonstrates the multidimensionality of crowding and gives support to the proposed mediating effects of social factors for the experience of crowding. High spatial density proved to be a stress-producing condition– a result that is in accordance with the results of other crowding experiments which have used different stress measures (e.g. Aiello, Epstein and Karlin, 1975; Paulus, Aunis, Seta, Schkade and Matthews, 1976; Saegert, 1973; Sherrod, 1974; and Worchel and Teddlie 1976).

Further could be shown, that the spatial inadequacies of the high density situation had been perceived by the participants and had led to negative evaluations of the room as being unpleasant, uncomfortable and not cozy. Also the other group members in high density are described to have been more crowded and participants felt more crowded themselves in the high as opposed to the low density conditions. Yet despite the found stress reactions and the descriptions of the potential stressful character of the experimental situation, no selfreports of subjectively experienced stress could be noted nor could any signs of social conflict be observed. Furthermore, when possibilities for group formation were given, high density even seemed to have positive effects on social relations. Of course, this does not mean that density can be assumed not to have any detrimental effects on social behavior at all or that interference is a variable that is irrelevant to crowding. It has already been mentioned that the operationalisation of interference seemed a little weak and it might well be that the participants did not consider the situational restrictions, the given tasks and the induced interferences to be really relevant for their own personal needs and thought the experimental situation to be just a game. Such an attitude towards the experiment could, of course, account for the absence of conflicts and severe stress reactions. This problem will be investigated further.

Nevertheless, the results demonstrate that there is no unitary response to crowding that the different antecedent conditions of crowding stress do have differential effects on reactions to crowding and that non-spatial, especially social factors can be important mediators of crowding. Further, empirical research on the role of the non-spatial factors in the crowding syndrome seems to be necessary in order to be able to progress from a merely "interactionist perspective" (Schopler and Stokols, 1976) to a interactionistic model of crowding that allows accurate predictions under what conditions crowding stress can be expected and how it can be prevented.

As has been mentioned, high density environments are commonly viewed to have serious deleterious effects on social organisation and behavior. Considering the steadily growing number of high density environments in our civilisation such a view of crowding leads to a dark prognosis about the quality of future living. The present research together with other studies in this field offers a more optimistic view of the problem of crowding in assuming that crowding is not only related to spatial but also to nonspatial variables, including social variables.

It is proposed that many negative effects of high density environments can be minimized through appropriate social and architectural planning.

Social processes can be influenced, modified and shaped. While it may sometimes be difficult to change the spatial variables of high density environments, it should be possible to introduce modes of social interaction that can reduce interpersonal tension and conflict. Baum and Koman (1976) have already shown the importance of social structuring for the experience of crowding. It is assumed that by structuring social processes and making social interaction patterns more transparent and more predictable for those involved, it should be possible to reduce the stress potential of high density situations (e.g. in bureaucratic institutions). Sherrod (1974) has shown that the experience of situational control is a crucial variable in determining whether high density is experienced as stressful.

Because it has been shown that the formation of cohesive groups make people feel less crowded, buildings should account for this fact and be designed in such a way that they foster communication and encourage group formation. This could be done by either building smaller apartment houses or breaking them down to smaller units, creating functionally flexible recreational areas that can serve the need for social gathering as well as for privacy, leaving open spaces and places (like miniature market places) which are assigned no specific function, adding points of artistic or cultural interest that give the environment an unique character etc. Well-known neighborhoods, friendship formation, the development of a sense of community make any environment a better place to live and add to the prevention of crowding stress, since crowding is not only a matter of spatial density.

CROWDING AND ENVIRONMENTAL CONTROL

Carl I. Greenberg

University of Nebraska at Omaha

Nebraska, U.S.A.

ABSTRACT: The research on the causes, consequences and mediating adaptive processes of crowding is viewed from a perceived control perspective. It is asserted that crowding is a result of uncontrollable inputs from others. Furthermore, exposure to crowded environments may result in a loss of perceived control over behavior-outcome contingencies, i.e., learned helplessness. In addition, loss of perceived control may result in the employment of ineffective coping mechanisms in individuals exposed to crowded environments; consequently, strengthening individuals' loss of control perceptions. A pilot study provided some support for these assertions.

CROWDING PERSPECTIVES

To date, a variety of theoretical perspectives pertaining to the causes, mediating adaptive processes and consequences of crowding have appeared in the literature. In a recent integrative theoretical article (Stokols, 1976) these perspectives were viewed as falling into one of three categories: (1) stimulus overload, (2) behavioral constraint, and (3) ecological.

The stimulus overload perspective asserts excessive stimulation as the critical antecedent condition for crowding to be experienced. Overload has been defined in general terms as the inability to cope with the environmental inputs impinging on an organism (Milgram, 1970). On the other hand, "environmental inputs" has been specified to a greater extent in later theories. For example, crowding is viewed as an overload of unwanted social encounters with others (Baum and Valins, 1977); an excess of social stimulation (Desor, 1972); an excess of intimacy cues from others (Kaplan and Greenberg, 1976); and failure to maintain a desired level of privacy (Altman, 1975).

Although each model has selected specific environmental inputs that cause overload, it can be assumed they are all describing the same antecendents of crowding, i.e., intrusive stimulation from other people that because of other mediating circumstances is seen as excessive. Thus high density environments –in particular, socially dense environments– have the potential to create excessive stimulation by decreasing the privacy of the setting for the individual.

When overload does exist, individuals are motivated to reduce the inputs into their system. The primary process for reduction of inputs is withdrawal. For example, Altman (1975) maintained that individuals seek an optimal level of privacy, or social stimulation. When there is an excess of environmental inputs people will be motivated to reduce them (i.e., withdraw from social interactions). In contrast, when there is a shortage of inputs from others, individuals will be motivated to increase these inputs (i.e., approach, or seek out social interaction).

Behavioral constraint perspectives have viewed crowding experiences as a function of reduced behavioral freedom. Proshansky, Ittelson and Rivlin (1970) and Stokols (1972) view behavioral restriction in similar ways. Both argue that environments that do not afford individuals with enough space, privacy, etc., are in effect restricting the choice behaviors open to them. Individuals presented with an environment that does not allow their needs to be met because of spatial considerations will, therefore, experience feelings of crowding.

In terms of the mediating adaptive processes, the behavioral constraint perspective would also predict that individuals experiencing crowding will be motivated to withdraw from social interaction. However, the underlying motivation for this withdrawal behavior is not to reduce overload, but to increase one's freedom of choice behaviors. This is not as discrepant from the overload model as one might think. If, for example, we adhere to Proshansky, et al.'s (1970) definition of privacy as the freedom to choose one's behavior, then, the behavior constraint model predicts that individuals are motivated to achieve a particular level of privacy. The overload model is arguing the same point –that a lack of privacy causes one to feel crowded. Consequently, the models are stating that privacy is a key component to experiencing crowding.

The ecological orientation towards crowding (e.g., Wicker, 1973) likens it with overmanned behavior settings (see Bechtel, 1977 for a complete description of behavior setting theory). In essence, this perspective takes the position that it is not density per se that facilitates crowding perceptions. Rather, it is the degree to which the setting is staffed vis-a-vis the variety of task activity. For example, in two behavior settings of equal social density (i.e., equal population size), the setting that has less tasks to be performed (i.e., overmanned) will be perceived by the participants as more crowded than the relatively undermanned setting.

The model as proposed by Wicker (1973) predicts that people who are in overmanned settings will also manifest withdrawal behaviors. However, since the model views crowding from the ecological unit rather than the individual unit, withdrawal behaviors would be to those outside the setting. Thus, for example, an overmanned behavior setting will implement stricter criteria for admission of outsiders into the setting. In a sense, then, participants of an overmanned setting are attempting to maintain the "privacy" of the setting from outside interference, or at least defend against making the setting less private than it already is.

Antecedents

Although the explanations as to why people feel crowded are diverse, all three perspectives do concur that perceptions of crowding should be conceptualized as something more than the objective factor of high density (Stokols, 1972; Zlutnick and Altman, 1973). High density is the objective availability of space while crowding is a more subjective state of psychological stress. Recent evidence on the distinction between high density and crowding has shown that crowding perceptions can also be facilitated by low density environments. For example, Greenberg and Firestone (1977) reported that interpersonal distancing violations by an interviewer increased feelings of crowding in the interviewee. Similar findings of interpersonal intrusion heightening crowding perceptions have also been documented (e.g., Sundstrom, 1975; Worchel and Teddlie, 1976). Hence, crowding does not appear to be synonymous with high density (see Rapoport, 1975 for a further discussion of this issue). Nevertheless, high density environments contribute significantly to perceptions of crowding, most likely because these environments promote interpersonal distancing violations. Moreover, these violations may be seen as creating environmental conditions of lower privacy levels than an individual desires (Altman, 1975).

In addition to closer interpersonal distances via high density, situational factors also influence crowding perceptions (Stokols, 1972; Zlutnick and Altman, 1973). For example, research has indicated that individuals involved in a task requiring interpersonal interaction reported greater feelings of crowding and discomfort compared to individuals involved in a non-interactive task (McClelland, 1974; Heller, Groff and Solomon, 1977); competitive tasks produce higher scores on crowding than cooperating tasks (Stokols, Rall, Pinner, Schopler, 1973); and anticipation of an unstructured group task produced greater degrees of crowding than the anticipation of a structured task (Schopler and Walton).

Similarly, personality variables heve been shown to affect crowding perceptions. Individuals with large personal space zones tend to have lower thresholds of crowding than individuals with small personal space zones (Cozby, 1973; Dooley, 1974), and Schopler and Walton (Note 1) reported that external locus of control individuals have lower thresholds of crowding than internal locus of control individuals.

In summary, crowding perceptions can be mediated or affected by other factors than high density. However, it is far from clear exactly which factors will intensify crowding perceptions under conditions of high density and which factors by themselves promote heightened perceptions of crowding.

Mediating Adaptive Processes

Regardless of the antecedent conditions that facilitate crowding perceptions, there appears to be consistent agreement as to various mechanisms individuals use to reduce or at least attempt to reduce stress due to crowding, i.e., withdraw from others around them. Generally, the studies that have investigated interpersonal behavior of "crowded" people have found: (1) lower frequencies of interaction with others (Hutt and Vaizey, 1966; Ittelson, Proshansky and Rivlin, 1972; Loo, 1972; Wolfe, 1975);

(2) decreased affiliative or intimacy behaviors (Argyle and Dean, 1965; Patterson, 1973), indicated by decreases in facial regard (Greenberg and Firestone, 1977; Ross, Erickson, Layton and Schopler, 1973; Sundstrom, 1975) and lower levels of self-disclosure (Greenberg and Firestone, 1977). Moreover, these withdrawal behaviors have been observed in situations where individuals expected to be crowded. Subjects "waiting for a crowd" exhibited less facial regard towards two confederates than subjects not waiting for a crowd to arrive (Baum and Greenberg, 1975; Baum and Koman, 1976; Greenberg and Baum, in press). In short, withdrawal from social interaction appears to be a process by which individuals attempt to cope with crowding, whether this condition was in existence or was forthcoming.

Consequences

The effects of crowding have also been observed with respect to task performance (see Freedman, 1975; Sundstrom, 1977). Decrements in task performance under brief exposure to high density have resulted in equivocal findings. Individuals exposed to high density are usually unaffected in their performance of simple tasks (e.g., Bergman, 1971; Evans, 1975; Freedman, Klevansky and Ehrlich, 1971; Sherrod, 1974; Stokols et al., 1973), although a few studies have found facilitating effects of high density on simple task performance (Freedman, Heshka, and Levy, 1975; Saegert, 1974). In summary, the evidence has not supported the notion that brief exposure to high density is detrimental to simple task performance.

On the other hand, studies that have used complex task performance have found poor task performance under high density environments (Evans, 1974; McClelland, 1974), while others have found no effects of room density (Dooley, 1974; Emiley).

Similarly, the aftereffects of exposure to high density have shown no decrement on simple task performance. Rather exposure to high density has increased simple task performance (Epstein and Karlin, 1975). However, studies using complex tasks (i.e., problem solving ability as a criterion measure) have found after effects due to high density (Sherrod, 1974). In short, the research tends to suggest that high density produces an inability to perform effective cognitive processing (i.e., cognitive deficit).

Summary

A lack of privacy rather than high density per se appears to be the central cause for one to feel crowded and subsequently provoke coping strategies to ameliorate crowding stress. If this stress is not effectively dealt with, negative consequences regarding one's task performance, interpersonal relations and general well-being might be observed (Sundstrom, 1977).

However, is it privacy per se that causes these negative consequences, or is it some aspect central to the notion of privacy that does? In other words, if privacy is the ability to control inputs and outputs from others and self, then, is it a general lack of control over socio-environmental stimuli that causes crowding? The remainder of this paper will be devoted to this issue.

ENVIRONMENTAL CONTROL

Altman (1975) considered privacy "as selective control of access to the self or to one's group" (p. 18). Similarly, Proshansky, Ittelson, and Rivlin (1970) stated that "the overall function of privacy thus is to increase the range of options open to an individual so that he can behave in ways appropriate to his particular purposes" (p. 178). In addition, Laufler, Proshansky and Wolfe (1970), Westin (1970) and Kelvin (1973) provided similar definitions and dimensions of privacy. That is, privacy is likened to control over the environment, particularly those stimuli originating from other people. Consequently, crowding can be conceptualized as a lack of control over socio-environmental stimuli.

Recent evidence suggests that exposure to living environments that, because of architectural design features do not allow for the regulation of potential social encounters with others, leads residents to feel crowded. Perceived lack of control over socio-environmental stimulation and a generalized perceived loss of the ability to control their behavioral outcomes may also be experienced (Baum and Valins, 1977). In an extensive series of studies, Baum and Valins (1977) reported the following with students living in long corridor-designed dormitories (34-40 residents) around undifferentiated shared living spaces versus students living in short corridor-design dormitories (of equal density; 6-20 residents) around differentiated shared living spaces: (1) reported greater feelings of lack of control over their environment; (2) reported greater perceptions of crowdedness; (3) performed less well on experimental tasks; and (4) withdrew more from interactions with strangers. In short, long corridor students displayed symptoms of helplessness (Seligman 1975) to a greater extent than their short corridor counterparts because they could not control or predict the frequency and intensity of interactions with others.

Furthermore, manifestations of helplessness, i.e., perceived loss of control of one's outcomes, have been observed in children of high density households. Rodin (1976) reported that children from high density living environments were significantly more likely than children from less dense households to allow another person (an experimenter) to control available outcomes for them. In other words, children living in environments that do not afford adequate privacy, or control over socio-environmental stimuli may also generalize this lack of control to other situations.

Moreover, Rodin (1976) posited that if high density living is analogous to helplessness training (learning independent response-outcome contingencies); then children from high density homes should exhibit greater degrees of helplessness when exposed to experimental conditions of uncontrollable outcomes. To test this hypothesis Rodin exposed children from both high and low density living environments to solvable (controllable) and unsolvable (uncontrollable) problems. The results indicated in unsolvable conditions children from high density homes performed significantly poorer in learning subsequent solvable problems (i.e., greater cognitive learning deficit) than children from low density homes.

In essence, crowded living environments appear to teach residents that they have little, or no control over the socio-environmental stimuli of that setting. This perceived loss of control eventually will generalize to other settings and consequently residents of high density households will acquire maladaptive behavioral strategies in dealing with a variety of circumstances (see Seligman, 1975). High density living thus may cause residents to learn to become helpless (Rodin and Baum, 1977).

Perceived Control

Based on the aforementioned research and theory, the experience of crowding can be viewed as a situation of uncontrollable stimulation from others. Increasing perceived control over the inputs from others may therefore ameliorate crowding effects. Sherrod (1974) tested this assumption. In his experiment, subjects were either exposed to high or low density rooms or a high density room where they were given the option of leaving (perceived control). Sherrod found that subjects who had perceived control over crowding had greater frustration tolerance (measured by the number of attempts to solve two insoluble puzzles) than subjects exposed to crowding alone. Subjects who perceived they had control over the aversive stimulation from others had greater motivation to perform the task than subjects in the uncontrollable crowding conditions. Uncontrollable high density, therefore, can produce motivational deficits similar to those found in the helplessness studies (Overmeier and Seligman, 1967; Seligman and Maier, 1967) while perceived control can ameliorate these effects.

Similarly, if high density can facilitate decreased expectancies for behavior-outcome contingencies, then individuals predisposed for perceiving their outcomes as not in their control (i.e., external orientation; Rotter, 1966) should consequently perceive high density environments as more uncontrollable, and crowded compared to individuals with a predisposition for perceiving their outcomes as in their control (i.e., internal orientation). In a test of this hypothesis, Schopler and Walton had internal and external subjects [based on Rotter's (1966) Locus of Control Scale] interact in high and low density rooms. As predicted externals tended to rate high density rooms significantly more crowded than internals. This result implies that generalized reinforcement expectancies can effect perceptions of crowding.

Why should externals feel more crowded than internals? Is it simply because they perceive the high density environment as more uncontrollable than internals; or is it because externals are ineffective in coping with high density environments through their behavioral coping strategies?

In a recent pilot experiment, this author attempted to ascertain if the latter hypothesis was operative.

Employing the anticipation method developed by Baum and Greenberg (1975), internal (N=40) and external (N=40) locus of control subjects expected to be in a small laboratory room with either nine (crowded) or four (not crowded) other subjects. Arriving alone, subjects encountered the experimenter. Subjects were lead to believe through a series of salient cues (e.g., 10 or 5 clip boards and pencils, 10 or 5 names on a list, and a 10 x 10 or a 5 x 5 matrix on the wall) that they would be in a group bargaining experiment with nine or four other subjects. In addition, the experimenter mentioned to subjects that the experiment would begin as soon as other (nine or four) subjects arrived.

Subjects were then ushered into the experimental room where they encountered 10 chairs in a semi-circle. Subjects were instructed to take any seat they wished. After a 2 minute duration interval a second same-sex subject (actually a confederate) arrived. The experimenter gave the confederate the same information as the real subject. At this point, observers (through a one-way mirror) began measuring the facial regard with the subject and the confederate. After 2 minutes a second same-sex confederate arrived. Again, the experimenter presented this person with the same information as the two previous space people. The second confederate then took a seat 1 space away from the first confederate. Again, a 2 minute observation period of the subject's facial regard with both confederates was recorded. At the conclusion of this observation period the experimenter gave each person a questionnaire to complete. The questionnaire measured the subjects' perception of the environment, their perceived level and attraction toward the confederates. After the questionnaire was completed the experimenter informed the subject that the experiment was over and the subject was debriefed.

The results of this study are far from conclusive. Subjects did not perceive the high anticipated density condition as more crowded than the low anticipated density condition, $F(1,76)<1$, n.s. Similarly, no differences in the amount of facial regard between subjects and confedates was observed between the high and low anticipated density conditions, $F(1,76) <1$, n.s., or between internal and externals in either density conditions, $F(1,76)<1$, n.s. Moreover, none of the manipulations affected the degree to which subjects chose less central seats, $F's(1,76)<1$, n.s. (less central seat was considered an indication of withdrawal from social interaction). In short, the study failed to produce any significant differences as regards measures of withdrawal coping behaviors, perceived stress measures, and attraction towards confederates.

However, internal analysis of the data revealed an important finding. Under conditions of high anticipated density internals tended to take a less central seat position the more they perceived the room with 10 people in it would be crowded, $r(20)=.43$, $p<.03$. On the other hand, for external subjects, no relationship between seat position and perception of crowding was found, $r(20)=.03$, n.s. The difference between these correlations was significant, $z=1.67$, $p<.05$. Superficially, this result implies that internals do employ a predictable potentially effective coping strategy while externals do not.

Conclusions

A lack of perceived control over outcomes appears to interfere with the development of effective coping mechanisms for environmental stressors. Similarly, situational circumstances that provide uncontrollable outcomes may tend to decrease one's perceived control. High density is only one environmental stressor that facilitates this dynamic process. Noise has been shown to be another.

Studies by Glass and Singer (1972) demonstrated that noise has detrimental effects on task performance only when it is perceived to be uncontrollable. Likewise, studies on the effects of helplessness have demonstrated that uncontrollable aversive

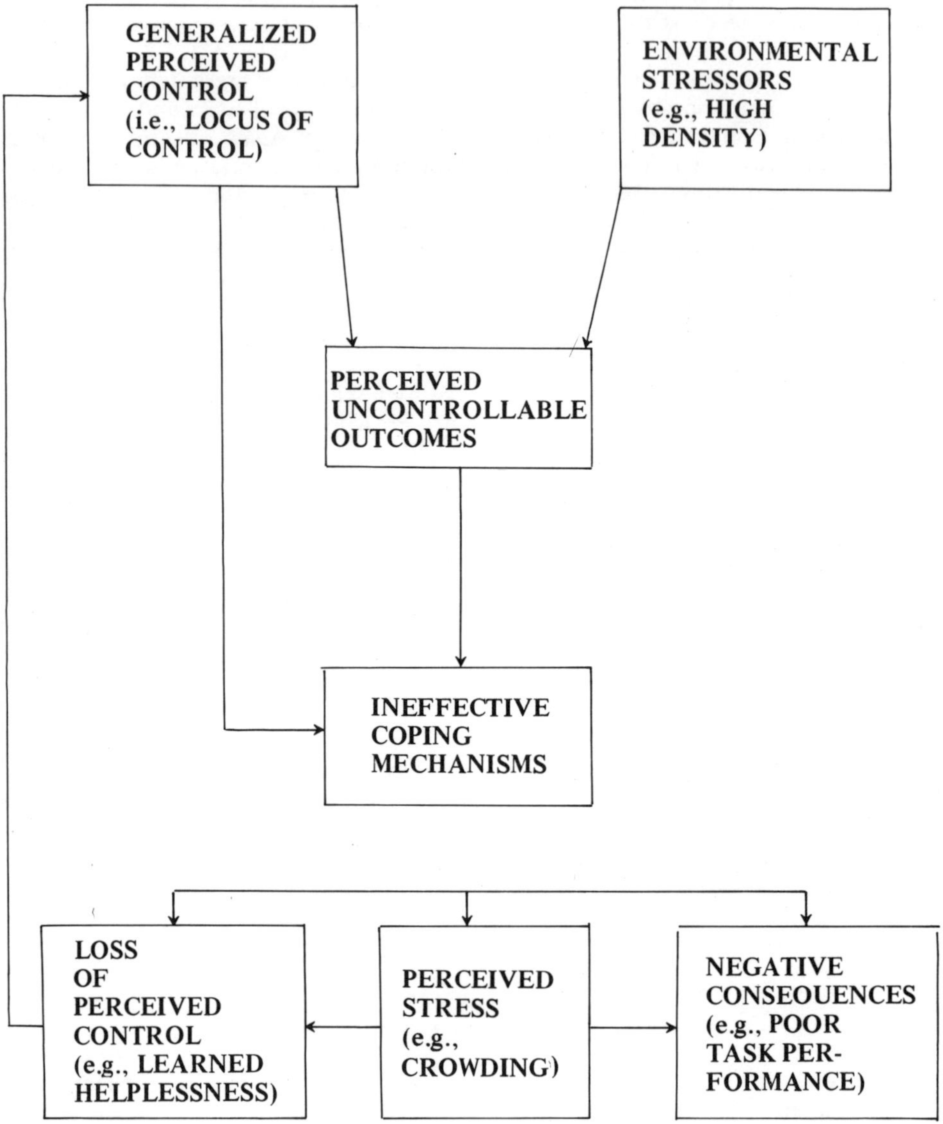

Figure 1: PROPOSED RELATIONSHIP BETWEEN PERCEIVED CONTROL AND CROWDING

noise stimulation can impair one's ability to learn response-outcome contingencies and create greater cognitive learning deficits (see Seligman, 1975).

As shown in Figure 1, perceived control is diminished when individuals are under the influence of uncontrollable outcomes. In addition, as one's perceived control decreases, the effectiveness of coping mechanisms also decreases, which leads to increased stress and greater loss of perceived control. Consequently, prolonged exposure to uncontrollable environmental circumstances may have detrimental effects on one's health and general well-being (Cohen, Glass, and Phillips, 1977). Those individuals who initially have a lower level of perceived control are more strongly influenced by environmental stressors and perceive them as more uncontrollable than those for whom initial perceived control is high. Nevertheless, perceived control and environmental stresses such as high density should be additive in producing negative consequences for the individual (Sherrod, 1974; Schopler and Walton).

Returning to the issue raised at the beginning of the paper, employing a control analysis appears to coalesce the three major perspectives of crowding. The overload and behavioral constraint perspectives fall right into this orientation. Overload theory views crowding as a state where one perceives the outcomes from others as uncontrollable and unpredictable. Only when these outcomes fall into the uncontrollable state does a perception of crowding emerge. Similarly, behavior constraint theorists would see uncontrollable outcomes as constraining the person from choosing behavioral responses. Again, crowding will be perceived only when the outcomes from others are seen as uncontrollable.

Even the overmanning orientation fits well in this scheme. As Wicker (1973) has stated when a setting has too many participants for the task activities a setting is then considered overmanned and therefore can be equated with crowding. The participants may then feel they have no control over the others in the setting and stress and social disorganization may emerge. However, as Strivastava (1974) points out, in conditions of extreme undermanning-the number of tasks is too great for the personnel in a setting– psychological dysfunctioning, i.e., stress, may also occur. It may therefore be that extreme undermanning environments are just as uncontrollable as overmanned and environments and crowding perceptions –although different in origin– may emerge in both settings.

A study by Wicker (1977) suggests this analysis might be operative. Using a slot car task, Wicker had four-person groups perform various job activities. In overmanned groups the number of job activities were less than in undermanned groups. Contrary to his predicted results, Wicker reported that the undermanned groups felt more crowded than the overmanned groups. As can be readily observed this result is comparable with the control analysis. Participants in undermanned groups might have felt that they could not control the outcomes of the task and consequently attributed their lack of control over the situation to crowding. The overmanned groups, however, may have not perceived the setting as uncontrollable and therefore did not feel crowded.

In sum, crowding as a result of uncontrollable stimulation from others appears to be a heuristic approach in ascertaining its effects on humans. The literature suggests that exposure to crowded environments is similar to helplessness training and the experimental effects of crowded environments on humans is similar to the effects found in learned helplessness experiments (Rodin and Baum, 1977). In contrast, reducing the uncontrollable outcomes of any environment, through stricter boundary controls and increased structure, may be a viable method of ameliorating crowding effects. Furthermore, by structuring living environments that promote learned competence opposed to learned helplessness it may be possible to increase the degree to which one perceives control and therefore decrease the negative consequences of crowding.

INDIVIDUAL AND CULTURAL DIFFERENCES IN COPING WITH CROWDS AND CROWDED CONDITIONS

James K. Mikawa

University of Nevada, Reno
Department of Psychology
Reno, Nevada, U.S.A.

Crowds of people have become a familiar pattern in almost all urban centers as population has increasingly congregated in cities. Crowding has had numerous effects on the economic, political, and social existence of individuals living in the cities. From the psychological viewpoint, it is generally assumed that crowding has serious negative effects on the behavior of humans. However, this assumption has not been adequately supported by scientific evidence (Baron, Byrne and Griffith 1974; Freedman, 1975).

Animal studies have demonstrated that under extreme conditions of crowding, disruptions are found in social relationships, reproduction, and physiological functions (Calhoun, 1962; Christian, 1963; and Hafaez, 1962). Comparable studies with human beings, of course, have not been done. Some correlations between behavior and crowded conditions have been found (Mitchell, 1971; and Marsella, Escudero, and Gordon, 1970). However, these correlations are contaminated by other factors present in high density areas such as poverty, poor housing and inadequate facilities. It is difficult to separate density influences from other significant factors which need to be considered as important variables.

Freedman (1975) states that high density does not generally have negative effects on people. He bases his conclusion on studies showing that when income, educational level, and ethnic background were controlled, density was not positively correlated with high crime rates. Schmitt (1957), however, found that in Honolulu density was associated with high crime rates. However, most of the studies in this area support Freedman's findings. Freedman (1975) also found that crowding in a room did not necessarily result in aggressive or negative behavior. Crowding tended to make people a more important stimulus and consequently intensified the typical reaction to them. These typical reactions could be both positive and negative, depending on the positive or negative characteristics of the situation.

Two elements, however, appear to be missing, or at least rarely mentioned in the crowding literature. One, how individuals differ with respect to their typical reactions to crowding across situations has not been investigated. People may differ in how sensitive they are to different crowded conditions and as a result, may cope differently with these situations. Two, the phenomenological view of persons toward crowded conditions has rarely been studied. How crowding is seen by individuals seems to be an important question.

In the present investigation, groups will be selected to the basis of their sensitivity to crowding. Persons who are high in sensitivity to crowding, and persons who are low in sensitivity to crowding will be interviewed in order to determine their negative phenomenological views of crowding, their coping styles, and their attitudes toward crowded conditions. It is expected that the high and low sensitivity to crowding groups will differ with respect to these variables.

Samples will be selected from areas in both the United States and Japan; cultural differences are expected. An interview format was used, recognizing difficulties in obtaining comparability of response on standardized scales developed in the United States. The exploratory nature of the study also dictated using a method which allowed relatively free responses.

Method

Subjects: College students from the United States and Japan served as subjects for the study. At the University of Nevada, Reno, twenty (20) subjects who scored high and twenty (20) subjects who scored low on the Globig Crowding Sensitivity Inventory were interviewed for their reactions to crowding. The (40) subjects were selected from an initial pool of 90 subjects who were given the Globig Crowding Sensitivity Inventory. In Japan, eleven (11) subjects who scored high and thirteen (13) subjects who scored low on the Globig Crowding Sensitivity Inventory were interviewed. The 24 subjects were selected from an initial pool of 100 subjects. A total of 64 subjects were interviewed in both countries.

Subjects in Japan were students enrolled at Waseda University in Tokyo. Tokyo is one of the highest-density areas of the world. College students in the United States were selected from the University of Nevada, Reno. Subjects participating in the study were general psychology students, obtained through the use of the subject pool.

Sex differences have been found in the crowding literature. Consequently, approximately half of the subjects in each grouping were female and half were male.

Procedure: The initial step was the selection of subjects using the Globig Crowding Sensitivity Inventory. College students in Japan were given the translated version of the Globig Crowding Sensitivity Inventory. The high scorers were compared with the low scorers. A similar procedure was used in the U.S.A.

After the subjects were selected, each subject was interviewed to determine: 1) his or her positive and negative reactions to crowds and the situations in which these have occured; 2) what positive and negative consequences have occured in the individual's life as a perceived consequence of crowd interactions; 3) what positive and negative reactions were felt by the individual when opportunities to be away from crowded conditions occured: 4) what are the positive and negative benefits from living in crowded conditions; 5) how an individual learns to cope with crowded conditions, including behavioral changes or patterns; and 6) what skills or attitudes have individuals developed to minimize or maximize the impact of crowds. A structural interview format was used (Appendix B). Results of the interview data will be used as a basis for further systematic study.

The data were analyzed to compare responses of high and low scorers on the Globig Crowding Sensitivity Inventory. In addition, five associations to the state of being alone and five associations to the state of being crowded were obtained from each subject. Cross-cultural comparisons between Japan and U.S.A. also were made.

In Japan, Japanese interviewers were used. Translated versions of the instruments were utilized. Interviews were tape-recorded. Research practices normally adhered to in Japan were followed.

Results and Discussion

The presentations of results and discussion are integrated since quantitative analysis was not emphasized in the present study and results can be more clearly understood within the context of discussion.

Cross-cultural differences were found between Japanese and American subjects on association responses to the state of being alone and the state of being crowded. High and low scorers on the Globig Crowding Sensitivity Inventory, however, did not differ in their respective countries. High scorers on the Globig Crowding Sensitivity Inventory were seen to be adversely sensitive to high density conditions and low scorers were relatively unaffected by these conditions. With almost no exception (94% of responses) Japanese subjects responded to the association task with "place" responses or some external referent such as "subway", "train", and "shinjuku". In contrast, most Americans (84% of responses) responded with feeling statements such as "tired", "nervous", and "relaxed". Japanese responses are similar to those found by Tanaka-Matsumi and Marsella (1976) in their study of depression. Japanese subjects tended to give responses which indicated external referent or places rather than feeling statements to the concept of depression. Tanaka-Matsumi and Marsella (1976) discussed these responses in terms of the nature of the Japanese "self". They saw the Japanese "self" as part of a larger social context which surrounds the individual rather than the individualized concept of self found in western cultures. It would be expected, then, that Japanese subjects would take outer-directed rather than inner-directed channels and respond with external referents rather than internal and individualized referents.

Japanese subjects, however, were able to respond with feeling statements in interviews. They commented on being "disgusted", "angry", and "feeling lonely". It may be that specific feeling reactions have to be elicited from Japanese subjects and that requests for general associations may result in external referent responses.

It was interesting to note that many cross-cultural reactions to crowding were quite similar even though Japanese subjects in Tokyo were obviously in higher density conditions than those in Reno, Nevada. The largest difference found between Japanese and American subjects was in their responses indicating how they coped with crowded conditions. In the majority of situations listed by American subjects (53%), they indicated no particular way of coping with stress produced by crowded conditions. It appeared that in many situations listed, crowds were not stressful enough to warrant special ways of coping. The most common way of dealing with crowded conditions was to avoid them (42% of responses given). They rarely listed situations which required them to persist in crowded places despite the discomfort. On the other hand, Japanese subjects regularly listed situations which required endurance daily such as on subways, trains, campus, and popular downtown areas. In addition to avoiding extremely crowded conditions by taking early or late trains, for example, Japanese subjects used a number of coping devices to reduce the stress of high density conditions. The coping devices given by these subjects in situations where avoidance was not possible were: reading, withdrawing, staring at the scenery, looking at people, concentrating or inner thoughts, and being indifferent. It appears that, in general, these coping devices are efforts to deny the immediate situations and to objectify people into a mass rather than to relate to someone as an individual. Internal distance was emphasized in situations where external distance from people was not possible.

In the Japanese sample, both high and low scorers on the Globig Crowding Sensitivity Inventory viewed certain aspects of crowding in a negative manner. The situations of traveling on crowded trains and subways, for example, were uniformly seen as negative. In the American sample, in contrast, the crowded situations listed tended to be entertainment activities such as sporting events and parties (39% of all situations given). Japanese women, in addition, uniformly indicated that dealing with "perverts" or people who sexually molested them in crowded conditions was a problem. In trains and subways, they could not tell who was molesting them because of the crowded conditions.

Japanese high and low scorers on the Globig Crowding Sensitivity Inventory differed on whether they preferred to be alone or be in a crowd in most situations. Those scoring low preferred to be in groups (69%) rather than being alone (8%). The percentage of persons who were somewhat ambivalent was 23%. Those scoring high on the Globig Crowding Sensitivity Inventory preferred to be alone (27%) compared to being in groups (8%). However, 55% were ambivalent.

In the Japanese sample, it was impressive to note the powerful attraction of crowds, even for those scoring high on the Globig Crowding Sensitivity Inventory. Crowds were seen to attract by their excitement, curiosity value, energy, and solidarity. Crowds seem to allow people to meet someone, feel at ease, do things, promote group feeling, increase feelings of existence, relieve gloomy feelings, escape loneliness and alienation, and provide security. On the other hand, in specific instances, crowds were seen as a problem and negative reactions were manifested.

Along with their attraction to crowds, Japanese subjects expressed more negative feelings about being alone than did the Americans. Being alone is seen as causing feelings of uncomfortableness and feeling lonely. To be part of a group is seen as providing identity and being the same as others is viewed as desirable and a natural way of existing. They feel a unity of human beings for each other. Even subjects scoring high on the Globig Crowding Sensitivity Inventory comment on the dangers of isolation. However, they do not react as negatively to being alone as those scoring low on the Globig Crowding Sensitivity Inventory.

In the American sample, positive and negative reactions to crowds were more clearly differentiated among the high and low scorers on the Globig Crowding Sensitivity Inventory. High scorers expressed mostly negative reactions to crowds (74%), no positive reactions, and a few ambivalent responses (26%). In contrast, low scorers reacted much more positively (70%), had no negative reactions, and had some ambivalent responses (30%). High scorers had a number of specific negative comments on how crowds affected them. They saw crowds as resulting in feelings of revulsion, self-consciousness, being scary, being killed, treated as a number, increasing feelings of wanting to get away, and being tense. Low scorers viewed crowds in a more positive light and saw crowds as interesting, powerful, and exciting. Crowds allowed interaction among people, meeting people, and provided security. Crowds also were seen to "bring out" the individual and relieve tensions of individuals if something was bothering them.

Both American and Japanese subjects who scored high on the Globig Crowding Sensitivity Inventory emotionally expressed several adverse reactions to crowding which indicated extreme uncomfortableness in these situations. For example, Japanese subjects expressed feelings of "being blue", "irritation", "being very impatient", "hatred", "being upset", and "being nervous". They also indicate body discomfort such as "I can't breathe", "can't bear bodily contact", "headache", "dizzy", and "tired". It appears that those who score high on the Globig Crowding Sensitivity Inventory have strong emotional reactions to crowded conditions which border on claustrophobia. Their reactions appear not to be controlled adverse responses to crowded conditions. A more intensive analysis of high scorers on the Globig Crowding Sensitivity Inventory is indicated.

In summary, it appears that differences were found between high and low scorers on the Globig Crowding Sensitivity Inventory in how they viewed and responded to crowded conditions. In addition, cross-cultural differences were noted in the United States and Japan, particularly with respect to free association responses to concepts indicating the state of being alone and crowded; and the responses of Japanese subjects regarding the attractiveness of crowds and the negative aspects of being alone. It also was interesting to note that the phenomenological view of crowding was an important factor in how subjects responded rather than extensive experience in high density conditions. Many subjects in Nevada had similar reactions to crowding as subjects in Tokyo, Japan. It seems that in the future research on crowding, factors need to be considered such as phenomenological views of crowded conditions, individual differences in chronic reactions to crowding, and cross-cultural variables.

PART III

MULTIVARIATE STUDIES OF PERCEPTION OF THE CROWDED ENVIRONMENT

Unlike the experimental studies in Part II, the three papers in this section differ both in method and in objective. Rather than constructing crowded and uncrowded settings and studying subjects' reactions to them, these investigations have started with naturally-occurring descriptions of crowded behavior settings within the normal territorial range of their respondents and then asked for certain complex judgments from similar respondents within which the settings are compared. The eventual outcome, after treatment of the data with multivariate analysis, is a set of underlying dimensions describing the "meaning" of crowding in terms of the respondents' cognitive structure.

Schopler, et al. find that American respondents code descriptions of crowded settings in terms of three underlying dimensions, with individual differences playing a relatively small part in the process. Working with English respondents and settings in London, Stockdale, et al. find a four-factor solution acceptable and also report small influence from individual differences. The reader is left to judge the degree of similarity among the American and English dimensions. Edgüer, et al., on the other hand, start with three dimensions and study the perceptions of respondents in Ankara, Turkey, of local behavior settings on each of the three dimensions of meaning provided by the semantic differential. They find, among other things, that the meaning of crowding is partly a function of the ecological sector in which the crowded setting is found.

These three studies seem to be breaking new ground in both a methodological as well as a theoretical sense. With regard to the former area, a variety of innovations, such as the use of naturally-occurring settings and multivariate analysis, deserve consideration with regard to theoretical innovation, such features as the identification of underlying dimensions of meaning in the crowding process and the use of samples from other cultures should be considered. The integration of these investigations with those from more controlled situations remains a problem for the future.

A MULTIDIMENSIONAL ANALYSIS OF SUBJECTIVE CROWDING

Janet E. Stockdale, Laura S. Wittman, Lawrence E. Jones(*), Denise A. Greaves

London School of Economics, Houghton Street, London WC2A 2AE, England

ABSTRACT: In order to examine the psychological dimensions underlying the perception of crowding, similarity judgments were obtained for a set of 15 descriptions of subjectively crowded situations. Subjects also made unidimensional judgments of crowding, environmental attributes and affective reactions. The dissimilarity matrices for 32 subjects were used as input to an individual differences multidimensional scaling analysis (INDSCAL). The results indicated that crowded situations were perceived as multidimensional varying along four identifiable dimensions. Multiple regression analyses, relating unidimensional judgments to the multidimensional configurations, identified these dimensions as: interpersonal overload/interference, alienation, anger vs. claustrophobia/helplessness and stress plus negative affect and behavioral response. Individual differences did not relate to saliences of subjective crowding dimensions.

INTRODUCTION

Over the past decade it has been generally recognised that crowding is not a unitary concept which may be operationalized by objective measures of density but is a perceived and subjective state –an experience of individuals– with multiple determinants. Any analysis which aims to determine the consequences of crowding must initially focus on the task of identifying the stimulus components of subjective crowding. The identity of the stimuli which induce subjective crowding vary widely according to the particular theoretical approach adopted (Stockdale, in press). Whilst there is minimal concensus as to what constitute the defining conditions for inducing feelings of crowding, there is increasing evidence that there are a range of factors which must be considered potential determinants of crowding. These include properties of the group in which the individual finds himself, such as size, structure

----

(*) University of Illinois at Urbana-Champaign, Champaign, Illinois 61820

and membership qualities, and properties of the physical and social environments and their compatibility with the ongoing task activity. Furthermore the influence of these variables will be moderated by demographic and personality characteristics of the participants and by their ability to cope with the situation.

It is a relatively easy task to provide additional evidence for the importance of these and other variables in the induction of crowding but this is an unproductive approach. At some point it is necessary to determine what variables are of major importance to individuals in perceiving situations as crowded. Only when this has been achieved can there be any attempt to formulate an explicit relationship between the judgments used by individuals to discriminate among potentially crowded situations as and the behavioral implications of these situations. There are two questions about the nature of this relationship which are of theoretical interest and practical importance. First, what dimensions do individuals use in perceiving situations as crowded? Although previous research has suggested that key factors may be stimulus overload (Milgram, 1970), perceived inadequacy of space (Stokols, 1972), unwanted social interaction (Valins and Baum, 1973), inability to attain desired levels of (Altman, 1975), behavioral interference (Schopler and Stockdale, 1977) and lack of environmental control (Stockdale, in press) there has been no direct evidence that such factors form the basis for an individual's perception of crowding. The identification of the dimensions of subjective crowding necessitates a technique by which perceived similarities and differences among stimulus situations can be directly inferred. They must not be a function of the experimenters' preconceptions. Multidimensional scaling methodology offers techniques for identifying those dimensions which underly the perception of crowding. Whatever dimensions emerge as salient to judgments of subjective crowding they will provide much needed guidance in defining a taxonomy of crowded situations –a necessary preliminary step in the linking of setting characteristics to the potential transient and long-term consequences expericenced by the participants.

The second question is whether the major perceptual dimensions are equally salient for all individuals, and, if not, are the differences in saliency related to personality differences among individuals? It might be expected, for example, that individuals' scores on variables previously linked to crowding, such as Internality/ Externality (cf. Schopler, McCallum and Rusbult, 1977), preference for privacy (cf. Altman, 1975) and interpersonal space (cf. Cozby, 1973) would be correlated with the weights associated with the perceptual dimensions of crowding. The use of an individual differences multidimensional scaling analysis permits the estimation of salience weights for each of the identified dimensions which can then be related to selected personality measures. The present study was designed, therefore, primarily to identify the dimensions used by individuals in judging similarities and differences among subjectively crowded situations, and also to examine the extent and character of individual differences in perceptions of crowding.

The INDSCAL Model

Multidimensional scaling techniques are concerned with the spatial representation of relationships among stimuli. The implicit judgmental model assumes that: (1) the individual conceptualizes the stimuli as though they are points in space, (2) the perceived dissimilarities among the stimuli are 'psychological distances' that are related to the distances among the points which represent them, and (3) the dimensions of the space are the relevant psychological dimensions along which the stimuli are judged and compared (Nygren and Jones, 1977). Individual differences multidimensional scaling analyses, such as INDSCAL (Carroll and Chang, 1970) generalize the procedures for scaling similarity (or preference) judgments so as to account for individual differences. INDSCAL involves the construction of an n-dimensional Euclidean group stimulus space in which the dimensions underlying the perceptions of stimuli are assumed to be common to all individuals. An important property of INDSCAL is its 'dimensional uniqueness'. Dimensions are uniquely determined and cannot be rotated or transformed in any way without changing the fit of the solution to the data. The axes obtained are therefore interpreted directly. Although INDSCAL assumes that all subjects share a common or group space they are allowed to weight the dimensions of this space idiosyncratically. Thus INDSCAL takes individual differences into account by computing dimensional saliences or weights for each individual that reflect his perception of similarities and differences among stimuli. The perceived dissimilarity between any pair of stimuli, then, is assumed to be linearly related to the weighted Euclidean distance between their respective points in the space. Salience weights range from 0 to 1 and the importance of a particular dimension for any subject is indexed by the ratio of that dimension's weight to the sum of the other salience weights.

METHOD

The first stage of the study consisted of a stimulus selection procedure in which 15 descriptions were selected to represent a larger population of descriptions of crowded situations. In the second stage subjects completed three types of task: (1) similarity judgments of each of the 105 possible pairs of descriptions: (2) unidimensional judgments of crowding, associated environmental attributes and affective reactions and (3) personality questionnaires and personal space measurement. In the analysis stage the similarity data were used as input to INDSCAL in order to determine the dimensions along which stimuli are judged and compared and to assess individual differences in the importance of each dimension. Multiple regression analyses, relating unidimensional judgments to the multidimensional configurations, were used to identify the dimensions which underly perceptions of crowding.

Subjects

Thirty six subjects (18 male and 18 female) were recruited from amongst the undergraduate student population attending the London School of Economics during the Lent Term, 1977. To avoid potential cultural differences all subjects were white and British-born and raised. Subjects were paid at the rate of £ 1.00 per hour for participating in the study. Data from 32 subjects (16 male and 16 female) were used for analysis following intrasubject reliability checks.

Stimulus selection

The stimulus set comprised 15 descriptions selected from a pool of 100 descriptions generated in response to the following question: 'Please think of a time in your life when you felt very crowded. Describe the incident you have in mind and your feelings at the time in as much detail as possible.' Where necessary the resulting descriptions were carefully edited to remove extraneous material or to correct errors in grammar or spelling, and were typed in a standard format. On the basis of previous crowding research 20 unidimensional bipolar 7-point scales of the form I feel..../ I do not feel.... were constructed. Each of the 100 stimuli were then rated on each of the 20 scales by 4 experienced judges. Two Principal Component analyses were then carried out on the stimulus by scales data matrix, one with respect to scales, the other with respect to stimuli. The primary criterion for selection was that stimuli should represent the majority of high and low mean ratings on four composite factors derived from the Principal Component analysis of scales. One stimulus that was judged medium on all four scale factors was also included. A secondary criterion was that stimuli should exemplify the five major stimulus factors derived from the Principal Component analysis of stimulus ratings. The factorial structure of the perception of crowded situations is examined in more detail in Stockdale and Jones (1977).

Material

The stimulus set is described in abbreviated form in Table 1.

Procedural Tasks

Personal space measurement. Prior to completion of the rating task an estimate of each subjects' personal space was obtained using a direct behavioral measure (cf. Horowitz, Duff and Stratton, 1965). An individual's personal space is the mean distance over four trials involving the subject walking toward same sex and opposite sex confederates, and the same confederates walking toward the subject until the subject feels that any decrease in interpersonal distance would make them uncomfortable.

Similarities/dissimilarities ratings. Subjects were first asked to read through the set of stimulus descriptions several times to familiarize themselves with them. Then subjects made judgments of the overall similarity between all possible pairs of descriptions (n=105) on a 9-point scale, ranging from '1=very similar' to '9=very dissimilar'. A Ross ordering (Ross, 1939), which allows maximal spacing between items and so reduces presentation bias in paired comparison procedures, was used to obtain the presentation order for all subjects. Subjects were instructed to make similarities judgments in terms of 'the feelings implicitly or explicitly expressed'. All subjects received 15 practice trials and 15 stimulus pairs were repeated after completion of the Ross order to provide reliability estimates.

Unidimensional scale ratings. Subjects rated each of the stimulus descriptions on 18 9-point scales listed in Table 3. The attributes defined by these scales reflected

Table 1

Set of Stimulus Descriptions in Abbreviated Form

A.	INTERVIEW	-candidate apprehensively waited for university interview; bewildered and confused by college ritual and behavior of students and principal.
B.	HOTEL	-hotel room hot and humid, small and oppressive; alone but felt hemmed in.
C.	BOX-OFFICE	-large crowd pushing toward box-office; wanted to retrieve bag and leave but movement nearly impossible. Claustrophobic
D.	WEDDING	-family wedding reception 'invaded' by friends of groom. Did not know them, found them boring and unlikable. Resentful.
E.	A-LEVELS	-severe, inescapable academic and emotional pressure from examinations and family tension; accentuated by unsympathetic society
F.	OFFICE	-working in open-plan office; disliked colleagues; felt unable to get away from the presence of others and trapped in a fixed routine.
G.	FLAT	-living in a small flat with two small children and husband working at home; expressed hostility, resentment and lack of privacy.
H.	MOTOR SHOW	-many people but in a large area; restricted choice of direction but generally refined and orderly movement.
I.	OXFORD STREET	-busy shopping street; movement and purchases impeded by slow moving window-shoppers; annoyance.
J.	TUBE	-standing in a rush-hour tube train; oppressively hot and unable to move.
K.	PROMS	-realization that it would be impossible to leave promenade concert quickly; felt faint due to heat and others pressing against barrier.
L.	PUB	-with boyfriend in pub; felt oppressed by the number of people and noise and took no part in the conversation, so felt separated from friends; had to leave.
M.	FOOTBALL	-crushed in football crowd leaving match; feet lifted off the ground and frightened of being trampled.
N.	TRAIN	-physically cramped and claustrophobic on long train journey; solved by overcoming embarrassment and intertwining arms and legs to get more comfortable.
O.	PRISON	-mental and physical constriction; following solitary confinement continually and intrusively questioned by officials in small, cramped cell.

hypotheses about the dimensions of subjective crowding and were selected on the basis of previous research (cf. Rapoport, 1975; Schopler and Stockdale, 1977). Subjects were instructed to respond as they perceived the person in the description to feel.

Personality questionnaires. Each subject completed the 25-item North Carolina Internality-Externality scale (Schopler, Langmeyer, Stokols and Reisman, 1973) and a 13-item Preference for Privacy questionnaire adapted from two subscales of Marshall's Preference for Privacy Scale (Marshall, 1971).

ANALYSES

Reliability Analysis

An index of intrasubject reliability or judgmental consistency was obtained by computing a Pearson product-moment correlation between the two sets of judgments made by each subject in response to the 15 pairs of descriptions repeated as a reliability check. This measure provides a conservative estimate of judgmental consistency but does serve to identify those subjects who achieve a low degree of consistency relative to the remainder of the sample. The distribution of intrasubject correlations suggested a minimum acceptable value of $r=+.59$, ($p<.05$) and 4 of the 36 subjects who completed the experimental tasks were therefore discarded from subsequent analyses for failing to achieve satisfactory intrasubject correlations.

INDSCAL Analysis

INDSCAL solutions based on the dissimilarities matrices for 32 subjects were computed in one to seven dimensions. A number of features of these solutions were then examined to determine the most adequate and parsimonious representation of the data. Values of the goodness of fit measure for each solution are shown in Table 2. The magnitude of this measure increases less rapidly after the addition of the fourth dimension. Similarly, increases in the fit measure for individual solutions were minimal when more than four dimensions were considered. Furthermore the pattern of salience weights indicated that the five- and higher-dimensional solutions did not add more interpretable information. Examination of the correlations between the unidimensional scale judgments and the dimensions of the stimulus space for

Table 2

Goodness of Fit Statistic For INDSCAL Solutions in One to Seven Dimensions

No. dimensions	Goodness to fit
1	.644
2	.714
3	.745
4	.772
5	.793
6	.812
7	.824

Table 3

Correlations of INDSCAL Dimensions With Mean Ratings of Stimuli on Bipolar Scales[a]

	Scale	I	II	III	IV	Multiple Correlation
1	(People in my way)	.532	.510	.025	−.453	.843**
2	(Too many people for space available)	.555	.579	−.333	−.428	.813*
3	(People too close to me)	.629	.574	−.178	−.230	.775*
4	(Cope with too many people at one time)	.614	.245	.039	−.030	.654
5	(Claustrophobic)	−.074	.384	−.608	−.549	.836*
6	(Too much going on around me)	.480	.081	.112	−.423	.730
7	(Have to fight for what I want)	.098	.412	.078	−.739	.874**
8	(Want to escape)	−.244	.042	−.214	−.796	.844**
9	(Do not feel free to do as I choose)	.126	.251	−.288	−.333	.452
10	(Helpless)	−.031	.053	−.549	−.464	.739
11	(Too little privacy)	−.271	−.114	.380	−.356	.559
12	(Anxious)	−.246	−.150	−.262	−.654	.780*
13	(Hostile)	−.321	−.281	.547	−.729	.941***
14	(Stress)	−.439	−.106	−.083	−.834	.920***
15	(Angry)	−.007	−.164	.684	−.394	.833*
16	(Do not feel I belong)	−.136	−.750	.193	−.158	.815*
17	(Do not feel involved)	−.033	−.645	.380	−.195	.756
18	(Crowded)	.188	.488	−.476	−.648	.838*

[a] The significance levels attached to the multiple correlations are inflated by virtue of the high intercorrelations among stimulus values.

The signs of the simple correlations reflect the direction of the rating scales and are of no intrinsic interest.

* $p<.05$
** $p<.01$
*** $p<.001$

solutions of differing dimensionality revealed that the addition of the fifth and sixth dimensions led to small increases (⩾0.1) in the values of the multiple correlations for only 5 subjects and 1 subject respectively, and that the pattern of simple correlations essentially remained unchanged. For these reasons the four-dimensional solution was chosen as optimal and the remainder of the paper focuses on the interpretation and conceptualization of the four-dimensional solution.

Dimensions of Subjective Crowding

The mean ratings of each stimulus description on each of 18 scales were computed from the unidimensional rating data for 32 subjects. Multiple regression analyses were then carried out using the mean ratings for each stimulus on each scale, in turn, as the dependent variable and the coordinates of the stimuli in the four-dimensional solution as predictors. The simple and multiple correlations between the unidimensional scales and the INDSCAL dimensions are presented in Table 3.

A number of the unidimensional scales correlated highly with the INDSCAL dimensions. Dimension I had high correlations with the mean ratings on Scale 3: People too close to me (r=+.629) and Scale 4: Cope with too many people at one time (r=+.614). Dimension II correlated highly with both Scale 16: Do not feel I belong (r=-.750) and Scale 17: Do not feel involved (r=-.645). Dimension III correlated highly with Scale 5: Claustrophobic (r=-.608) and Scale 15: Angry (r= +.684). Dimension IV had high correlations with Scales 7: Have to fight for what I want (r=-.739), 8: Want to escape (r=-.796), 12: Anxious (r=-.654), 13: Hostile (r=-.729) and 14: stress (r=-.834).

The multiple regression procedure locates a vector corresponding to each unidimensional scale in the group stimulus space and the multiple correlation indexes the goodness of fit. If for a particular scale the simple correlations associated with the dimensions are low but the multiple correlation across all dimensions is high, then, even though the attribute defined by the scale cannot be used to identify the dimensions, the attribute can be accurately predicted by the solution dimensions. A high simple correlation between a particular scale and dimension of the solution implies that the associated vector lies close to that dimension and may give some insight into the identity of that dimension. In order to clarify the identity of the dimensions additional multiple regression analyses were carried out in which unidimensional scales were combined, with equal weighting, and mean ratings on these composite scales were used as dependent variables. The construction of these composite scales was based primarily on hypotheses derived from the factor structure of the original stimulus population and from previous research findings. Those scales which provided the best fit with the four-dimensional solution were used in interpreting the dimensions and the correlations between these composite scales and the INDSCAL dimensions are shown in Table 4. Vectors corresponding to each composite scale were located in the group stimulus space using the standardized Beta weights as direction cosines and are shown in Figures 1 to 3, with letters representing the positions of the stimulus descriptions(*)

(*) The given planes parsimoniously summarize the four-dimensional solution.

Table 4

Correlations of INDSCAL Dimensions with Mean Ratings of Stimuli on Composite Scales

	I	II	III	IV	R
Interpersonal overload/ interference	.546	.365	−.131	−.462	.765*
Alienation	−.084	−.716	.301	−.183	.795*
Claustrophobic/ Helpless/Angry	.042	−.279	.874	.240	.927***
Stress+negative affect	−.364	−.205	.238	−.874	.897***
Stress+negative affective/behavioral response	−.247	−.038	−.013	−.861	.770**

Dimension I was labelled "Interpersonal Overload/Interference". This dimension correlated (r=+.546) with mean ratings on a composite scale composed of the following unidimensional scales: - 1: People in my way, 2: Too many people for space available, 3: People too close to me, 4: Cope with too many people at one time and 6: Too much going on around me. Figure 1 shows a plot of the I x III plane of the four-dimensional stimulus space and the location of the Interpersonal Overload/Interference vector. Examination of the projections of the stimulus decriptions on this vector revealed that those situations involving large numbers of people in close physical proximity (e.g. football games, tube train, motor show) were typically judged as involving more overload and interference than those involving minimal interaction with others (e.g. being alone in a hotel room).

Dimension II was correlated (r=−.716) with ratings on a composite scale composed of two unidimensional scales:– 16: Do not feel I belong and 17: Do not feel involved. Dimension II was labelled alienation since it differentiates between those situations in which internal reasons for feelings of crowding were expressed and those in which feelings of not belonging and uninvolvement were not salient. The II x III plane of the stimulus space and the location of the alienation vector are shown in Figure 2. In general, situations which were judged highly alienating were instances of 'psychological' crowding whilst those situations lower in alienation were dominated by attributes of the physical environment. This finding suggested that a categorical structure may provide an appropriate representation of these stimulus judgments.

As pointed out earlier, Dimension III had high correlations with the mean ratings on Scales:– 5: Claustrophobic, 10: Helpless and 15: angry. Dimension III correlated most highly (r=+.874) with a composite scale derived from these scales (with appropriate reflections). The location of the vector corresponding to this composite scale is indicated in Figure 2. Dimension III was labelled anger vs. claustrophobia/helplessness because it differentiated between situations expressing or resulting in feelings of anger or resentment and those commonly described as 'claustrophobic' or 'oppressive', which were also characterized by severely restricted movement and consequent feelings of helplessness.The descriptions which are rated

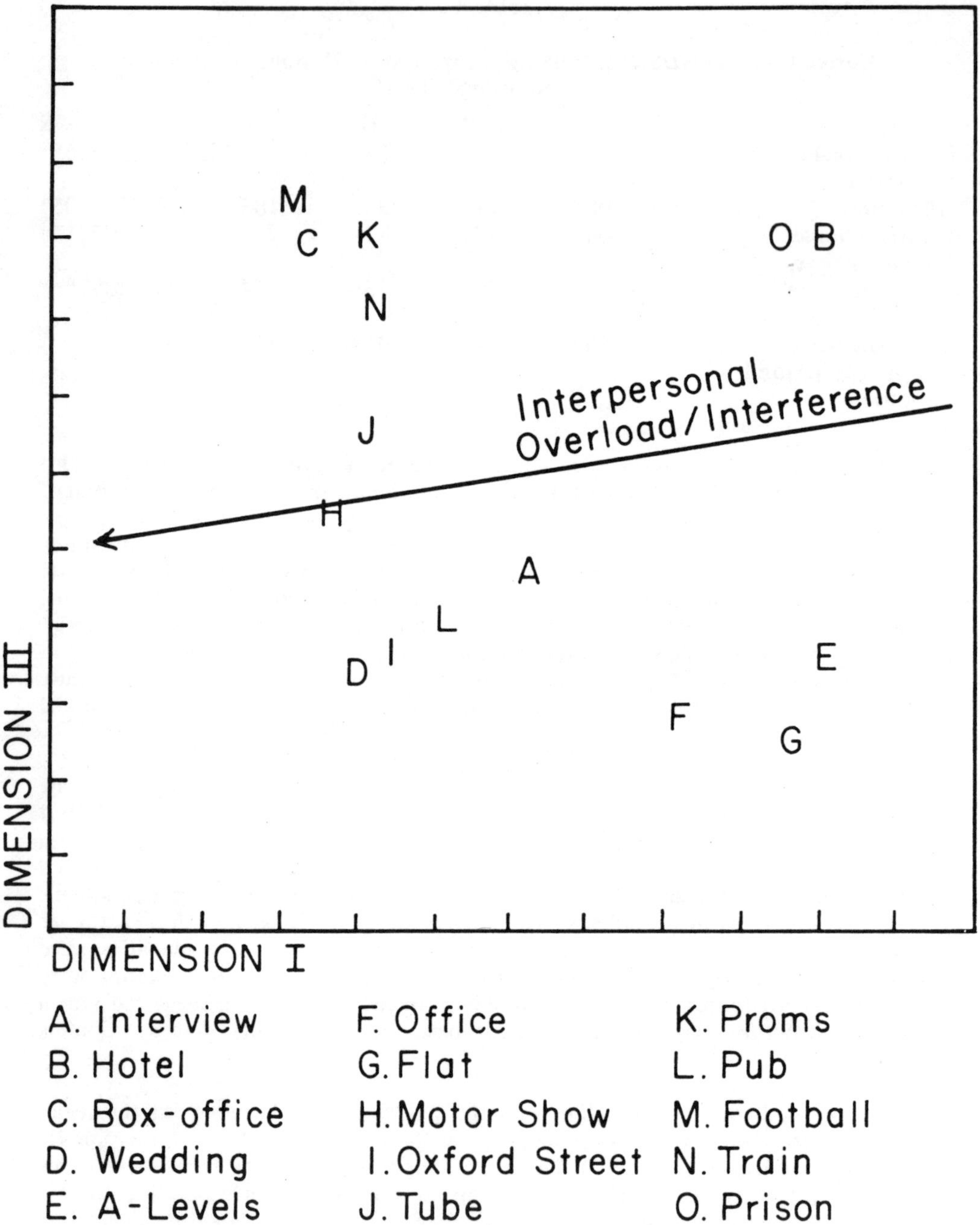

Figure 1. Dimensions I and III for the four-dimensional INDSCAL solution

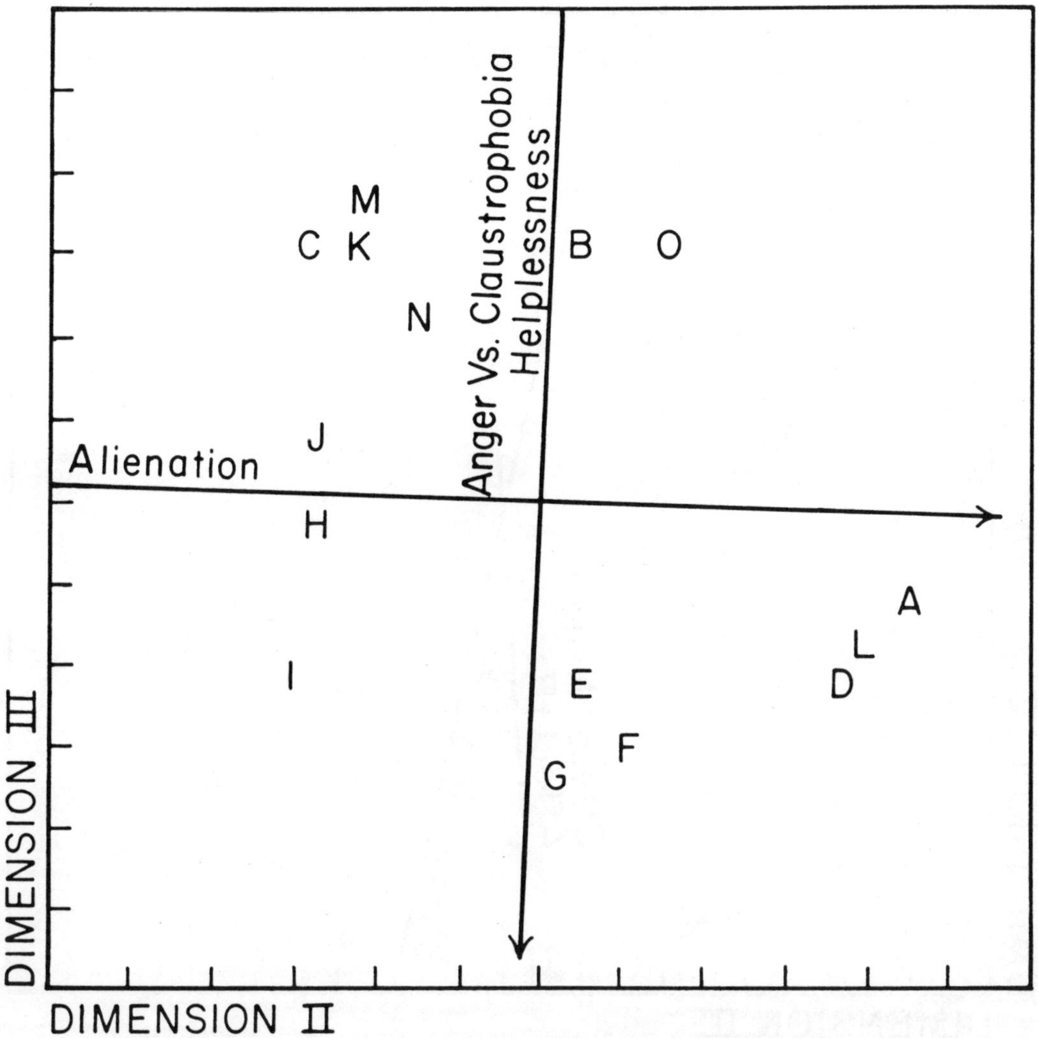

A. Interview
B. Hotel
C. Box-office
D. Wedding
E. A-Levels
F. Office
G. Flat
H. Motor Show
I. Oxford Street
J. Tube
K. Proms
L. Pub
M. Football
N. Train
O. Prison

Figure 2. Dimensions II and III for the four-dimensional INDSCAL solution.

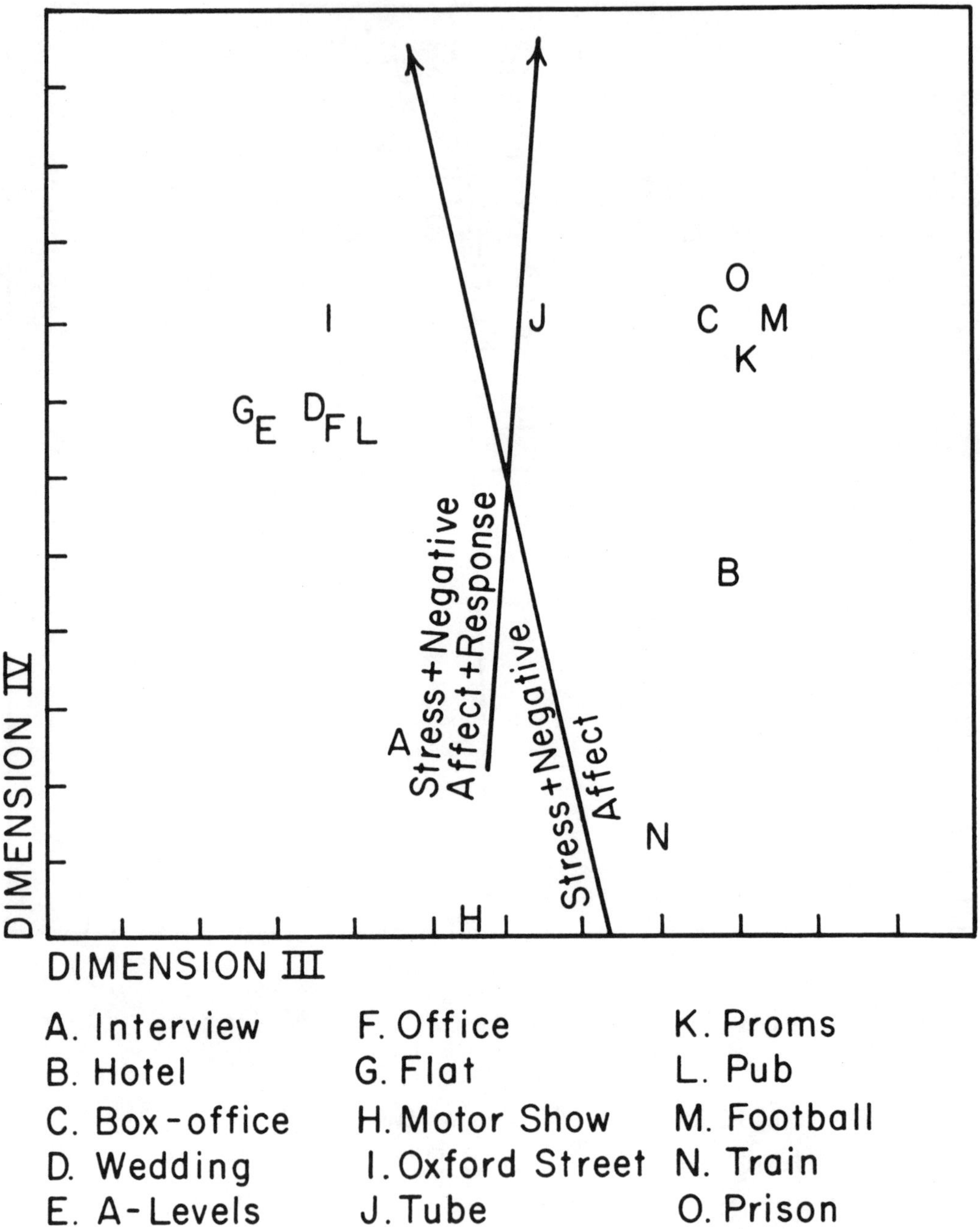

Figure 3. Dimensions III and IV for the four-dimensional INDSCAL solution

highly claustrophobic and induced feelings of helplessness included the majority of those judged to involve 'physical' crowding. One situation which was characterized by close physical contact with others (H: Motor Show) was not judged to induce feelings of claustrophobia but this description explicitly stated that the situation was 'not claustrophobic'.

Among the unidimensional ratings the highest correlation with any of the four INDSCAL dimensions was that between Scale 14: Stress and Dimension IV (r=–.834). Dimension IV was also highly correlated with other scales which either expressed negative affect or encompassed feelings or response toward desired but inhibited action. A composite scale labelled Stress plus Negative Affect, comprising Scales 14: Stress, 12: Anxiety, 13: Hostility and 15: Anger, correlated highly (r= -.874) with Dimension IV. The location of the corresponding vector is shown in Figure 3. There was also a high correlation (r=–.861) between Dimension IV and a second composite scale which included the following additional scales:– 7: Have to fight for what I want, 8: Want to escape, 9: Do not feel free to do as I choose and 10: Helpless. This resulted in the best-fitting vector and its location is shown in Figure 3. Dimension IV was therefore labelled Stress plus negative affective/behavioral response.

Only four situations were rated relatively low on stress and the negative implications of crowding, and there are interesting differences between these and the remainder of the stimulus set. For example, stimulus H: Motor Show, which expressed a strong feeling of personal control and orderliness within the setting, received the lowest stress rating. The description of travelling by train (Stimulus N) differed in that the original sense of discomfort was ultimately resolved to the satisfaction of everyone involved. Similarly, the negative feelings created by the University interview procedure (Stimulus A) were finally resolved with an initially bewildering situation becoming meaningful. The remaining eleven situations were largely disorderly or even chaotic and there was no expressed resolution to the problems, and consequent stress, created by the physical or social environment.

The distribution of stimulus situations along Dimension II suggested it would be fruitful to determine whether there is an underlying categorical structure. In order to examine the hypothesis that people perceive the same stimuli in the same way the standardized aggregate dissimilarities matrix was submitted to Johnson's Hierarchical Clustering Analysis (Diamter Method). This revealed that the stimuli clearly divide into two main clustres which are differentiated by whether feelings of crowding are generated by physical or psychological factors. A dendogram showing the clustering of the stimulus situations and their inclusion levels is shown in Figure 4. Those situations which were judged to involve 'physical' crowding were typically short-term , secondary environments in which there were a large number of strangers. Physical density and physical discomfort were factors in all these descriptions. Crowding was easily attributable to external environmental attributes which were not within the control of the participant. Within this cluster there was one major sub-division. Situations which evoked anxiety about personal safety or personal integrity were differentiated from situations, which, although dominated by the physical impact of a large number of people, were also characterized by lack of fear and a definite purposeful attitude on the part of the actor (Motor Show and Oxford Street).

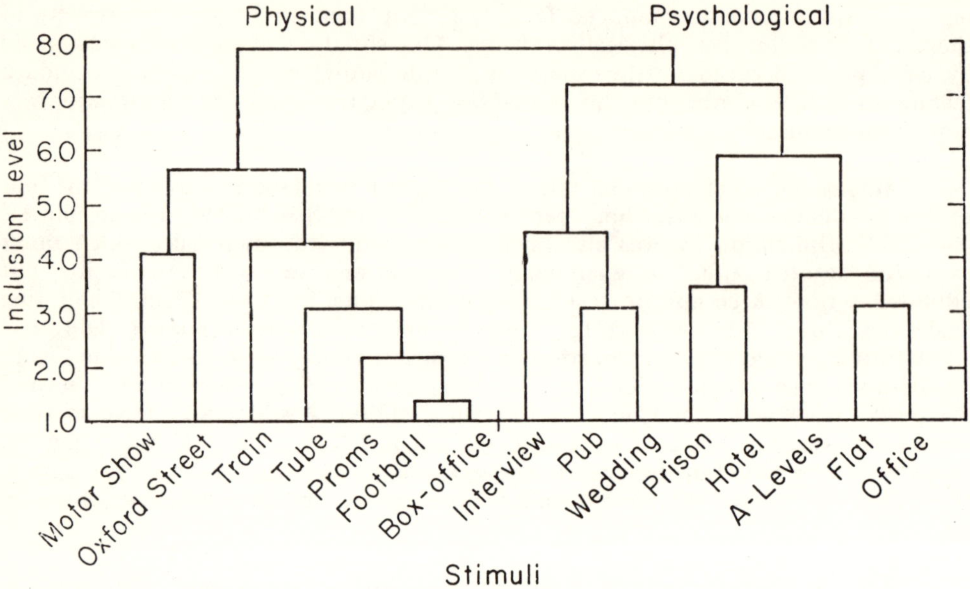

Figure 4. Dendogram showing clustering of stimulus descriptions and their inclusion levels

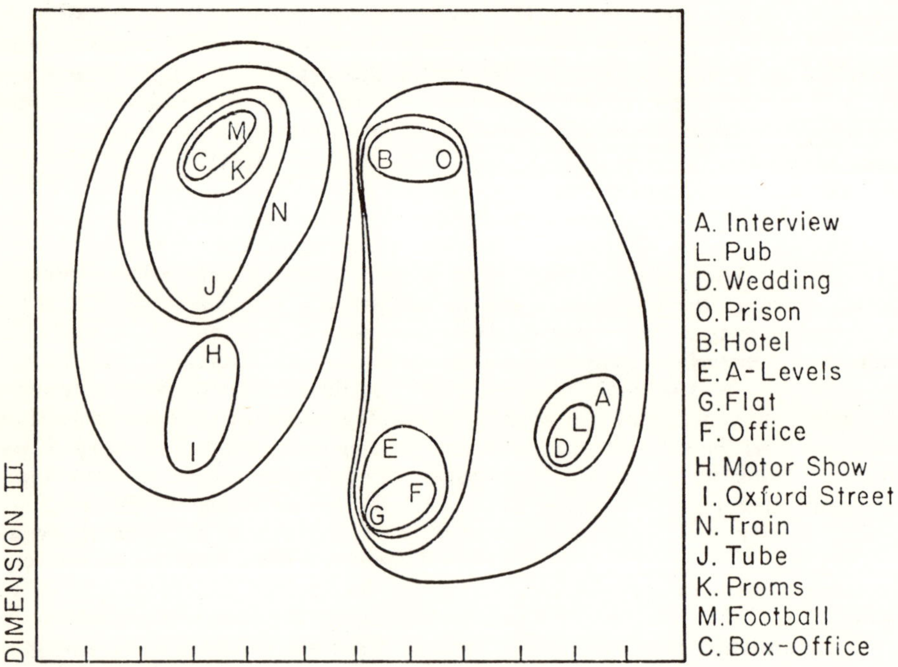

Figure 5. Dimensions II and III for the four-dimensional INDSCAL solution showing the clustering of stimulus descriptions within Dimension II.

In contrast with the first stimulus cluster, those environments in the second cluster, which were characterized by 'psychological' crowding, generally contained fewer people in physically smaller settings. The attribution of crowding was obviously mediated by variables other than simply physical density or number of people. The socio-emotional environment was prepotent and the crowding stress experienced appeared to be due to the type and amount of involvement with others. Within this cluster descriptions of relatively transient social settings (University Interview, Pub, and Wedding) were distinguished from those which referred to living and working environments - essentially primary environments. Figure 5 shows the clustering of the stimulus situations relative to Dimension II. A qualitative structure, offers an alternative interpretation to that offered by the multidimensional scaling solution and, whilst it is possible that this merely reflects the selection of the stimulus sample, it is likely that individuals do construe crowding in terms of whether the source is physical or psychological. A primary feature of those situations in which crowding is attributed to psychological pressure is a feeling of alienation from others.

Individual differences

Examination of the salience weight matrix resulting from INDSCAL revealed that Dimension I was most salient for 63% (N=20) of the subjects, Dimension II was most salient for 19% (N=6) with Dimensions III and IV each being most salient dimension for 9% (N=3) of the subjects. The pattern of salience weights was similar for males and females (Table 5) and there were no significant differences between the weights estimated for males and females for any of the four dimensions (Mann-Whitney U Test). To determine whether the INDSCAL dimensions were related to personality variables as measured, a series of multiple regression analyses were carried out, using scores on the personality scales as independent variables and individual salience weight ratios, for each dimension in turn, as the dependent variable. Table 6 shows the correlations of the salience weight ratios for each dimension with the personality measures. Neither the Externality/Internality nor Preference for Privacy scores proved to be significant predictors of dimension salience and, although personal space was the single best predictor of dimension salience, at no point did the simple or partial correlation approach significance.

DISCUSSION

The results of the present multidimensional scaling analysis suggest that individuals utilize at least four dimensions in making judgments of crowded

Table 5

Mean Salience Weights for each INDSCAL Dimension for Males and Females

	I	II	III	IV
Males	.518	.318	.243	.115
Females	.458	.330	.284	.201

Table 6

Correlations of the Salience Weight Ratios For Each Dimension With Personality Measures

Dimension	Internality Externality	Preference for Privacy	Personal Space	R	P
I	.008	–.096	–.198	.306	.424
II	–.085	.014	–.181	.217	.712
III	–.033	–.071	–.124	.202	.756
IV	–.071	.0155	–.192	.224	.691

situations, and confirm that crowding cannot be conceptualized simply in terms of properties of the physical environment. The first dimension reflected the impact of interpersonal interaction and the second the degree of alienation. One end of the third dimension was defined by feelings of anger, the other by feelings of claustrophobia and helplessness. The fourth dimension reflected the degree of stress and negative effect experienced and the extent to which it was resolved. It is of interest to examine the structure of the four-dimensional space and the distribution of stimuli along several of the component dimensions as a first step toward evolving a taxonomy of crowding situations. Each description was classified by reference to its position relative to the origin of the stimulus space(*). The categorization of the stimulus set across all four INDSCAL dimensions is shown in Table 7.

A major question which must be considered is whether the procedure used to select the stimulus set biased the subjects' similarity judgments, such that these judgments must be regarded as indirectly dependent on the scales by which stimuli were selected. The stimuli were selected to represent a larger population of stimuli and a major selection criterion was that they should reflect combinations of high and low loadings on four factors derived from unidimensional ratings of the total stimulus population. Inevitably the stimulus set did not represent the frequency with which descriptions of a particular type were generated. The preselection analysis was based on independent data from different subjects and although the analysis generated four major factors their identity was nat directly congruent with those which emerged from the multidimensional scaling analysis. The methodology does not demand that the results of the two procedures should be comparable in that individuals are quiet capable of making judgments along dimensions that they do not use in construing their environment.

A major feature of the stimulus classification shown in Table 7 is that no descriptions are judged low relative to both dimensions I and II. Stimuli typified by a low level of interpersonal interaction all involved strong feelings of alienation and psychological pressure. This pattern of stimuli is understandable in logical pressure and alienation are not dependent on interaction with a large number of

----

(*) This classification scheme is arbitrary to the extent that it fails to identify those stimuli that lie close to the origin.

Table 7

Categorization of Stimuli Across All Four INDSCAL Dimensions

I:	Overload/ Interference (Hi vs. Lo)		Hi				Lo			
II:	Alineation) (Hi vs. Lo)		Hi		Lo		Hi		Lo	
III:	(Anger vs. Claustrophobia (HiA vs. HiC)		HiA	HiC	HiA	HiC	HiA	HiC	HiA	HiC
IV:	Stress (Hi vs. Lo)	Hi	D,L	–	I	M,C,J,K	G,E,F	0	–	–
		Lo	–	–	H	N	A	B	–	–

others, whereas physical crowding situations are usually characterized by physical proximity and behavioral interference. The matrix also suggests that intense social interaction and feelings of claustrophobia and helplessness are features of situations involving physical crowding and minimal feelings of alienation , and do not characterize situations involving strong feelings of alienation and psychological pressure. The perceptual structure represented by the matrix also indicates the lack of stress-reducing mechanisms in the context of intense interaction and feelings of alienation and anger. This analysis identifies some of the boundaries that might characterize a taxonomy of crowding situations, and could be of direct use in designing manipulative research. Effort should be directed toward identifying compatible and realistic combinations of values on the dimensions underlying the perception of crowding and determining whether perceived differences relate to the behavioral consequences of both short- and long-term involvement in these situations. For example, situations in which different characteristics are salient may offer different options for reducing subjective crowding and so correspond to different coping strategies.

It is encouraging that the results of a multidimensional scaling study by Schopler, Rusbult and McCallum, (1977) lend some support to the generality of the dimensions identified in this study. Using a different scaling technique, different stimuli and subject populations, Schopler et al. identified the dimensions of subjective crowding as physical-psychological, familiar-unfamiliar and resultant stress. These three dimensions have much in common with Dimensions II and IV identified in the present study.

The stress model has commonly been used to describe the etiology of crowding and the results of this analysis to identify stress as a major component of crowding. Contrary to recent conceptualizations of crowding neither lack of privacy (cf. Altman, 1975) nor lack of control (cf. Stokols, 1976; Stockdale, in press) emerged as dimensions of subjective crowding. The unidimensional scale crowding did not identify any of the dimensions but was associated with three out of the four dimensions. This result implies that the feeling of "crowding", as understood and used by the subjects is not identifiable with any one aspect of crowding, but reflects a number of different aspects, which may vary across individuals and may indeed differ for a given individual at different times. A preliminary step in determining whether the term crowding shows systematic differences in meaning across subjects would be to examine the factorial structure of crowding ratings derived from a subjects by subjects correlation matrix. To the extent that the term crowding reflects individual differences in usage, so that it is impossible to know how subjects are using the scale, then judgments of crowding, such as those used to provide manipulation checks in experimental studies of crowding, are ambiguous.

The present study also aimed to examine the contribution of three personality variables to individual differences in the perception of crowded settings. However, the individual differences scaling analysis did not reveal any of the personality variables measured to be significantly related to dimension salience. Clearly individuals differ in the importance they attach to the perceptual dimensions but this study was unsuccessful in determining what factors contribute to these individual differences. Further studies should explore the variables underlying differences in dimension salience and their potential behavioral consequences.

The methodology employed in this study represents a radically different approach to the problem of characterizing crowded situations. A primary difficulty with past studies is that they reflect how crowding theorists rather than individuals construe crowded environments. Although the current approach offers no way of unequivocally identifying the perceptual dimensions as antecedents, concomitants or resultants of crowded settings, the dimensions are in no way predetermined or imposed on the subjects but are derived by analysis. Moos, (1976) argues that it is important to identify similar underlying dimensions along which very different social environments can be characterized. This view is particularly applicable to the variety of environments which induce feelings of crowding, which in the past has generated so much conceptual confusion. The identification of the dimensions underlying the perception of crowding is the first step in analysing the contribution of environmental, personal and cultural variables to crowding and assessing the full impact of crowding on the individual and society.

CONCEPTUAL DIMENSIONS OF CROWDING:

A MULTIDIMENSIONAL SCALING ANALYSIS

John Schopler, Caryl E. Rusbult and Richard McCallum

University of North Carolina, Chapel Hill, N.C., U.S.A.

ABSTRACT: In order to assess the extent to which the conceptual distinctions used by crowding theorists are contained in peoples' perceptions of crowded situations, a three-part multidimensional scaling experiment was undertaken. In Phase I, free descriptions of crowding instances were collected. Phase 2 involved a new set of respondents who made similarity judgments of the descriptions. The analysis revealed that respondents discriminated among the stimuli on three dimensions and suggested that women and men use different salience weights. Phase 3 sought to find objective measures of the attributes defining the three-dimensional space and found that the dimensions could be termed physical-psychological, familiar-unfamiliar and resultant stress. The implications of the results for crowding research are discussed.

Several years ago we asked the simple question, "What are the conceptual dimensions that determine peoples' perceptions of being crowded?" The predominant view in the crowding literature portrays crowding as an experiential state whose antecedent conditions are not solely defined by high density. Despite this consensus, there is wide divergence about what constitutes the necessary and sufficient conditions for inducing the state of feeling crowded (cf. Schopler and Stokols, 1976; Stockdale, in press). As an initial step toward evaluating these various conceptions, it was thought important to identify how people, rather than crowding theorists, construe the domain of being crowded. Information about the extent of match between the conceptual dimensions used by people and theorists would be useful information for gauging the applicability of the ideas employed by the former group to explain the behavior of the latter group. Furthermore, two subsidiary benefits might also be gained from research pursuing the initial question. Identification of the major dimensions could be used (1) to see whether particular groups of individuals (e.g., women vs. men, internals vs. externals, urban vs. rural) differed in the weights they typically assigned to any of the dimensions and (2) to evolve a taxonomy of situations producing crowding. The focus of the present paper will be on our efforts to answer the initial question, which proved to be sufficiently difficult to delay progress on the two subsidiary goals.

The burgeoning literature on crowding offers an impressive array of suggestions concerning the best way to define the stimulus conditions producing crowdedness. These include such definitions as perceived inadequacy of space (Stokols, 1972), excessive stimulation from social sources (Desor, 1972), exposure to too many unwanted social interactions (Valins and Baum, 1973), inability to attain desired levels of privacy (Altman, 1975), or interference from others (Schopler and Stockdale, 1977). The literature, however, provides no guidance concerning which, if any, of these dimensions are actually operative in the minds of individuals. Identifying such dimensions could be achieved by travelling along any one of several research paths. After some reflection, we decided to embark on a multidimensional scaling route. Multidimensional scaling is a family of data processing procedures developed explicitly to identify conceptual dimensions. It lends itself to data collection that is shielded from the inadvertent introduction of the researcher's own preconceived ideas, with respect both to generating an initial pool of stimuli and to responding to those stimuli. Furthermore, it identifies dimensions that need not be registered in the awareness of respondents. This seems especially important in light of a growing doubt about people's ability to report accurately on their cognitive processes (Nisbett and Wilson, 1977). Finally, it should be noted that we were not unmindful of the proximal presence of Forrest Young, one of our colleagues, who is knowledgeable about the procedure and is willing to give freely of his time as a consultant. Because multidimensional scaling may not be widely known, we will first provide some general background before turning to a description of the research.

OVERVIEW OF MULTIDIMENSIONAL SCALING

Multidimensional scaling (MDS) procedures enable the identification of the dimensions, or factors, that individuals perceive and employ in discriminating among a set of stimuli. The user begins with a set of stimuli that represents the topic of interest. The stimuli may be of any form (e.g., colors, words, names of countries, paragraphs) and should represent an adequate sample from the population being studied. Respondents are then asked to state their beliefs concerning the similarities/dissimilarities among the stimuli comprising the set. This portion of the procedure may take many forms, a common one being to require respondents to rank order all stimuli in terms of their similarity to each single stimulus. For example, if the stimulus set consisted of 10 stimulus objects, each person would be asked to perform 10 similarity rank orderings —one ordering relative to each of the 10 stimuli. For larger stimulus sets, where the collection of a complete set of rank orderings would be extremely time-consuming, a partial set of orderings may be obtained from each respondent. The collection of a portion of the full rank ordering set enables the computation of the full set, with little loss of accuracy. These data, the similarities matrix, are the input for standard MDS programs. The reader is referred to Shephard, Romney, and Nerlove (1972) and Young (1975) for a full description of the mathematics underlying these procedures.

With most MDS computer programs the individual is free to request the computation of multidimensional solutions in any number of dimensions. The choice of one solution over others depends upon the "difficulty" of fitting the data to a space of that dimensionality. A measure of the difficulty is the "stress" of that solution. Naturally, as the solution allows for more dimensions, the stress of the fit decreases. Assuming that the most appropriate solution is the one that promises to most accurately and parsimoniously represent the relationships among the stimuli, one observes increases in solution dimensionality and looks for large decreases in stress followed by a "leveling off" of the stress values. The solution that produces the greatest stress decrease before the "leveling off" is generally the correct solution.

After the relationships among the stimuli are determined, the characteristics, or dimensions, that define the multidimensional configuration must be identified. A set of potential defining attributes are selected, and respondents are required to judge the extent to which each of the initial stimuli possesses each of the potential attributes. Assuming that the true defining attributes are included in this set, the characteristics that best fit the actual multidimensional configuration are considered to be the dimensions that define the space. These are dimensions that individuals employ in discriminating among the stimuli.

A final issue concerns individual differences in the weighting of these dimensions. The respondent population may be divided into several parts, representing distinct groups of individuals (or single individuals). Some MDS programs enable the user to request that weights be computed for each group (or individual) for each dimension. The assumption implicit in this procedure is that although all individuals perceive the stimuli in terms of the same set of dimensions, these dimensions are differentially important across groups (or individuals). Observation of the relative importance of each dimension for various groups (or individuals) is thus possible. The reader is referred to Carroll (1972) and Takane, Young and deLeeuw (1977) for a detailed description of the theory underlying individual differences procedures in MDS.

METHOD

The study consisted of three phases. The first phase involved the collection of paragraphs which comprised the crowding stimulus set. In Phase 2 respondents were required to produce several rank-orderings of these stimuli. These data enabled the identification of the multidimensional space which represented respondents' perceptions of the relationships among the crowding stimuli. In order to determine which characteristics represented the major dimensions in the obtained multidimensional space, Phase 3 respondents were asked to rank order the crowding stimuli relative to a number of conceptual dimensions. The procedures employed will be described separately for each of these phases.

Phase 1

Seventy undergraduates from the University of North Carolina were asked to respond to the following statement:

Please think of a time in your life when you felt very crowded. Decribe the incident you have in mind, and your feelings at the time, in as much detail as possible. Each person was given one sheet of 8 1/2 by 11 inch lined paper on which to write an essay. Respondents were allowed to write for as long a period of time as they desired. The resultant essays were typed verbatim on individual sheets of paper. Those essays that were difficult to comprehend (due to serious errors in grammar or spelling) were removed from the set, as were duplicate descriptions. The remaining 46 essays comprised the crowding stimulus set, and ranged from five to twenty-one lines in length with an average of 11.4 lines per essay.

Phase 2

Twenty-one males and 23 females participated in Phase 2 of the study in partial fulfillment of the requirements for an introductory psychology course. Eight groups consisting of from four to six, same-sex persons were asked to complete the experimental task. Two respondents were unable to complete the task because of time limitations, and one respondent completed the rankings incorrectly. Data from these individuals were excluded from the experimental analyses.

Upon arrival at the experiment each respondent was assigned to a separate table and was asked to complete the North Carolina Internal-External Scale (Schopler, Langmeyer, Stokols and Reisman, 1973). Internality-externality assignments were determined for all subjects on the basis of the mean value obtained in the validation sample. All those scoring higher than the validation mean were defined as externals, and all those scoring below the validation mean were defined as internals.

The experimenter explained that he/she was interested in discovering the similarities and differences among a set of descriptions. Respondents were to familiarize themselves with the descriptions (which were the crowding stimuli obtained in Phase 1) and then to rank order them in terms of their similarity to one of six "target" descriptions.

Although the use of any six targets would enable the identification of as many as five dimensions in the multidimensional stimulus space, the six targets selected were those that seemed to represent distinct definitions of the concept "crowding," as contained in the initial stimulus set. The six categories and a summary of their respective target stories are provided in Table 1.

The experimenter stated that respondents were to rank the stimuli in terms of similarity to each of the targets with respect to the experience of being crowded. After explaining the ranking procedure in greater detail and answering questions concerning the task, the experimenter provided each respondent with the set of 46 crowding stimuli, the first of six target descriptions, and a coding sheet on which rankings were to be recorded. Respondents completed rankings at their leisure and exchanged their completed targets and recording sheets for new ones approximately every half hour. Order of targets was randomized over respondents. After completing the six rank orderings subjects were thoroughly debriefed.

Phase 3

In order to facilitate the task of labeling the crucial dimensions defining the experience of subjective crowding, the data obtained in Phase 2 were analyzed via

Table 1

The Six Preliminary Crowding Categories and Their Target Items

Category	Item Number	Content of Target Story
(1) Large Group size.	10	Author is separated from her family at state fair, feels confined in flow of crowd.
(2) Close, physical proximity	5	Waitress is constantly hindered by coworkers' activities.
(3) Stimulus Overload.	21	Author felt pressure from exams and social demands.
(4) Lack of personal control	4	Author finds it unbearable to be home for summer under parents' authority.
(5) Outsider to ongoing activities.	2	Author attends football game at girlfriends' school, feels he is in a strange place with strange people.
(6) Being in an unfamiliar situation	34	Author in a class of 450 students, who seemed to know each other, but not him.

the program ALSCAL (Takane, Young and de Leeuw,1977). The program, utilizing a least-squares model, identified three dimensions (or factors) underlying the 46 stimuli. Considering both these data and a number of important theoretical factors (Milgram, 1970; Schopler and Stockdale, 1977; Proshansky, Ittelson, and Rivlin, 1972), several hypothetical factors were delineated as potential defining dimensions. The 14 factors chosen for the Phase 3 attributes ranking task were the following:

1) the number of people present in the situation;
2) the amount of physical density present in the situation;
3) the extent to which the author's feelings are caused by the presence of others;
4) the extent to which the author feels like an outsider to the other people described;
5) the extent to which the author feels distressed or unhappy;
6) the extent to which the ending described is pleasant;
7) the extent to which the author is familiar with the situation (setting, people, etc.) described;
8) the extent to which the author's feelings are determined by others' intentional actions aimed at him rather than by actions which are unintentional or unavoidable;
9) the extent to which others interfere, block, or thwart the author;
10) the extent to which the author is in control of events in the situation;
11) the extent to which the author's feelings in the situation are unique, i.e., not shared by others in the situation;
12) the extent to which the author's feelings are due to himself/herself rather

than to features of the situation;

13) the extent to which the author perceives that he/she is confronted with too many inputs (noise, social demands, etc.);

14) the extent to which the author is concerned with "psychological" factors as opposed to the "physical" aspects of the situation.

Seventeen male and 18 female introductory psychology students participated in the experiment for course credit. One male and three female respondents failed to complete the ranking task, and their data were not included in the analyses.

Upon arrival at the experimental session each person was assigned to a separate table. The experimenter explained that the purpose of the experiment was to identify the characteristics that define the experience of being crowded. The respondents' task was to rank order 27 descriptions (a subset of the crowding stimuli) with respect to a number of attributes (the 14 potential factors).

In order to simplify the complex attributes ranking procedure somewhat, preliminary ALSCAL analyses were employed in order to split the crowding stimulus set into two equivalent halves. Observing the obtained distances between all pairs of stimuli, those pairs of stimuli that were closest to one another were designated pairs, and one member of each pair was randomly assigned to Subset or to Subset B. Those stimuli whose closest seconds were at a distance greater than .25 were included in both subsets. The correlations of the average locations of the resulting 27 stimuli for the two subsets were quite high on all 14 rankings. The average correlation across all 14 factors was .57, and 11 of 14 correlations were significant beyond the .05 level. Thus, the two subsets indeed appear to have been equivalent.

The experimenter distributed booklets containing 14 sheets, each of which listed a factor and provided spaces for respondents to record their rank orderings. These 14 sheets were randomly ordered in the booklets across respondents. Each person was also given a set of index cards on which were printed the 27 stimuli which comprised Subset A or Subset B. One-half of the males and one-half of the females received each subset.

After distributing the experimental materials the experimenter explained that the task required 14 separate rank orderings of the 27 stimuli (one for each of the attributes). After describing the procedure in some detail, the experimenter answered questions and the respondents proceeded to complete their orderings, a task which entailed approximately two hours of work. At the end of the two hour session all respondents were thoroughly debriefed.

RESULTS

Multidimensional Configuration of the Data

The data obtained in Phase 2 of the experiment were used to identify the multidimensional space that represented respondents' perceptions of the crowding stimuli. The mean ranking of each stimulus relative to each target was computed, the dissimilarities matrix was developed, and a nonmetric multidimensional scaling program, ALSCAL (Takane, Young and deLeeuw, 1977) was applied to these data. Solutions were obtained in one, two, three, and four dimensional Euclidean space.

The stress values for the four solutions are presented in Table 2. To determine which of the four solutions was the most accurate and parsimonious representation of the stimuli, the reduction in stress that resulted from the addition of each new dimension was considered. The four-dimensional solution did not appear to reduce the resultant stress to an appreciable degree, so the three-dimensional solution was considered to be the most adequate. Figure1 portrays the three-dimensional space that represents the relationships among the crowding stimuli. It should be noted that ALSCAL orders dimensions in accordance to the relative spread of the stimuli on the dimensions.

Dimensions of Crowding - The Attributes Data

The ranking task completed by subjects in Phase 3 provided the attribute vectors necessary to determine which of 14 hypothesized characteristics best described the obtained stimulus space.

The attribute vectors were related to the obtained three-dimensional solution via the computer program PREFMAP (Carroll and Chang, 1970). Employing a multidimensional unfolding procedure (Carroll, 1972), solutions were obtained for three separate models. Although each model attempts to find for each attribute an ideal point (or vector) in the three-dimensional space, the assumptions underlying the models differ somewhat. Models one through three are point-fitting methods, and differ with respect to several important assumptions. The first model allows each attribute its own orientation and pattern of weights. The second model allows for the differential weighting of dimensions by each attribute, and the third model assumes that all attributes possess the same orientation and weighting of dimensions. Model four finds the vector (rather than the point) that best represents each attribute's position in the three-dimensional space. Model one was considered inappropriate for the data in the present study, so it was not employed as a means of analysis.

The root mean squares of the three models do not seem to differ greatly (the respective values for models 2, 3 and 4 were .840, .809 and .792) but because the vector model (model four) yielded larger F-values (see Table 3) than models two or

Table 2

Stress Values Resulting from Derived Solutions in One, Two, Three, and Four Dimensions

Dimensionality	Stress in Distances
1	0.262
2	0.165
3	0.124
4	0.087

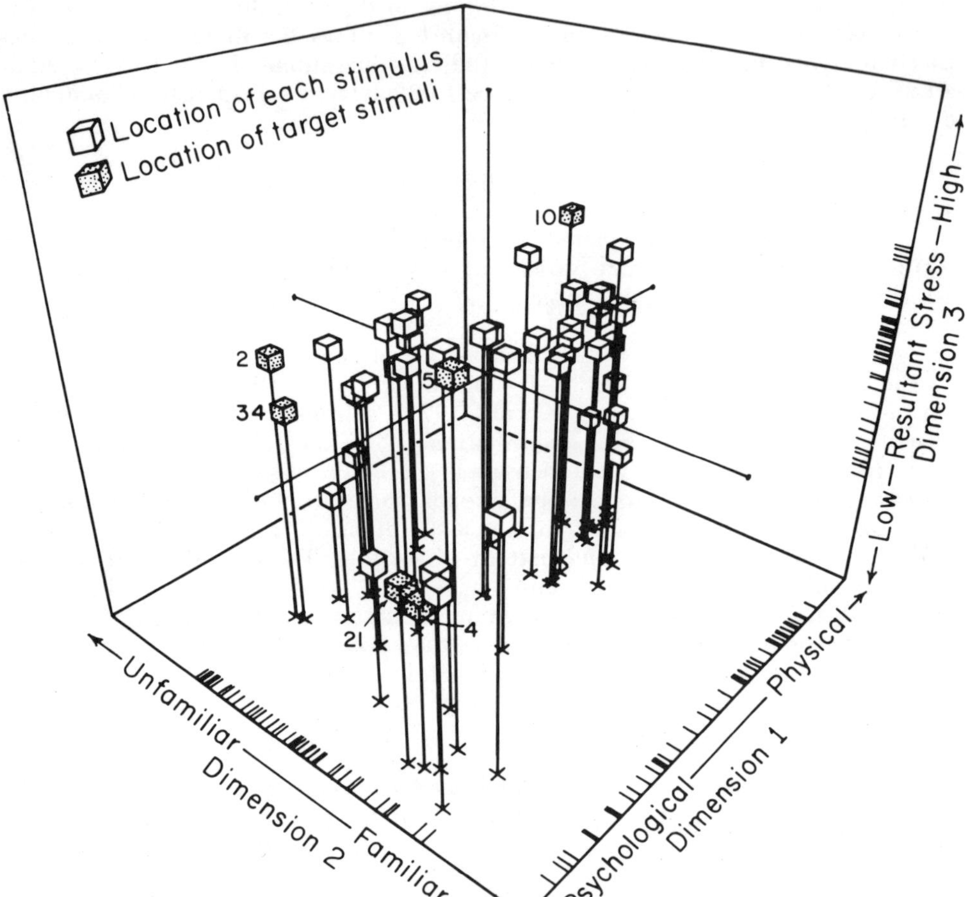

Figure 1. Three-dimensional location of 46 crowding stimuli

three for 12 of 14 attributes, model four was selected as the appropriate description of the fit between the attributes data and the derived three-dimensional space.

The correlations between the obtained squared differences and the mean rankings of the stimuli for each attribute are also presented in Table 3. The statistics in Table 3 describe the goodness of fit of each attribute vector with the existing three-dimensional configuration. As the analyses were conducted on the assumption that the data were ordinal rather than interval, the correlations and F-ratios should be considered descriptive rather than inferential. Because the probability values associated with the statistics are not exact, they are not presented. Large F-ratios and correlations should be viewed as significant, i.e., not due to chance. The direction cosines of the attribute vectors are displayed in Table 4.

The attribute vector aligned most closely with dimension one of the three-dimensional space is attribute 14 (cosine=-.98), representing the hypothesized "psychological-physical" characteristic of the crowding stimuli. The correlation between the ranking values for attribute 14 and the obtained squared distances was .93, and the F-value (df=3,42) was 96.51. Although the "psychological-physical" variable appeared most closely to approximate dimension one, several additional attributes were relatively closely aligned with this dimension. Attribute 12, representing the "personal-situational" variable, closely approximated dimension one (cosine= -.86, r=.78, F(3,42)=21.23), as did attribute 2, the "physical density" factor (cosine=.81, r=.95, F(3,42)=118.14). The "personal control" factor, attribute 10, was well aligned with dimension one (cosine=-.85), although the fit of this vector to the derived configuration was not as good as that achieved by attributes 14, 12, and 2 (r=.70, F(3,42)=13.38).

Table 3

F-Ratios and Correlations Between Obtained Squared Distances and Mean Rankings of Stimuli for Attributes One Through Fourteen (Model Four)

Attribute Number	F-Ratio	Correlation
01	78.70	.92
02	118.14	.95
03	5.64	.54
04	127.20	.95
05	35.94	.85
06	29.33	.82
07	16.60	.74
08	9.84	.64
09	14.98	.72
10	13.38	.70
11	16.33	.73
12	21.23	.78
13	13.35	.70
14	96.51	.93

Table 4

Direction Cosines of Ideal Attribute Vectors

Attribute Number	Dimension 1	Dimension 2	Dimension 3
01	.59	.79	.19
02	.81	.49	.31
03	.12	−.35	.93
04	−.26	.93	.28
05	−.34	−.13	.93
06	.15	.14	−.98
07	.06	−.99	.03
08	−.75	−.65	.11
09	.20	−.60	.78
10	−.85	−.28	−.44
11	−.63	.19	.75
12	−.86	.08	−.51
13	−.64	.27	.72
14	−.98	.07	−.19

Dimension 2 of the derived configuration appeared to be best represented by the "familiarity with the situation" factor, attribute 7(cosine=-.99), and attribute 4, the "outsider" factor (cosine=.93). Although attribute 7 was more closely aligned with dimension 2 than was attribute 4, the latter attribute provided a better fit to the data (for attribute 7, $r=.74$, $F(3,42)=16.60$); for attribute 4, $r=.95$, $F(3,42)=127,20$).

The third dimension of the stimulus space was most closely approximated by attributes 6 and 5, the "pleasant ending" (cosine=-.98) and "distressed author" (cosine=.93) variables. The correlations between the ranking values and the squared distances were high for both attribute 6 ($r=.82$) and attribute 5 ($r=.85$), as were the F-values (for attribute 6, $F(3,42)=29.33$; for attribute 5, $F(3,42)=35.94$). A third attribute, the "feelings caused by others" factor, aligned well with the derived configuration (cosine=.93), but did not provide as good a data fit as did attributes 6 and 5 ($r=.54$, $F(3,42)=5.64$).

No attribute vector direction cosines other than those mentioned above surpassed a value of .80,the designated cutoff point for consideration as representative of the three derived dimensions of crowding.

Internality-Externality and Sex Differences in Weighting of Dimensions

In order to assess the extent to which respondents' internality-externality and sex influenced their perceptions of the crowding stimuli, these factors were examined through the use of a multidimensional scaling program, which allows for differential within-group scaling of the data (Takane, Young and deLeeuw, 1977). Employing

Table 5

Dimension Weights for Males, Females, Internals, and Externals (ALSCAL)

Group	Dimension 1	Dimension 2	Dimension 3
Males	.10	.16	.11
Females	.20	.08	.09
Internals	.23	.10	.04
Externals	.20	.08	.10

Phase 2 rankings of the crowding stimuli, separate dissimiliarities matrices were computed for males and females and for internals and externals. ALSCAL was used to analyze these data in order to discover the differences between internals and externals and between males and females in the weighting of the three dimensions. The respective weights for each of the three dimensions are displayed in Table 5. It should be noted that these salience weights are not scaled to reflect the variance accounted for, and their small magnitude is of no import. Internals and externals did not appear to differ greatly in their weighting of the three dimensions. Division of the subject population into males and females, however, revealed differences in weightings of the three dimensions. While females relied more heavily on dimension 1, the "psychological-physical" variable, in discriminating among the crowding stimuli, males' appeared to attach greater importance to the second dimension, the "familiarity with the situation" factor. For dimension 1, the males' weight was .10 and the females' weight was .20, and for dimension 2 the respective weights were .16 and .08. Males and females did not seem to differ strongly with respect to their use of dimension three in distinguishing among descriptions of crowded situations.

DISCUSSION

Although the steps undertaken in this research are somewhat complex, the final results are fairly simple and straightforward. We began by collecting personal descriptions of instances of crowding and selected 46 representative stories. These stimuli were presented to a new set of respondents who provided similarity judgments in a form suitable for multidimensional scaling using ALSCAL 1. The analysis revealed that the respondents discriminated among the stimuli on three dimensions. In order to identify the meaning of the dimensions another new set of respondents judged the degree to which the 46 stimuli possessed each of 14 attributes. This analysis showed that each of the three dimensions were closely aligned to a different set of attributes. Because there is no inherent necessity in these procedures for obtaining such close alignment, we feel some confidence in asserting that peoples' perceptions of crowding are organized around at least three meaningful, basic dimensions.

The first dimension is anchored in a salience of physical, situational characteristics, at one end, and a salience of psychological characteristics at the other end. It is closely aligned with the attributes of whether the situation contains high or low density, whether the feelings of the actor appear to be generated by the situation or by the actor, as well as whether the actor is perceived as having little or much control over events. The three most extreme examples of crowding instances on the physical end of the dimensions, stories 20, 24, and 26, all involved attendance at rock concerts. The three most extreme examples illustrating the psychological end of the dimension are stories 4, 30, and 46, which all involve the necessity of resolving a cognitive conflict, e.g., parental desires versus the actor's desire about choice of a university. It seems quite reasonable to have the physical-psychological dimension define a cluster of attributes that essentially involve perceiving situational causation for behavior, on the one hand, and personal causation for behavior, on the other. Reactions to the former kinds of situations are dominated by the physical arrangements of the setting and the structural organization of the participants. The psychological situations require a uniquely personal resolution.

The second dimension can be termed unfamiliarity. It spans lack of familiarity with people/settings to familiarity with people/settings. It is anchored at one end by descriptions like story 19, in which the actor attended a huge military dance where she knew "no one," and at the other end by such instances as story 13, in which the actor was disturbed by a childish, spoiled rommmate in a cramped dormitory room. The alienation dimension was something of a surprise. At the outset of the research we would not have given it much priority. Although the presence of alienation from other people or unfamiliarity with settings can easily be understood as a component of crowding, the opposite end of the dimension is somewhat puzzling. Familiarity with people/settings does not seem to constitute a factor that contributes to crowding. Indeed, it appears that in the presence of familiar people/settings, additional components, such as high density or constraints, must exist for crowding to occur. The presence of unfamiliarity suggests the experience of crowding is unique to the actor(s) who feel(s) unfamiliar. It may well be that lack of familiarity, in itself, converts some ordinary situations into crowded situations. For example, Stokols, Smith, and Prostor (1975) introduced screening devices into a naturalistic setting and found the surprising result that feelings of crowding were augmented. If the participants were acquainted with the setting, their results may merely reflect the impact of unfamiliarity.

The third dimension involves the stress experienced by the actor at the end of the depicted sequence. It has been termed resultant stress. At one end of the dimension the stress is low, while at the other end it is high. The ends of the dimension are illustrated, respectively, by story 7, where the actor describes a large, competitive chemistry class in which he eventually feels less lonely, and story 5, where, in her role of waitress, the actor is constantly hindered by a coworker. The information needed to judge this dimension requires knowledge of the outcomes experienced by the actor. It is not contained in particular attributes of the setting and may not be available to an outside observer.

Peoples' perceptions of crowding appear to be adequately encompassed by the three dimensions. It should be noted, in passing, that the 46 stimuli are not evenly distributed over this three-dimensional space. Although the resultant stress dimension divides the stimuli into two equivalent sets, the combination of the first two dimensions do not. Of the 24 stories scaled on the physical side of the first dimension, only 7 are scaled toward unfamiliarity, while 17 are scaled toward familiarity. For the 22 stories scaled toward the psychological end, 13 are scaled toward unfamiliarity and 9 toward familiarity. To the extent to which the 46 stimuli are representative exemplars of crowding instances, it appears that whether a situation is classified as physical or psychological interacts with whether it is unfamiliar or familiar. It is entirely plausible that situations containing physical determinants of crowding are also likely to be ones with which the actor is familiar, rather than unfamiliar, with the people/setting.

Salience weights for each of the dimensions were calculated for internals and externals, as well as for men and women. A firm conclusion about these results is not possible because, at the time of data analysis, ALSCAL did not permit testing the significance of the salience weights. Our data set was simply too large. It does appear, however, that standing on internality-externality does not affect the salience weights for any dimension, whereas sex of respondent does affect use of dimensions 1 and 2. Compared to men, women place twice as much importance on dimension 1 and only half as much on dimension 2. Women appear to be more sensitive to the physical-psychological aspects of situations, while men are more sensitive to unfamiliarity. The possibility exists that the strategic problems for dealing with crowding are somewhat different for women than for men. Women might be more prone to react to the physical-psychological components, while with men, unfamiliarity might weigh more heavily.

We believe the evidence favoring the existence of the three dimensions is quite strong. As with any single study, it is of critical importance to know whether the results generalize beyond the procedures employed and the sample of people studied. It is, therefore, fortunate that another paper in this symposium has addressed the same problem, but with a different multidimensional scaling procedure and a different sample of respondents (Stockdale, Wittman, Jones, and Greaves). The broad outlines of their results correspond closely to ours. They also identified a resultant stress dimension, an alienation dimension, similar to our unfamiliarity dimension, and two distinct clusters corresponding to our physical-psychological dimension. It is reassuring to see that the present results are not restricted to stimuli generated and judged by the American college sophomore.

The beginning made by the present research points to several directions for future research. It will be necessary to determine if the taxonomy of crowding situations created by the three dimensions makes a functional difference for reactions to particular situations, especially for combinations of the first two dimensions. The possible existence of salience weights should be further explored both with respect to their presence or absence in different groups and for their behavioral consequences in reacting to crowded situations. Finally, it would also be informative to extend the present procedures to discover how crowded situations differ from situations that are not crowded. This information would complement the present results and provide a comprehensive picture of the cognitive dimensions people use to structure situations.

MULTIVARIATE DIMENSIONS IN THE PERCEPTION OF PUBLIC AREAS BY TURKISH UNIVERSITY STUDENTS

Nükte Edgüer and Ayhan LeCompte

Department of Psychology, Hacettepe University, Ankara, Turkey

ABSTRACT: 46 Turkish university students viewed and rated their reactions to 18 color slides of familiar public behavior settings. Photographs had been taken of these scenes under crowded and uncrowded conditions. Each slide was rated on four validity scales and nine semantic differential scales. Respondents rated crowded and cluttered settings significantly different than they did uncrowded and uncluttered settings in each of three different types of settings, thus establishing a validity measure for the original selection of slides. Results for the semantic differential ratings showed that slides with children were rated significantly more positively, regardless of the level of crowding, but that slides depicting objects and transportation scenes were both judged to be more positive under uncrowded conditions. Ratings of scales on the activity dimension were more related to the type of setting than to the degree of crowding, with slides of children and youth rated as more active, and slides of objects as more passive. The potency dimension was not affected by either crowding or the types of settings used in the study. It was concluded that this methodology can be used to "bridge the gap" between antecedent ecological variables and consequent perceptual reactions.

INTRODUCTION

The study of crowding, as a multidisciplinary phenomenon, proceeds at many different levels of analysis. The presently reported study is an attempt to map out a methodology for investigating the meaning of crowding as a perceptual-individual event in a person's life.

In Turkey, for various demographic and historical reasons, most urban public behavior settings can be considered crowded. That is, they have an "overmanned" quality about them, in the sense that, at least at peak times, the number of available people exceeds the capacity of the behavior setting to absorb or process them without a delay period. Before proceeding to sketch out the underlying logic of the present study, however, it is necessary to digress and establish a few objective definitions.

At an ecological level, the present study is concerned with the perception of public behavior settings. A public behavior setting is an objectively defined unit, identified by Roger Barker (1968), which has, at minimum, three defining characteristics: It possesses 1) an ongoing social pattern of behavior with 2) definite spatial and temporal boundaries and 3) a congruent relationship between the appropriate behavior and the surrounding environment. As defining attributes, these characteristics can be identified in any setting.

Consider, for example, public transportation. In Turkish cities, a wide variety of vehicles exists for the purpose of moving goods and people from one point to another. They have a standard behavior pattern (locomotion), follow a prescribed route at predictable intervals and specify a certain relationship between the social behavior of passengers and the surrounding and supporting physical milieu (for example not smoking, sitting quietly and not slamming doors in dolmuşes). These ecological requirements of public behavior settings interact with persons as individuals who inhabit behavior settings in ways that are presently not understood.

The present study then, is an attempt to develop a methodology for generating and analyzing data on individual perception of crowding in public behavior settings. As such, it has two basic conceptual difficulties to surmount. First, it must define a stimulus that stands for a public (ecological) event. Secondly, it must create a response medium that can reflect the person's perception of the event, but do so in an objective manner. These requirements, as vague and general as they are, do help in designing an adequate study by narrowing down the range of possibilities. For example, consider the possibility of a questionnaire study in which respondents are asked to visualize a location of a certain type with which they are familiar and answer certain questions about it. Such a study, as useful as it may be for other purposes, utterly fails to bridge the gap between ecology and perception on at least two grounds. First, it leaves the stimulus at the level of a private, psychological event with no guarantee that two respondents may be thinking of similar events at the time that they are generating data. Secondly, such a study provides no objective structure to analyze the responses that are given.

Of course, given a certain amount of control over the setting, one can overcome these problems directly. Many experiments conducted in a laboratory are precisely such attempts. The stimulus is admirably defined and producable at a moment's notice. The response medium is highly objective, often reduced to a choice between two buttons. Unfortunately, such control, while it seems to solve the problem of defining a stimulus objectively and creating a stable response medium, involves at least two other problems of equal methodological importance. The first difficulty with such studies lies in the artificiality of the setting that is created. In the attempt to produce a highly controllable setting, the investigator may emerge with an environment so unusual that the subject is at a loss as to appropriate behavior and must turn experimenter for cues. This produces a remarkably tractable subject, but leaves the investigator with data of doubtful value. The second difficulty with such high control settings is an out-growth of the first; it relates to what Campbell and Stanley (1963) have called "external validity" and delineates the degree to which it is possible to generalize from the findings of a study. Understanding nothing about the important parameters of a phenomenon, an investigator seeking laboratory control may easily abstract the wrong features for production in a more simplified form in an experiment, thus producing ungeneralizeable results.

The present study is an attempt to study the linkage between crowding as an objective, ecological fact and the psychological perception of it on the part of the individual. It has been designed in an effort to avoid both the Scylla of fuzzy, unanalyzable data collected from natural settings and the Charybdis of overprecise, unrepresentative results from laboratory setting. It does this through presenting color slides of recognizeably familiar behavior settings to unselected respondents, thus achieving control over the stimulus and collecting ratings on previously factored scales to measure the degree of reaction. In this design, the previous experience of the respondent in the setting displayed or in a similar setting becomes a part of the perceptual system under investigation. Analysis of the ratings of respondents to types of public behavior settings under crowded and uncrowded conditions then provides the data base to allow some understanding of the affective meaning of crowding to urban Turks.

METHOD

In this first attempt to investigate Turkish perceptions regarding crowded and uncrowded public behavior settings, a number of limiting operations were necessary in order to permit quantative evaluation of data. Such steps are perhaps inevitable in a scientific investigation, but seem more dramatically restricting when done for the first time in a previously uninvestigated area. In the following sections, these limiting steps will be described, particularly with respect to the degree to which they influence the generalizeability of results.

Selections of stimuli

From a large number of existing photographs and situations 18 color slides were chosen to be used as the stimuli in the present study. To be included, each scene had to 1) be a recognizeably Turkish behavior setting, 2) depict a dense or sparse collection of people or objects in that setting and 3) provide an example of children and/or youth, transportation or objects. The latter three categories were used because they seemed intuitively to represent types of community behavior settings in which large variations of density commonly occurred. For example, transportation settings can be easily varied from almost empty to extremely full, merely by varying the hour of the day in which they are observed. No attempt was made to sample other types of community behavior settings, such as government offices or restaurants, mainly because of limitations in subject time and scarcity of other resources.

Selections of rating scales

Each of the slides was rated after exposure on a set of 13 different seven point graphic rating scales. The scales were each anchored at the right and left sides of the page by a pair of polar opposite terms or phrases and bound into a booklet with one

Table 1

The Rating Scales Used in the Study

Series 1

Slide Number:

Tıkış tıkıs Sıkışık (Tight-squeeze)	1	2	3	4	5	6	7	(Wide-open) Ferah
Tenha (spacious)	1	2	3	4	5	6	7	(Not spacious) Tenha değil
Boş (Empty)	1	2	3	4	5	6	7	(Full) Dolu
İyi (Good)	1	2	3	4	5	6	7	(Bad) Kötü
Büyük (Large)	1	2	3	4	5	6	7	(Small) Küçük
Canlı (Lively)	1	2	3	4	5	6	7	(Dead) Cansız
Nahoş (Not Beautiful)	1	2	3	4	5	6	7	(Beautiful) Hoş
Hafif (Light)	1	2	3	4	5	6	7	(Heavy) Ağır
Yaşlı (Old)	1	2	3	4	5	6	7	(Young) Genç
Zevkli (Pleasant)	1	2	3	4	5	6	7	(Unpleasant) Zevksiz
Yüksek (High)	1	2	3	4	5	6	7	(Low) Alçak
Hızlı (Fast)	1	2	3	4	5	6	7	(Slow) Yavaş
Kalabalık (Crowded)	1	2	3	4	5	6	7	(Uncrowded) Kalabalık değil

page for each of the 18 slides. Table 1 displays a facsimile of one page of the response booklet.

In Table 1, free English translations have been provided in parentheses and the scales have been numbered consecutively. In all other respects, the table is a faithful rendering of one page of the response booklet used in the study.

Scales 1, 2, 3, and 13 were provided as a validity test of the degree to which unselected respondents actually saw crowding and non-crowding in the slides in the manner intended by the investigators. In addition, the analysis of these scales was designed to study the similarity between human and non-human density in the stimuli. They were counterbalanced in direction and arranged randomly on the mimeographed stencil.

The remaining 9 scales are taken from a recently completed, extensive cross-cultural investigation of affective meaning (Osgood, May and Miron, 1975). According to the Turkish data presented in this study, scales 4, 7 and 10 measure the evaluative dimension of meaning, scales 5, 8 and 11 the potency and scales 6, 9, and 12 the activity dimension. These three dimensions presumably exhaust the mathematically defined affective meaning space and the 9 scales presented in Table 1 represent the best measures of these dimensions across a wide variety of stimuli. Clearly, the results of the present study are limited by the choice to include these particular response scales and only further research can determine the degree of representativeness of the three underlying dimensions to describe the perceptual meaning of crowding in Turkey. On the other hand, however, similar dimensions have been found in many cultures across a wide variety of stimuli; thus, the use of these particular scales increases the possibility of intercultural comparisons tremendously.

Respondents and Procedure

A total of 46 respondents participated in the study in two separate testing sessions, with 23 respondents in each session. Respondents were unselected university students who were recruited from Psychology classes at Hacettepe University. Data were collected in the Spring of 1977 and the procedure for the two sessions was identical. In both cases, respondents were ushered into a large room and seated at the tables facing a screen. Booklets were available at the tables and as the instructions were read, the respondents were told to examine each scale carefully and make sure that the words and the rating scales were clearly understood. At this time, any questions were thoroughly discussed. The instructions emphasized that the respondents were to imagine themselves in each slide and to rate how they would feel. After the instructions were read, the lights at the front of the room were dimmed and the first slide was shown. Each slide was preceded by a slide number which respondents were instructed to write at the top of the page. Slides were exposed for approximately ten seconds piece, with respondents filling out their ratings immediately afterward.

The purpose of the two testing sessions was to counterbalance the order of stimuli in order to determine whether or not such factors as practice with the scale, or fatigue of the respondent had any systematic effects on the rating of the slides. On the second session, the slides were presented in reverse order, but in all other respects the procedure was the same as in the first session. Order of the slides was determined randomly, with the exception that within each set of three slides, one each of crowded, uncrowded and cluttered (high object density) slide was shown in order to break up any incipient sets.

Design of the study

The basic design for data collection consisted of two levels of order (forward and backward) with male and female respondents nested under order and running across ratings and type of setting (children+youth, transportation and object density). Table 2 presents a diagrammatic representation of this design.

This type of design leads, more or less naturally, into a repeated measures analysis of variance performed separately for each of the Evaluation, Potency, and Activity dimensions and for the validity scales. For each group of slides, the appropriate scales were summed to make a single dependent variable score for each respondent.

In addition to the analysis of each of the dependent measures, a multivariate analysis across the 18 stimuli was also performed. The purpose of this analysis was to determine what underlying dimensions accounted for the variance of the ratings.

RESULTS AND DISCUSSION

Analysis of sex differences

Although no predictions had been made regarding sex differences, each rating scale was evaluated on every slide to determine whether or not systematic differences did, in fact, exist in the data, Of the 468 comparisons that were generated by this procedure (i.e., 2 orders x 13 ratings x 18 slides), 5 proved to be significant at the 1 percent level by the t test, two in favor of males and three in favor of females. As this proportion is almost exactly one percent of the possible comparisons, the simplest explanation of these observed differences is that of a random or chance event. This is not to say that sex differences in the perception of crowded public behavior settings do not exist, but merely that the present sample of settings and rating scales failed noticeably to elicit them.

Analysis of dependent variable measures

For all four dependent variables the same procedure was followed; first, the separate ratings were summed into a single score for each slide, then the ratings for the three slides in each level nest were summed. For example, slides 5, 11 and 14 constituted the level nest for Uncrowded, Children and Youth. Ratings on these three

Table 2

DESIGN OF THE STUDY					
AREA 1 CHILDREN and YOUTH		AREA 2 TRANSPORTATION		AREA 3 OBJECTS	
3 Uncrowded Settings	3 Crowded Settings	3 Uncrowded Settings	3 Crowded Settings	3 Uncrowded Settings	3 Crowded Settings
ORDER ONE : MALE RESPONDENTS					
ORDER ONE : FEMALE RESPONDENTS					
ORDER TWO : MALE RESPONDENTS					
ORDER TWO : FEMALE RESPONDENTS					

Table 3

Summary of Four Univariate Analyses of Variance of Ratings

Source	df	Validity Scales	Evaluation Scales	Potency Scales	Activity Scales
Between Ss.	3				
A(Order)	1	7.65	91.69	0.04	23.46
Ss Within grps	2	8.37	6.25	3.74	15.29
Within Ss.	20				
B(Location)	2	362.86**	331.60**	25.32	307.27**
AB	2	32.59	20.34	.94	1.36
BxSs within grps	4	15.15	7.30	2.09	5.40
C(Crowding)	1	7728.99**	522.20**	15.89	8.04**
AC	1	7.15	5.13	.00	2.95
CxSs within grps	2	15.52	.64	7.47	.22
BC	2	292.66**	106.13**	7.29	84.01**
ABC	2	23.61	11.94	.22	13.41
BCxSs within grps	4	3.04	4.83	1.47	2.94

**$p<.01$

slides were summed and the mean scores for females and males within each order was found. Since no sex differences had been found, mean scores for males and females were treated as replicates. Thus, the data were analyzed within a three factor, repeated measures design with mean scores for males and females constituting a single replication. This design permits a determination of the significance of Order (forward, vs. backward), Type of Setting (children and youth, transportation and object scenes) and Crowdedness (crowded vs. uncrowded), as well as the evaluation of the significance of their various interactions (Winer, 1962). By using a single replication (mean scores for males and females), a conservative bias was introduced into the analysis, which was judged to be a desirable thing in view of the previously untried methodology of the study.

The results of these analyses with each of the four dependent variables is summarized in Table 3.

The first thing to note about the results as displayed in Table 3 is the complete absence of significance of the Order factor. In none of the four analyses is Order significant, nor is it found to be significantly interacting with any of the other factors. Apparently the arrangement of the ratings in the booklet and that of the slides in the series was effective in reducing any tendency of the respondents to change their set over time and rate differently.

Inspection of the pattern of significance in Table 3 indicates that in each case

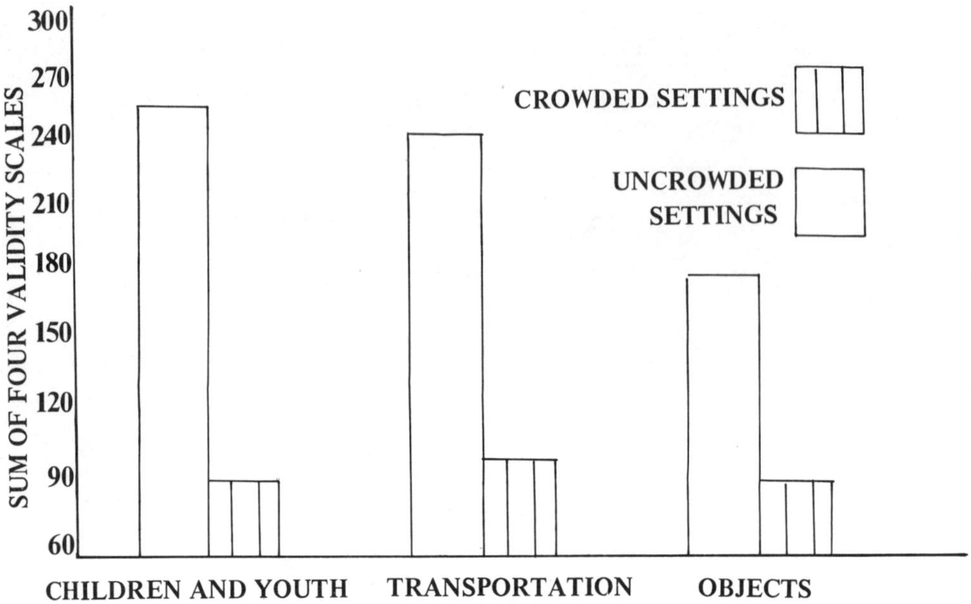

Figure 1: DIFFERENCES IN MEAN RATINGS ON FOUR VALIDITY SCALES DEPICTING PUBLIC BEHAVIOR SETTINGS IN THREE AREAS

but that of the potency scales, Type of Setting and Crowding are significant and that these two factors interact significantly with each other. All of these findings are significant at the 1 percent level and nothing else even approaches an acceptable level of statistical significance. Because of the presence of significant interaction terms in each analysis, the results cannot be simply described or summarized. For this reason, the four dependent variables will be separately treated and described.

Figure 1 displays the six means in the significant Crowding by Type of Setting interactions. Inspection of the figure shows that in each case, the ratings for the crowded/cluttered slides are dramatically different from their counterparts in the uncrowded/uncluttered conditions. Statistical analysis supports this impression and a test of multiple comparison among the means in the set indicates that the interaction occurs because of the lower means for uncluttered objects and the slightly elevated mean for crowded transportation settings. In these two cases, the contrasts between crowded/cluttered and uncrowded/uncluttered slides were not seen by respondents to be quite as great as in the slides of children and youth settings. However, the vastly greater percentage of the variance clearly occurs along the crowded/uncrowded dimension and indicates that the slides used in the study were indeed perceived as crowded/cluttered and uncrowded/uncluttered. This factor controls 83 percent of the total variance in comparison to 8 percent for the type of slide factor and 6 percent for their interaction (Hays, 1973).

Figure 2 displays the set of means in the crowding in the crowding type of setting interaction for the three evaluation scales. Statistical analysis of the differences

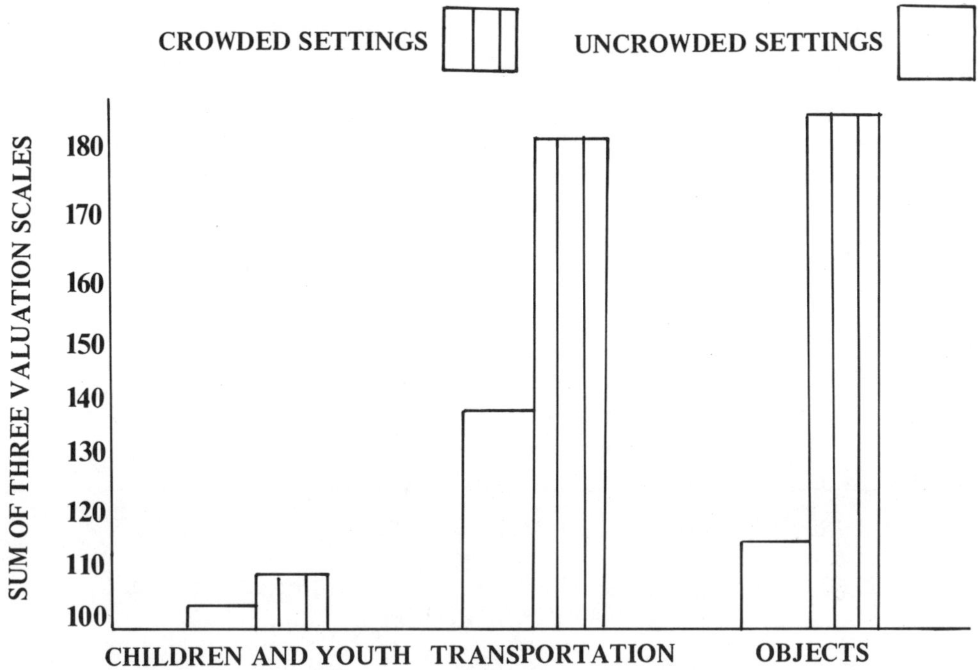

Figure 2: EVALUATION DIFFERENCES IN THREE TYPES OF BEHAVIOR SETTINGS

among the six means in this set reveals that the greatest contrast occurs between slides showing cluttered and uncluttered objects. Transportation settings tend to be rated as negative in emotional tone, whether crowded or uncrowded, while slides showing children and youth are overwhelmingly rated as positive, regardless of the degree of crowding. In terms of the degree of perceived pleasantness, slides showing children are rated highest, followed by uncluttered objects and uncrowded cars and buses in that order. The most unpleasant ratings were given equally to both crowded transportation and cluttered object scenes.

In contrast to the evaluation dimension and the results with the validity scales, the analysis of the three scales on the potency dimension shows minimal differences. The set of six means in the type of slide by crowding interaction is plotted in Figure 3 as a contrast to demonstrate the fact that not all ratings are sensitive to the variables that have been manipulated in the present study. As would be expected from the negative results of the analysis of variance shown in Table 3, the means are all very similar. Apparently the potency dimension is not involved in the judgment of the qualities of slides of urban behavior settings.

The results for the activity dimension are displayed in Figure 4. Statistical analysis of the differences among the set of six means indicates that slides of children and youth are seen as more active than either transportation or object scenes, but that

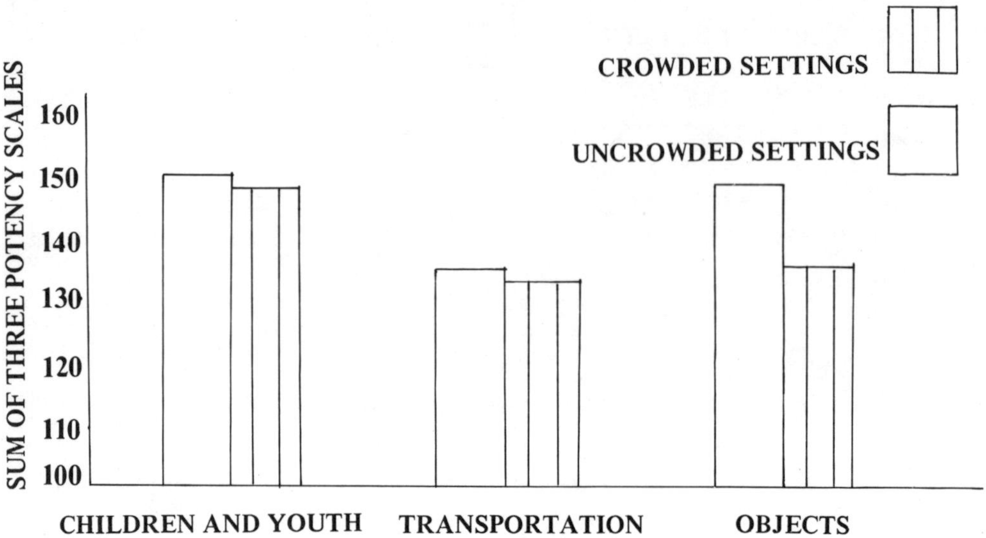

Figure 3: POTENCY SCALE MEANS FOR THREE DIFFERENT TYPES OF BEHAVIOR SETTINGS

crowded settings are seen as more active than uncrowded settings in the children and youth category. Slides showing a high degree of object density (clutteredness) are rated as significantly more passive than are either uncluttered object scenes or transportation scenes. Unlike either the analysis of the validity ratings or that of the evaluation scales, the majority of the treatment variance in the analysis of activity scales piles up along the Type of Setting Factor rather than the Crowding factor or their interaction. 69 percent of the treatment variance is accounted for by the Type of Setting factor alone, whereas crowding accounts for less than 1 percent and the interaction for 22 percent.

Multivariate analyses

Two different procedures were employed in an attempt to tease out the underlying structure of the data. The means for each rating scale on each of the slides were intercorrelated and the resulting correlation matrix was submitted to a principle-axis factor analysis program. Next, the same intercorrelation matrix was subjected to a cluster analysis routine in which the original matrix was reduced to a zero-one matrix from which characteristic roots and vectors were extracted. Results of these procedures produced an encouraging degree of agreement. In both cases, three underlying factors or clusters were identified. The third factor/cluster could be clearly identified as a "natural environment" dimension since all of the scenes in which a background involving non-urban structure (e.g., beaches, water, grassy hills) are loaded here. Factors one and two both involve types of crowding but the differences between them is not a simple one since the factor loadings are equally distributed among the remaining slides.

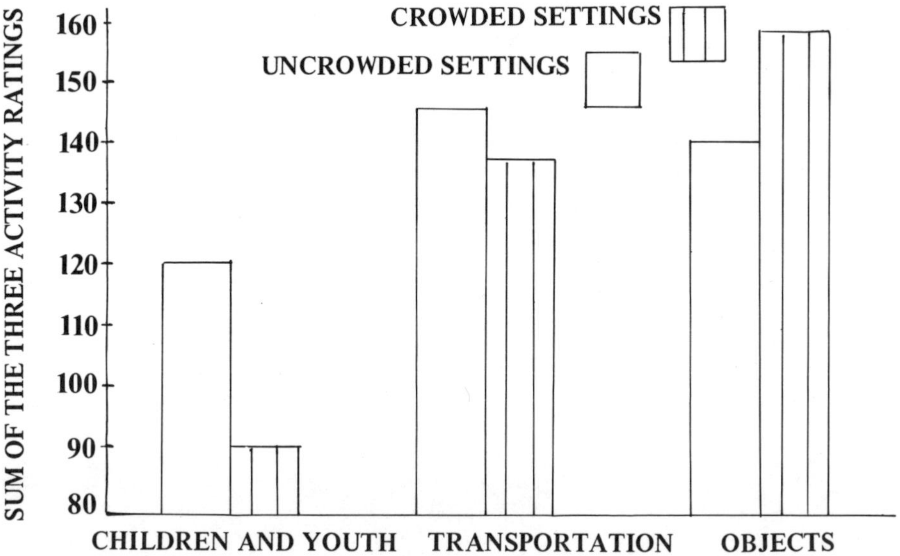

Figure 4: DIFFERENCES IN MEAN ACTIVITY SCALE RATINGS FOR THREE DIFFERENT TYPES OF BEHAVIOR SETTINGS

CONCLUDING COMMENTS

Clearly, the methodology developed in the presently reported study has shown promise. Naturally occurring public behavior settings have been sampled and respondents have shown by their ratings that they perceive crowded settings differently depending on the presence of other factors such as the presence of objects or children. The perceived pleasantness of the setting is a combined function of the presence of children and the absence of crowded or cluttered conditions. On the other hand, the perceived activeness of the setting seems to be more or less a linear function of the type of location than of the presence or absence crowded or cluttered conditions, with children, transportation and non-peopled object scenes perceived as progresssively more passive in that order.

There seem to be at least three fruitful directions in which the present findings can be extended. First, the presently utilized sample of public behavior settings can be extended in a variety of ways, such as including other types of settings in the public domain (restaurants, stores and parks, for example) or by including family settings (living room, mealtime and bedrooms). Secondly, other types of response measures can be brought into use, such as adjective check-lists or other types of rating scales on which sex differences may be more apparent.

Perhaps the most exciting extension of this type of methodology, however, is in the area of cross-cultural investigation. The use of Semantic Differential scales as response measures and color slides as stimuli would seem to greatly facilitate the application of a similar method in another country. If American respondents, for example, given the same or similar stimuli and response scales, would produce a different pattern of results, then the cornerstone for a truly objective study of the intercultural perception of crowding will have been established.

PART IV

CASE STUDIES OF CROWDING CONSEQUENCES

To put flesh on the bones of abstract, quantative models of crowding and its human consequences would seem to be a highly desirable thing. Without the concrete immediacy of the actual phenomenon, the experimentalist might mistake his laboratory for the world and the theoretician, his dreams for reality. One function of this final section, then, is to confront the reader who may have found some new ideas in previous sections, with a chance to apply them. Another no less important function is to sensitize oneself in unexpected, heuristically valuable directions.

The paper by Sidney Brower seems to fulfill both functions admirably. He shows how the use of public spaces such as streets and playgrounds changes as the density of the surrounding neighborhood increases. Derek Hall provides an excellent example of the way in which people conceptualize their community and how their subjective perception of their social and built environment can be represented. Two papers focus on family crowding in Turkey. Bayazit, et al. and Ipek Gürkaynak very convincingly show the relevance of this area, on which the other papers in this collection have been regrettably silent. Another area of great relevance relates to crowding in the squatter communities in developing countries. Goethert, et al. have studied this problem in Turkey as well as elsewhere, and contribute a valuable paper on their experiences. Finally, Mark Glazer, a cultural anthropologist, uses his participant observation training to compare two Jewish communities, one embedded in American, the other in Turkish culture.

SHARED SPACES IN CROWDED AREAS

Sidney N.Brower

City of Baltimore, Department of Planning

Maryland, U.S.A.

ABSTRACT: A study of high density residential areas in the City of Baltimore showed that while there was intensive outdoor recreational activity, little of this activity happened in the parks. The study explored the reasons for the reluctance to use parks and concluded that a basic problem was the absence of management. Management of shared spaces is an important consideration in high density areas. The paper argues that where management resources are limited, public open spaces should be designed to make the most effective use of whatever resources there are. Guidelines are offered for the design of manageable spaces.

In crowded areas, where land is a scarce commodity, people spend a great deal of their leisure time in public outdoor spaces like parks, playgrounds, plazas and streets. It is not surprising then, that in plans for improving older, high density neighborhoods, we place so much stress on the need for new and better public open spaces. This paper is about a study of recreational open spaces in several high density, low income, residential neighborhoods in the City of Baltimore. The study tried to find out how these spaces are used, how people feel about them and what can be learned from them. The people who live in these neighborhoods are predominantly black and have low incomes. Relatively few residents own their homes and the transiency rate is high. The outside housing-unit density (persons per acre) and the inside housing-unit density (persons per room) are both considerably higher than the city average. The houses are two and three storey rowhouses. Each has a rear yard. Many of the houses have been converted into apartments.

Baltimore provides an interesting case study because of the many small neighborhood parks that are being built: over fifty have been completed since 1960

Note: The study on which this paper is based was funded through a grant from the Center for Studies of Metropolitan Problems of the National Institute of Mental Health. The study was conducted jointly by the Department of Planning and the Johns Hopkins University.

and many more are planned. In one forty-block area known as Harlem Park, there is a park in 3 out of every 4 blocks. The findings of the study have corroborated some early suspicions: while there is intensive outdoor recreational activity throughout the area, little of this activity happens in the parks. In systematic observations of 12 park blocks in Harlem Park over a three month period, more than 19,000 people were recorded (this includes all the people seen out of doors except for those in automobiles). Almost 80% of them were sitting and talking, playing games or involved in some other form of recreational activity. (The other 20% consisted of pedestrians and people who were working.) Few of the recreators were in the parks. Most of them were on the sidewalks. Along the streetfronts there were almost three people recreating to every pedestrian, and there were three times as many people recreating on the streetfronts as in the parks. The parks were used mostly by children, but there were twice as many children on the streetfronts. Adults hardly used the parks at all.

The Harlem Park parks are all smaller than one acre in size. They are located in the center of blocks, surrounded by the backs and rear yards of houses, accessible and visible to the street only through narrow alley entrances. The inner block parks in Harlem Park are the prototype for many parks now being built in other sections of Baltimore. They have the advantage that they can be built on land otherwise occupied by outbuildings, stables, warehouses and relatively few occupied housing units. The findings in Harlem Park do not, however, apply only to inner block parks; they point to a number of themes that come up again and again in studies of over 40 parks in a variety of locations in several inner city neighborhoods over a period of five years.

Not only are the parks underused, but maintenance of the parks is a serious problem. City agencies are unable to keep up with the trash, debris, broken glass, abandoned sofas, mattresses, refrigerators and automobile parts that accumulate in the parks. Local residents are equally concerned about the kinds of people that collect in the parks - like winos, drug addicts and unruly teenagers. Many residents are afraid to go into the parks especially at night, and they will not let their children play there.

In spite of these problems, inner city residents continue to ask for more neighborhood parks and the city responds by continuing to build them. The feeling on the part of both residents and the agencies seems to be that the problems are not caused by the parks but by the people who misuse the parks. They reason that parks in themselves are improvements to the physical environment providing opportunities for safe, pleasant recreation: it is unfortunate that residents cannot now take full advantage of these opportunities, but the problem is not one of park design, it is one of social reform. I shall not argue with the need for social reform, but I hope to show that parks can be designed to serve present-day residents without short changing the needs of future, more fortunate generations. Nor shall I claim that the park problems can be solved by more careful selection or arrangement of interior spaces and equipment. What really happens is determined less by these physical elements than by the actions of the strongest and most insistent user groups. The parks were originally designed with particular recreational uses and particular user groups in mind (climbers were for children to climb on, sandboxes for tots to dig in, benches for adults to sit on, etc.), but because nobody really exercised control over the park, children and most adults could easily be displaced by small, aggressive, antisocial groups that were much in evidence in the community. The official managing body was the city government, and its representatives included policemen, health enforcement officers, sanitation collection crews and park supervisors. Because the city has limited resources, however, these managers were seldom there. The city felt that residents who live around the park should share the management responsibility, but the resident associations were loosely structured and fragile and could not be relied upon to cope with troublesome and dangerous elements in the park.

The result was that nobody really controlled the parks. The city had inadequate resources and resident associations were unskilled. The parks became unsuitable places for general recreation and came to be used for activities that preferred to be away from the public eye.

The solution to the major problems in the parks is to provide proper management. If only we could provide paid managers in all the parks! To be realistic however, we must accept the fact that public resources are severely limited and will be for the foreseeable future, and that we will have to make more effective use of the resources that we have now. One way to do this is to make the parks more manageable. They should make lighter demands upon managers, increase the chances that management efforts will be successful, and provide opportunities for increasing management skills. I believe that we can design parks to be more manageable if we are sensitive to the prevailing customs, and the circumstances that support them.

Plans for the improvement of existing neighborhoods do not of course have to provide for the perpetuation of existing customs; but many customs are tenacious, even if people do not feel strongly about them, because they are convenient and have become generally accepted. When we develop a plan that assumes people will do things they don't do now, or will stop doing things they do do now, that plan will depend for its success on the existence of management. The greater the difference between desired uses and present day uses, the greater the need for management, or else the more unpredictable the outcome.

In Harlem Park for example, we found the following customs apply to the use of space.

Most people spend most of their recreation time in the immediate vicinity of home. The spaces that are most frequently used tend to be chosen not for their intrinsic qualities so much as for their location: they are convenient to home. For adults being close to home means that they do not have to leave the house and their family, they can listen out for the telephone or the baby, attend to the front door, keep an eye on the stove. For children it means that mother is on hand to hear calls for aid and to respond to complaints. In low density areas, most residents have private land attached to their dwellings, but in high density areas most residents have neither yards nor balconies, and so they use the public spaces that are closest to home.

The most common form of recreation is sitting and talking, and this is usually tied to socializing. Socializing depends largely upon chance meetings with neighbors and passers-by. The streetfront, the locus of all arrivals and departures and the scene of frequent and varied activities-like those of street hawkers, mailmen, social workers, etc.- is the best place for this kind of casual socializing. The most frequent reasons given for sitting out front are: habit, custom, where one meets one's friends and where the action is.

Community organizations have a spatial component: people are far more likely to associate with residents they can see from their front steps than with residents they can see from their back yards. The street is the focus of the neighborhood. Parks at the back are, as residents see them, at the junction between neighborhoods. This means that responsibility for supervision and maintenance of the parks in inner block locations do not automatically fall upon any one of the natural community groups. In addition, cooperative energies are less likely to be directed toward maintaining a park in the back (a space that has the drawback of being associated with the service side of the house, and the place where household garbage and trash are stored and collected) than toward maintaining the streetfront, the public face of the neighborhood. Residents' maintenance and embellishment of the streetfront is,in fact, characteristic of many study blocks.

Security considerations have a strong influence on residents' choice of recreation space. For most residents, the places where they feel most secure are places that are close to home. They are very apprehensive about the presence of outsiders (people they don't know and have no control over) in these places. The streetfront is safer than the park because it is open, visible, lighted at night, patrolled by the police, and in an way an extension of the house. The fact that others residents recreate on the streetfront, that many sweep and wash the section of sidewalk in front of their individual houses, furnish it with seats and tables, and decorate it with plants and window decorations, adds to the sense of security: outsiders and troublemakers will be discouraged by the presence of people who obviously have a social investment in the area, and who know one another and so are likely to help one another in case of need.

Many adult residents feel that children are more likely to get into fights if they are concentrated in a park than if they play in a dispersed fashion along the streetfront. The chances of fights in the parks are all the greater because of the presence of strange children from other neighborhoods, and because of teenagers who tend to "take over" from the younger children. Children also get hurt on the equipment and on broken glass. They rely heavily on their parents to intercede in case of trouble. Parents seldom accompany children to the park; they can usually be found sitting out on the streetfront.

In addition to the troubles that keep children out of the parks, there are certain advantages to playing on the streetfront. Many children's games are especially suited to the streetfront. Most of these games are traditional and self-regulating, requiring no supervision and a minimum of equipment. Many are linear in nature, using the park as part of a track that extends through connected yards, alleys, sidewalks and streets. Many involve ball playing.

Parks are particularly good for structured team games because they are the largest open spaces in the neighborhood. The equipment tends to get in the way, though, and most parks have "no ball playing" signs.

Individual households appropriate the section of sidewalk immediately in front of the house. They play an active role in its maintenance and management.

The park seems to belong to nobody. There is no one to control what happens there, and this greatly reduces its usefulness for general recreation.

Many play activities, those of boys especially, involve interactions with the adult world – selling newspapers, running errands, conversing with older people, going to the store, visiting the firehouse. There is frequently more excitement and challenge to be found in the general living environment–in dark alleys, vacant buildings, strange neighborhoods – than in any piece of play equipment. There is also danger from fast moving automobiles.

The attitudes and choices that bear upon the use of space in Harlem Park are not unique. They can be found in the studies of other people in other places (see for example, Jacobs, Yancey, Newman, Appleyard, Department of the Environment). Comparisons show that the forms of behavior in Harlem Park are generally characteristic of lower and working class environments. There is one problem however, that residents share with all people who live at high density: how can a particular space be made to serve the different needs of many different users. Incongruous activities that in low density areas choose to disperse, must find a way of co-existing in a single space. Management is necessary to resolve conflicts. If there are any principles that can guide us in designing parks that contribute to the effectiveness of management, then these principles can be applied in other high density areas.

I should like to suggest several such principles. Parks should be designed to express the identity of the managing group, to establish their authority and to keep demands upon them within practicable limits.

If a park is to be managed by a public agency and it is necessary to rely on the services of roving officials, then the park should be placed where it is fully visible from the public right of way. The manager's job will be made easier if the park is actively used, if it does not provide an attractive hideaway for antisocial groups, does not have multiple entries that offer easy escape for troublemakers, and can be controlled at night with special lighting, and fences if necessary. Official signs can proclaim public management (although the absence of restrictions will usually be a more telling indication). Familiar uniforms and badges of office will identify individuals who act as agency representatives.

If a park is to be managed by a local community organization, it should be placed within the social territory of the group. In Harlem Park for example, it will be easier to get residents to assume the management of a space in the front of the house than at the back. The park will be identified with the managing group rather than the general public, if members of the group have control over who uses it. This can be done on a permanent basis (the park is made a private facility), or temporarily (the park is reserved for local residents during the most trouble-prone times, generally at night). Control of access not only establishes the rights of the managing group, it makes it easier for troublemakers to be identified and dealt with. The introduction of equipment and facilities desired by but not available to surrounding communities, can prove to be a powerful attraction to outsiders and can add greatly to management problems.

The tendency of each resident in Harlem Park to manage the section of sidewalk in front of his own house, suggests a third alternative for managing public parks – the independent actions of individual residents.The space should be treated as a collection of smaller spaces, and individual parcels should be identified with individual homes. Objects associated with private users should be emphasized so as to establish a visible claim to each of the parcels. Elements associated with public parks, and large-scale equipment designed to accommodate large groups of people should be kept to a minimum.

Whatever its identity, the managing group will have an easier job if the park is designed so that it can accommodate activities that are popular. A park that caters to the special needs of special user groups is going to depend heavily upon management if it is located in an area where open spaces are scarce, and many people look for a convenient location rather than a specialized facility. It is more important that a park be designed to discourage uses that the managing group particularly wishes to exclude, than that it be tailor-made for a narrow range of selected activities. A park should not be thought if as a stage set, but rather as a stage equipped for mounting a variety of different performances.

I have mentioned three different management strategies. There are of course others, like management by a church group or a social club or by a single strong community leader who assumes complete charge of the park. There are examples of all of these in the study blocks. The most viable arrangement, however, seems to be strong management by resident organizations, served and assisted by public agencies.

Where resident organizations are weak, as in Harlem Park, it is particularly important that parks be attractivenotonly to children but also to supervising adults. There are certain play activities in which adults are more likely to participate than others; playing ball games, say, rather than climbing on a jungle gym. Features could be installed that depend upon the presence of adult operators; control of night lighting for example, or operation of a spray pool, or storage and erection of moveable pieces of equipment. The parks could be designed so that resident-initiated maintenance and improvement efforts have highly visible results. In low income areas these results must be able to be achieved with little cash outlay; for example by cleaning, polishing, painting or planting. In all this, residents should be able to rely upon public agencies for information, advice, education and assistance.

In addition to suggesting guidelines for the design of parks, the Baltimore study clearly shows that parks, no matter how well designed, cannot satisfy all the recreational needs of residents in high density areas. Planners who concern themselves only with the provision of parks, neglect the places where people spend most of their recreational time – within the neighborhood itself. In Harlem Park, for example, many recreators are not likely to be diverted from the streetfront into the parks, because the streetfront remains the outdoor space that is most convenient for home-based recreation. Many children will continue to play in the streets because their games cannot be contained within a park, or because many things outside the park are more fun than those inside. Recreation should be recognized as a legitimate use of the streetfront and steps should be taken to reduce the dangers of automobile traffic. The impact of automobile traffic is especially severe in crowded areas.

One approach is to exclude automobile traffic on residential streets, either permanently (which in existing areas may require traffic rerouting and changes in street layout), or for certain hours of the day (using, for example, moveable barricades). Findings in Baltimore suggest that it may be advantageous to retain automobile traffic because of convenience and security (lights at night and patrolling police cars), and also because elimination of the noise and hazard of the automobile may be replaced by the noise and dangers associated with overactive play. Turning the street into a park can interfere with pedestrian movement, attract outsiders, result in property damage and hurt (especially to young children and elderly people), be a source of loud and constant noise and generally constitute a serious nuisance for abutting residents. Because there are no private outdoor spaces associated with most of the dwelling units, residents tend to see the sidewalk and other shared spaces immediately next to home, as an extension of their domestic space. The use of these spaces by outsiders, especially if the use is continuous, active and noisy, can be very distressing to residents: an invasion of their territory. For this reason, spaces immediately next to houses should be seen as improvements to the housing environment, suitable for home-based recreation and subject to the control of abutting residents, and not as elements in the public open space system.

Pedestrian safety may well require reducing the volume and speed of automobile traffic, but as with parks, proper management is the key to the success of the streets. This suggests that what is needed in high density areas is not just park management but management of open spaces. If some of the ideas for designing manageable parks are applied to the streetfront, then it suggests that facilities for concentrated play and features that attract or accommodate outsiders—benches tables, play equipment, etc.—should be kept to a minimum (residents being encouraged to bring out their private furniture and equipment when needed for their own use), steps should be made suitable for playing or sitting on, streetfront elements should be designed to withstand recreational use (for example windows protected against stray balls), and adequate provision should be made for trash storage and trash collection.

What we learn from the Baltimore study is that social problems and conflicts can greatly diminish the usefulness of public places. This is especially likely to happen in crowded areas and is a particular problem in residential areas where public open spaces serve as extensions of the home. To ensure that public open spaces are usable, they must be managed, and the most effective form of management is one that relies heavily on the involvement of local residents. A major objective for designers of public open spaces should be to create spaces that residents are able to manage.

This demands of residents living in crowded residential areas, that they demonstrate their social responsibility far more frequently than their counterparts who live in the suburbs. If this involves any lessening of personal privacy, it must be balanced against the advantages of being able to share in the benefits of a richer and more satisfying public environment.

COMMUNITY AND BEHAVIOR IN URBAN ENVIRONMENTS

Derek R.Hall

Sunderland Polytechnic, England

ABSTRACT: This paper relates to the conference theme by focussing upon perceptual and behavioural responses to the social, political and built environment in urban 'communities' of varying density and character. It analyses the spatial, social, political, economic and semantic dimensions of 'community' as perceived by a large sample (587) of residents of a medium-sized southern English city (Portsmouth, Hampshire) both in high and low density environments within its boundaries. It is suggested in this paper that residents' images of the built environment and the salience of 'community' both as an image and as a function of the socio-morphological environment, are functionally most meaningful and interrelated in high density contexts.

POSTULATES

In analysing some of the responses to the questionnaire survey, and in relating them to various aspects of Portsmouth's spatial structure and to previous survey findings, four basic postulates were employed:

(i) that the proportion of sampled residents articulating a community area in Portsmouth, by virtue of the functional value attached to the concept in this survey, will be considerably less than those recognising a 'home area' or comparable concept in previous surveys elsewhere in Britain (Research Services, Ltd., 1969a, 1969b; Hampton, 1970; Baker and Young, 1971, 1973; Bryant and Hall, 1971);

(ii) that due to conceptual (Goodchild, 1974; Baril, 1971) and methodological (Spencer, 1973) filter processes, lower socio-economic groups will be less likely to articulate a community area than higher group respondents;

(iii) that within the wider context of non and less-place urban realms (Everitt, 1976), general assumptions about working-class propinquity as against middle-class

TABLE 1

Local area articulation/recognition of a local community area*
"An area round here .... where you feel at home" + ++

	Portsmouth*		Sheffield+	Country boros' 60-250,000++
	No.	%	%	%
Able to conceptualise	253	43.1	85	75
Unable to conceptualise	327	55.7	14	25
Other answer	7	1.2	–	–
Totals	n=587	100.0	100	100

Sources * writer's survey,
+ Hampton (1970), 101,
++ Research Services, Ltd., (1969a), 12.

Portsmouth*	Sheffield+	County boros' 60-250,000++	CBS over ¹]²√

mobility (McClenahan, 1929, 1945; Janowitz, 1951; Webber, 1964a, 1964b; Doherty, 1969) lead to an expectation of different class interpretations of local community, with emphasis on interactional factors by the former and physical and aesthetic factors by the latter groups.

'COMMUNITY AREA' ARTICULATION

Of the ultimate total of 587 respondents interviewed in the questionnaire survey, just over 43% (43.1%, n=253) were able to recognise and cartographically articulate a local community. It was postulated that due to the slightly different, implicitly emotionally and functionally stronger terms used, the positive response to this question would be considerably less than in previous surveys using such concepts as 'home area'. Table 1 confirms this hypothesis by comparing such responses to surveys of large English towns.

In terms of the size of these areas, however, Table 2 supports previous research which suggested that in built up areas the psychological area which people associate and take an interest in, often extends over a number of streets but incorporates a population of no more than about 10,000 (cf. Lee, 1968).

As can be seen from Table 2, by far the largest proportion of articulated community areas in Portsmouth (10.8%) fell within the group larger than a single street but smaller than the equivalent of a ward. The implicit and explicit nature of the survey question, however:–

"Is there an area around here in which you live, that you can call a 'community area'?"

would appear to preclude the possibility of articulating relatively large areas, a bias, albeit relevant perhaps only in a handful of cases, which must be acknowledged.

Fig. 1 represents generalisations of such areas articulated by at least 10% of the area sample interviewed. Emerging from this map are large areas in the city where little recognition of community area was apparent in the survey. Other blank areas are those where the sample respondents interviewed was too small for valid inclusion. Fig. 2, in attempting to overcome the problem of subjectively imposing 'average' boundaries, distinguishes areas of high and low community area articulation by contour surface representing, in absolute terms, the number of sampled cognitive maps containing that part of the city. By contrast, Fig. 3 presents community articulation in terms of area sample percentages.

'Residual' articulated areas have been represented in their own right (Fig. 4). These can be seen as eccentric in that while their residents feel them to be part of a wider local community area, the inhabitants of the rest of this community area do not. It was often found to be the case, as in Highbury and Old Wymering, that the eccentric area contained often high density municipal housing in a district of otherwise relatively low density private sector dwellings. Also represented on Fig. 4 are 'isolated' articulated community areas, that is, those articulated by one or two persons or less than 10% of an area sample. These were of two types: those articulated in relative isolation (e.g. South Parade), and smaller, second tier areas falling within already defined community areas (e.g. Portsbridge within Highbury).

BOUNDARIES AND FRONTIER AREAS

Styles (1972) has suggested that physical boundaries are used to delimit community areas by those whose who are not familiar with an area. On the other hand, a bend in a road, or a house are employed by local residents to mark the boundaries of areas considered to be of different status and character. While such micro features of spatial cognitions were lost in the various filter processes involved

TABLE 2

Approximate size of conceptualised areas

	Portsmouth*		Sheffield+	County boros' 60-250,000++	CBS over 1/2m++
	No.	%	%	%	%
Larger than a ward area	3	1.2	15)	62	36
Ward area equivalent	16	6.3	12)		
Group of stress	179	70.8)	64	33	55
One street or smaller	55	21.7)		5	9

Sources * writers's survey,
+ Hampton (1970), 106,
++ Research Services, Ltd, (1969a), 15.

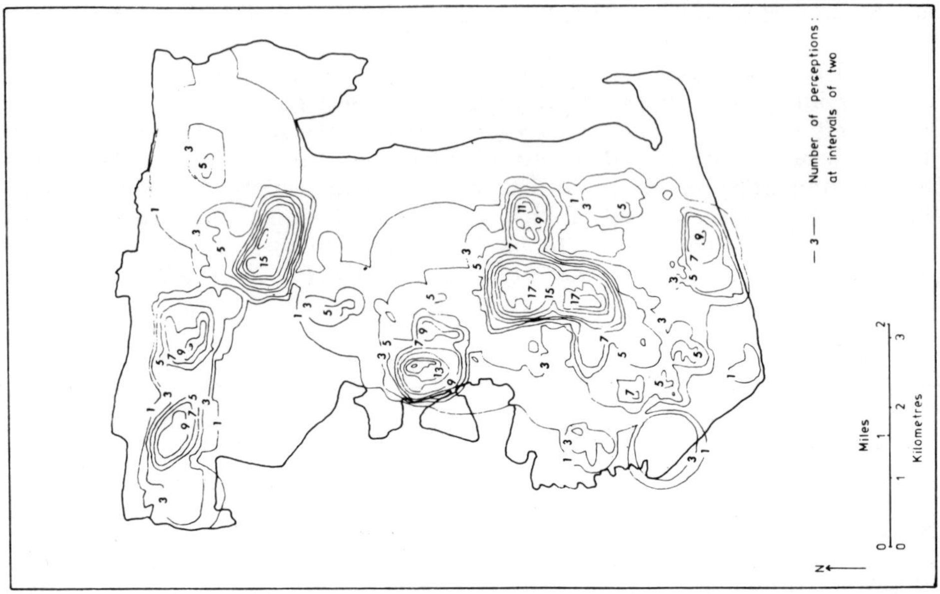

Fig. 2 Community area surface map

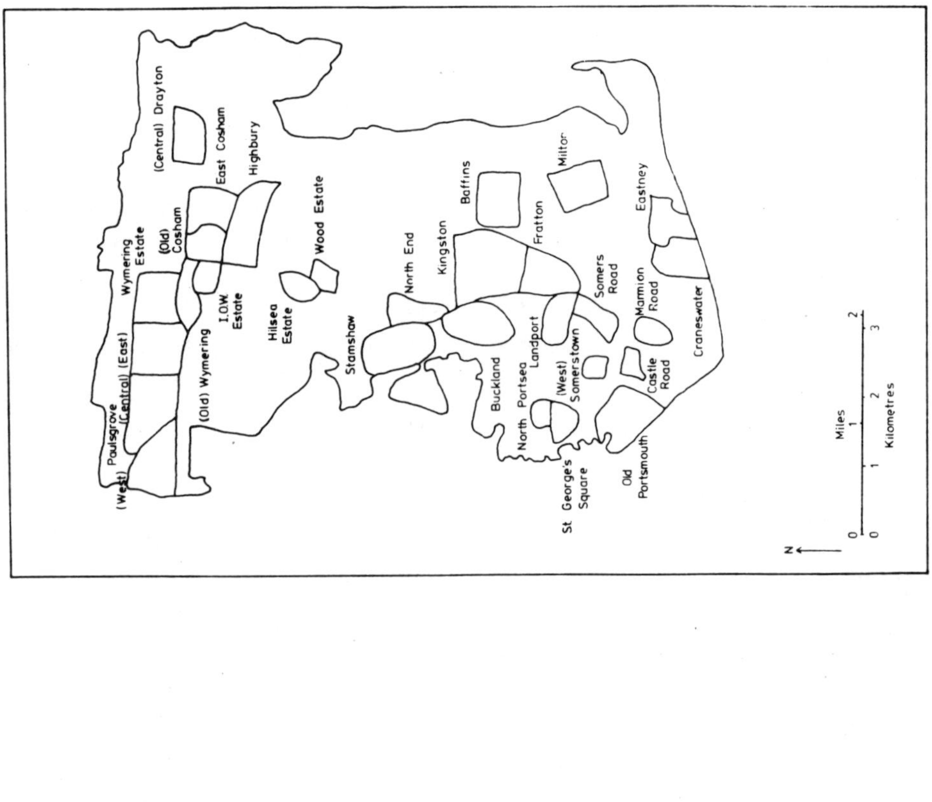

Fig. 1 Generalised community areas.

in their cartographic representation, an important duality is reflected in Fig. 2's general surface representation of Portsmouth's articulated community areas. Some areas' boundaries appeared to be closely and 'accurately" defined by a large proportion of their respondents —reflecting the existence of widely held and easily recognisable psycho-spatial barriers acting as boundaries to social areas (Fig. 5). Other areas, however, were defined with a far greater degree of imprecision and a lack of spatial correspondence with each other. This latter type is here termed a 'frontier' since such features would appear to represent relatively gentle gradients rather than distinctly sharp boundaries between areas. (Comparison can be made with the arguments for using the terms 'boundary' and 'frontier' at higher, politically defined spatial levels: Prescott, 1967; Kasperson and Minghi, 1969). Such 'frontier' areas in Portsmouth arising out of the survey, included cemeteries, parks, sports grounds, a hospital and the upper slopes of Portsdown Hill overlooking Paulsgrove and Wymering.

(i) where public access from different adjacent social areas is available simultaneously (e.g. local parts);

(ii) where public access from both local inhabitants and non-local area inhabitants is an implicit feature (e.g. cemeteries);

(iii) where public access is both non-locally based and for a specific purpose (e.g., hospitals, Service's sports grounds);

(iv) where large-scale demolition and extended temporary dereliction have apparently rendered an area devoid of any overt physical or psycho-spatial function, where public access is thus relatively irrelevant.

Thus emphasised in such 'frontier' areas is a lack of territorial specificity, whereby areas are 'shared' by residents of different territorial origins. Such a context can be seen as the antithesis of what Boal (1969) on Belfast has termed 'activity segregation', whereby different groups not only do not live in the same area, but also do not interact with each other.

RESIDENTIAL STRUCTURE AND COMMUNITY AREA ARTICULATION

Distinctive and well bounded areas of the 'natural' or built environment do not necessarily induce local interaction and involvement, as early attempts at architectural determinism in British new towns revealed (Heraud, 1966, 1968a, 1968b; Broady, 1968). In this respect, three general types of area in Portsmouth failed to articulate a community consciousness in the questionnaire survey. These can be classified as:

(i) high status, low density areas of high personal social and physical mobility where 'community' had very little meaning within the locally defined area —reflecting Janowitz's (1951) concept of a 'community of limited liability'. Much of the Upper Cosham-Dragton-Farlington area, a rather sprawling, high status mainland area of

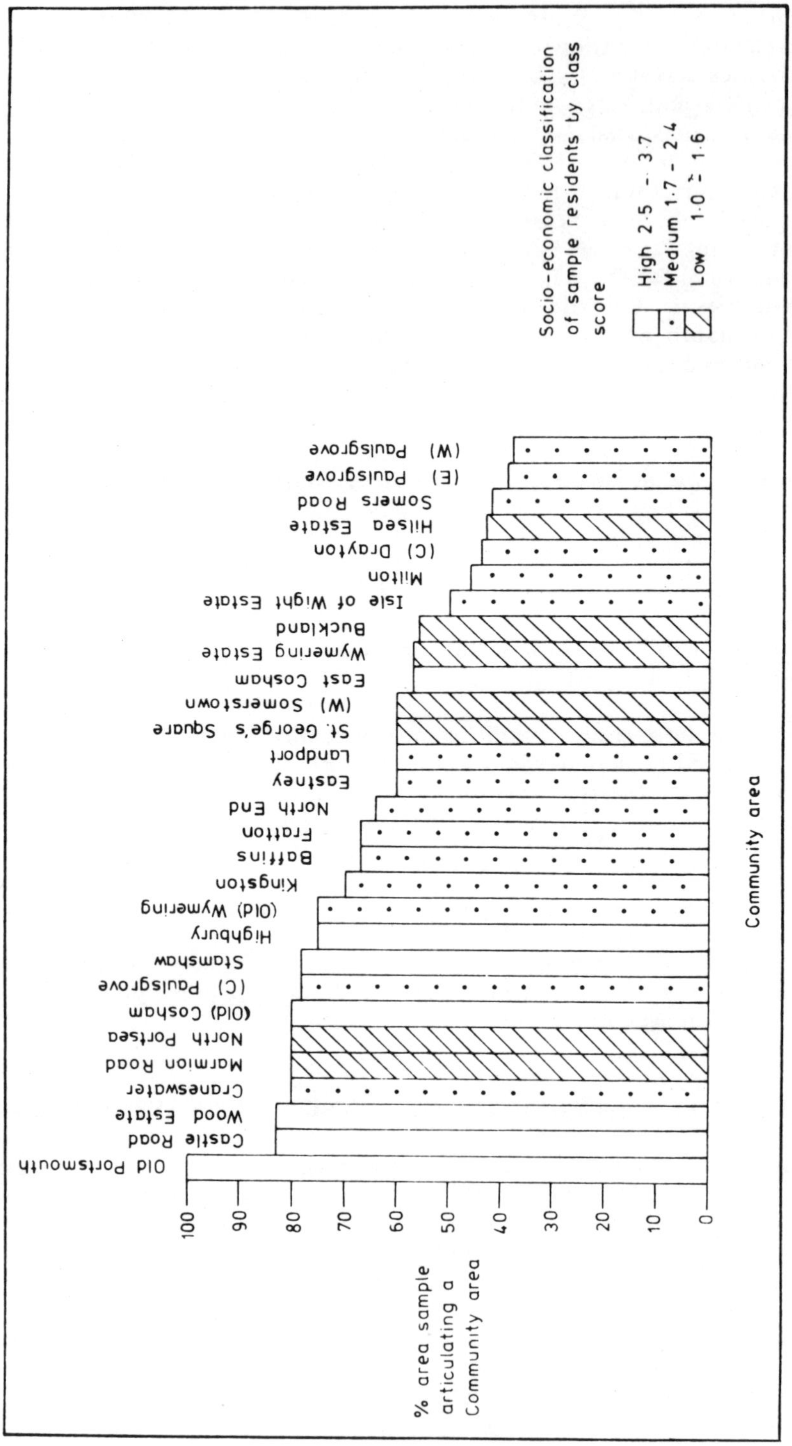

Fig. 3 Community area articulation

large, generally interwar private housing, exemplified this type:

(ii) high density areas of subdivided dwellings experiencing a high turnover of population: especially relevant here was the presence of accommodation catering to holiday-makers during the summer months and to the large student population during the rest of the year, in much of Southsea;

(iii) medium density areas predominantly inhabited by 'upper working' and 'lower middle class' families comprising terraces of 4/6-to-a-block units built between the wars and lacking many of the recognised interactional features of residential areas such as small shops and pubs (e.g. Copnor, North End, Hilsea).

The patterns derived from Figs. 1 and 2 suggest that articulation of a local community area in Portsmouth as derived from the questionnaire survey, would seem to have been strongest in four general types of morphological area:

(i) middle to high density middle class urban 'villages' – Craneswater, Old Portsmouth– where values of morphological distinctiveness lending character to a status area, fashionabilty, and relative social homogeneity within a well-defined area were seen to be important;

(ii) old, high density, working-class areas of closely packed terraces - Stamshow, Fratton, Kingston - where time and social similarity had induced valued interaction in areas not yet threatened by redevelopment;

(iii) certain medium to high density municipal estates where the provision of social facilities, e.g. community centres, together with relative social and morphological homogeneity were important;

(iv) medium density areas of well-defined physical boundaries coupled with a degree of internal cohesion, such as Highbury, bounded by railways on two sides and main roads on the other two, and possessing an active parish church.

Fig.3 and Table 3 reflect the degree of articulated 'community area consciousness' of the survey sample in each of the recognised community areas. Superficially, at the aggregate level, it would appear from Fig. 3 that little discernable relationship existed between an area's socio-economic structure and the degree to which the survey sample were able to articulate a community area. It can be claimed, however, that social groups possess, and may see themselves possessing, different and often conflicting interests, as well as articulating these interests differently. Baril (1971) has suggested for example that in appreciating differing experiences of the real world, analysis should first be undertaken of the language used to express these appreciations. From this premise, Goodchild (1974) applied the work of Bernstein (1958, 1959) on class differences in language to differences in environmental images. He pointed to Bernstein's distinction between 'public' language (short, grammatically simple sentences of poor syntax) used by most people, and 'formal' language, accurately formalised grammatically and syntactically, used only by managerial and professional status groups, the difference being primarily a function of differing levels of conceptual ability. Thus it is argued that those restricted to the 'public' code are less likely to conceptualise aesthetic aspects of the environment, but are

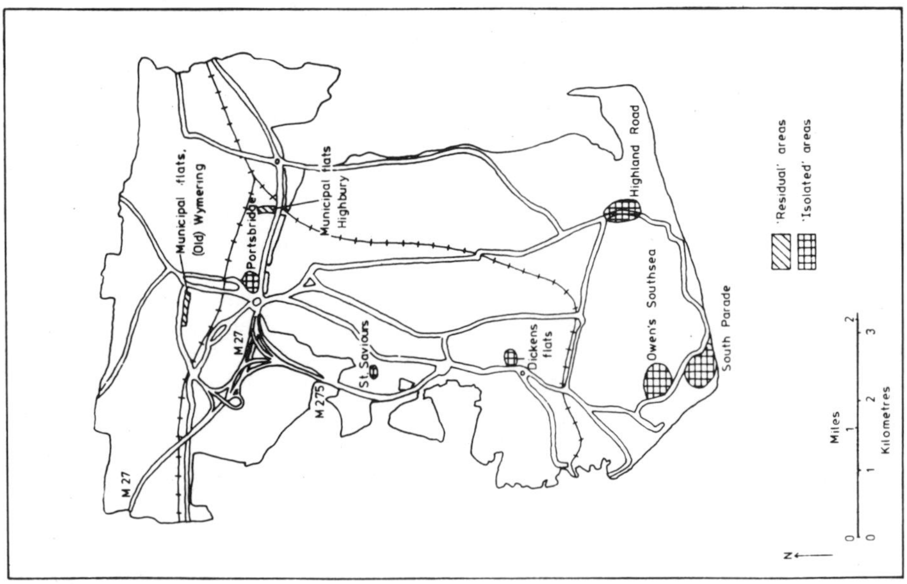

Fig. 4 'Residual' and 'isolated' articulated community areas

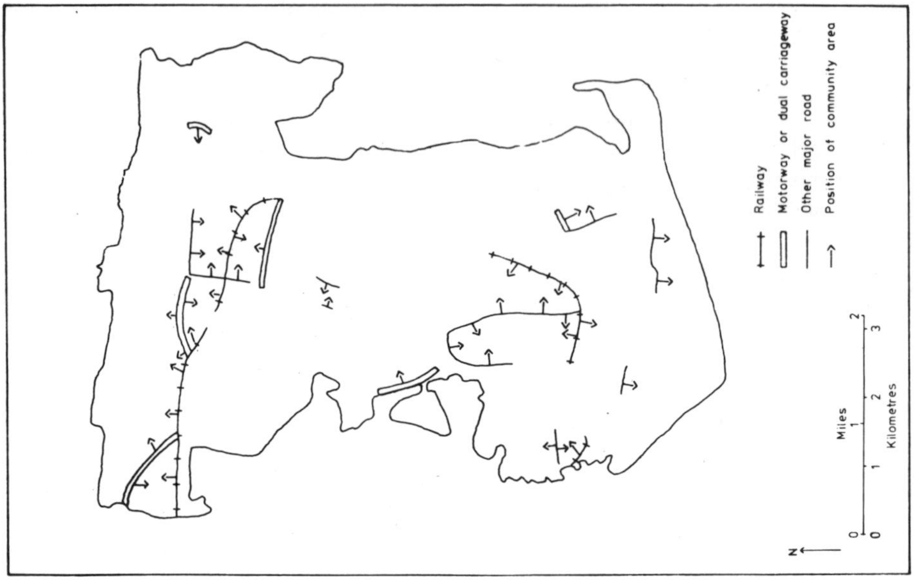

Fig. 5 Major linear boundaries of articulated community areas

TABLE 3

Degree of community area articulation (% area sample articulating a community area)

1.	Old Portsmouth	100	16.	Eastney	60
2.	Castle Road	83	17.	Landport	60
3.	Wood Estate	83	18.	St. George's Squ.	60
4.	Craneswater	80	19.	(West) Somerstown	60
5.	Marmion Road	80	20.	East Cosham	57
6.	North Portsea	80	21.	Wymering Estate	57
7.	Old Cosham	80	22.	Buckland	56
8.	(Central) Paulsgrove	78	23.	I.o.W. Estate	50
9.	Stamshaw	78	24.	Milton	46
10.	Highbury	75	25.	(Central) Drayton	44
11.	Old Wymering	75	26.	Hilsea Estate	43
12.	Kingston	70	27.	Somers Road	42
13.	Baffins	67	28.	(East) Paulsgrove	39
14.	Fratton	67	29.	(West) Paulsgrove	38
15.	North End	64			

essentially concerned with basic needs of day-to-day life. Such a conceptual dichotomy could be claimed to be one of the filter processes operating to support the argument (Spencer, 1973) that higher socio-economic groups are more likely to reflect community area recognition if only because they can more readily understand and articulate the concepts involved. However, a lack of statistical significance between employment group and community area articulation was found in the survey, thereby confounding the second postulate put forward at the beginning of the paper. The third postulate, however, was borne out in that there was seen to be an increasing likelihood of community articulation with increasing length of residence in one particular place. A very significant relationship, at the 0.1% level, was revealed between length of residence and local community area articulation ($x^2 = 28.01$, df=4). While research is still ongoing, particularly in relation to urban population and built environment densities, initial findings tend to support the argument that due to a combination of class composition and historical legacy of the built environment, higher community area recognition in areas of long residence also corresponds to relatively high density residential contexts. Higher population mobility in higher socio-economic low density areas and relatively new low socio-economic medium density areas would not appear to reflect such a relationship, given the constraints of historical evolution. Exceptions did, however, exist in Portsmouth. Certainly further detailed research, both in time and space, is needed, and it remains to be seen whether density *per se* (cf. Lee, 1968) can be isolated as an important variable.

THE IMPACT OF SOCIAL ENVIRONMENTAL FACTORS ON FERTILITY BEHAVIOR IN TURKEY AND IN THE UNITED STATES AND ITS IMPLICATIONS FOR CROWDING

Ipek Soykan Gürkaynak

Faculty of Education, Ankara University, Ankara, Turkey

ABSTRACT: After a general introduction to social psychological, cultural and immediate environmental factors affecting fertility, a research project designed to demonstrate the influence of the immediate environment on American and Turkish women's desire to use contraceptive devices and commitment to limiting family size is reported.

Both cross-cultural and intra-cultural results showed that because of the actual/ perceived closeness of parents and the availability of hired child-care help there exists a help-support culture in Turkey which influences fertility behavior in the direction of more children, fewer acceptable birth control methods and more good reasons to have a large family. Effects of women's employment in Turkey are also discussed.

INTRODUCTION

One purpose of this project was to begin some cross-cultural and intra-cultural survey research on some as yet unstudied factors influencing fertility desires, birth control attitudes, and family size, and the overall implications of these on crowding. Two virtually identical studies, conducted in Kansas, USA and Ankara, Turkey, on the impact of social psychological and environmental factors on fertility behavior of women,will be reported.

Although it has been defined in a variety of ways, crowding in general implies an interpersonal process where there is the subjective feeling of lack of significant personal space and more interaction than desired (Altman, 1975).

Contradictory evidence in studies on the effects of human crowding abounds; for example, in terms of the specific relationship between crowding and different forms of social and psychological pathology (Galle, et al., 1972; Gillis, 1973; Carnahan, 1974; Booth and Edwards, 1976).

Thus, Ausley Coale, the then President of the Population Association of America, said in 1968 that the problems of pollution, juvenile delinquency, etc., should be attacked directly rather than through population reduction since at the present rate of population growth in America, the density would not be very high even if the population rose to a billion (Coale, 1968).

As mentioned above, it is true that we do not know the exact extent to which population increase actually contributes to a society's social ills (e.g. juvenile delinquency, crime, drug addiction, etc.). However, we do know, even through intuition and common sense, that everywhere there are limits to the capacity of the natural environment to support an expanding population, public facilities such as education and transportation become overloaded, and access to them becomes more difficult as numbers of people grow. We also know that noise, intrusion, impersonality and inactivity increase as the pressure of population increases (Milgram, 1970; Altman, 1975).

In short, then, uncontrolled population growth is a severe threat to the physical and emotional well-being of the people. And the people are increasingly aware of this. A Gallup Poll of city dwellers, suburbanites and rural residents in 1966 revealed that 22% of the respondents preferred to live in the cities. This percentage was found to have dropped to 18% in 1968. Another Gallup Poll of city dwellers in 1970 indicated that only 13% wanted to remain in the cities.

In view of the foregoing, the population problem in general and fertility in specific is discussed in this paper as the major cause of crowding. It is a cause behind which lie certain social psychological and environmental factors, some of which have been treated in the study to be reported.

Although much information has been accumulated on birth control techniques, on their availability to and usage by women of different personality make-ups, different social strata and different cultures, we know comparatively little about the specific factors in the psychological environment which bring about differential effects upon the birth rate within different cultural groups (Miles, 1970).

If middle class women consider three children to be the ideal (or desired) family size, for instance, we would have difficulty in attributing the cause of that apparent attitude only to values concerning children, or even contraceptive availability as opposed to situational factors such as the size of potential children's bedrooms, availability of child-care help, and other kinds of aid which may reduce the psychological costs of having children, as well as social norms influencing people to have children.

Social Psychological Factors Influencing Fertility Behavior

One meeting ground for psychology and population is the study of the social psychological factors that affect women's views on family size, contraception, value of children, work outside the home, etc.

In this respect, studies identifying the elements of motivation which may be subject to external influence and therefore are in some degree controllable might prove of high utility (Miles, 1970). Stycos (1955), in his Puerto Rico study, used intrafamilial variables to explain the discrepancy between stated small family ideals and the actually achieved large families. He showed that these vague values of a small family were overshadowed by "...aspects of conjugal family structure that interfered with actions necessary to strengthen or realize these values" (Fawcett, 1970, p.41).

One may generalize from this study for underdeveloped countries in general where social pressure, where it exists, is usually on the side of having as many children as one may raise --, the "necessity" of a sister for a boy or a brother for a girl, the fact that one can never know full satisfaction until one is a mother; in short, the assertation that it is only "normal" for a woman to want and to have children. In dealing with such motives, couples should somehow be persuaded to act more in line with their own economic self-interest, to obtain psychological satisfactions through sources other than large families, to modify their basic positive attitudes toward large families and achieve a societal consensus to stop population growth within a limited time span so that a major thrust towards enhancing the quality of life becomes possible.

Since many studies show an inverse relationship between a woman's fertility and her level of occupational aspiration (Tangri, 1969; Farley, 1970; Waite and Stolzenberg, 1976), one way of lowering family size desires may be to open up other avenues for adult satisfaction, i.e., creating non-familial roles which offer significant competition to the familial ones in such a way that women will become significantly less involved in the latter kind of role and find other forms of self-realization (Blake, 1965; Miles, 1970).

"Social changes which make marriage and motherhood less of the sine qua non of adult status for women should drastically alter the numbers of mothers and children (Tangri, 1972, p.2)." Therefore, "...policies expressly related to family roles, and opportunities for legitimate alternative satisfactions and activities, constitute the crux of future reduction in family size because they directly assault the motivational framework of reproduction (Blake, 1965, p.68)."

In summary, then, it is believed that further knowledge of the above social psychological factors (such as the life style, role and status of women, family attitudes, institutionalization of reproduction, values of children, social pressures, ideas of "normalcy", etc.) that affect fertility and fertility desires would strongly be needed, including a more thorough investigation of the specific cultural and immediate environmental factors which may be of vast importance. These last two factors will now be discussed in turn.

Cultural Norms Influencing Fertility Behavior

Assuming that reproduction is very important to the family and to the society at large, its level should be very much determined by cultural norms related to child bearing behavior. The norms about these matters, however, are themselves very much related to the nature of the society that they are a part of. One can demonstrate this with an example: The general assumption today is that high birth rates in developing countries are primarily a result of unwanted babies. This assumption is based on the fact that research shows desired parity to be practically always lower than actual parity. In other words, women desire a smaller family size than they are likely to have by the end of their reproductive years (Mauldin, 1969; Fawcett, 1970). This view completely overlooks the institutionalization of reproduction in these countries and individual desires for large families, and consequently, places heavy emphasis on intensive planned parenthood programs. Not only have such programs been unable to lead to marked reductions in birth rates among the illiterate masses, but they also fail to consider cultural and psychological aspects of the issue. Blake (1965) asserts that family planning programs are faced with a dilemma: On the one hand they have to introduce contraceptive means, and, on the other, they have to lead to more modest family size desires, since, although desired parity is lower than actual parity, it is not low enough to reduce population growth substantially.

The creation of more modest family size desires is not an objective easy to accomplish, mainly because of the cultural norms that may induce women to have larger families even if birth control methods are readily accessible.

In Turkey, for example, there are cultural norms that (a) work to actually induce women to want large families, and (b) work to reduce the desire to have a small family. In the first category are the factors such as the horrifying thought of not having any children. Barrenness is regarded as a "shameful disgrace" at worst and as a "pitiful misfortune" at best (Timur, 1956). A childless woman is usually considered to be sterile because not wanting any children is quite incomprehensible if not inhuman. As a matter of fact, women who don't want children or want only one child are generally considered selfish and caring for their own comfort only. Another factor in this category is the high prevalence of the desire for at least one male child.

The list of cultural norms can go on indefinitely, including the differential socialization of the two sexes, the widely accepted subservient role for the woman, religious or supersititious beliefs that absolve one of economic and other considerations since "Allah will take care of the children because Allah gives them" (Timur, 1956).

The second category above is based mainly on the fact that the culture is generally very child oriented: Social settings welcome children, a mother can take her child to the hairdresser's as easily as she can take her or him to a meeting or a social gathering. Children are well tolerated even when they are not welcome, and they are not thought of as a nuisance.

The present study was designed partly to compare, where possible, such child related norms in the United States and in Turkey, although it was only the former study which was particularly designed to focus on norms. The main focus of the Turkish part of the study was to examine the effects of the immediate environment on fertility behavior. We now turn to this problem.

The Immediate Environment Influencing Fertility Behavior: The Case of Turkey and the United States.

It is an oversimplification to attribute the great differences in underdeveloped countries between desired and actual parity to the lack of birth control techniques. Specifically, the differences between desired and actual parity in Turkey cannot be attributed to this factor since both abortion and contraceptive devices are available (Tuncer, 1968; Tuncer, M., 1971).(*)

Obviously, the dissemination of contraceptive knowledge and techniques does not necessarily lead to a stable population; otherwise, all countries where this kind of information is available would have achieved it. The issue of the general social context of family life, home environment, and cultural environment in the sense of what these have to offer to family planning or to the absence of or objections to it, should be examined seriously as an empirical problem.

As will be explained shortly, the home environment (but not, in general, the attitudes) of Turkish middle class women differs in a few important respects from that of the American middle class women.

In Turkey, there are two important immediate environmental variables which may make additional pregnancies less unpleasant than they would for an American woman, and thereby reduce the desire to limit family size. One of these factors is the availability of servants, (i.e., child-care helpers), the other is the possibility and the common practice of calling on parents and relatives for help with the children. With respect to this latter practice, one can say that in the American family, the relations between parents and married children are governed by different norms. Upon marriage, children establish themselves in neo-local residence, and "especially in the middle class it is believed that parents thereafter are freed from further responsibility for them; they are on their own and must fend for themselves" (Sussman, 1953, p.22).

To the best of the author's knowledge the only piece of research done in the

----

(*) In Turkey, the laws dating from the early 1900's and prohibiting the import, sale and distribution of contraceptives were repealed in 1963, in view of the fact that rapid population growth can result in serious problems in the near future. IUD's and oral contraceptives are readily available now. Support has been provided by a $ 3.6 million AID Development Loan and technical aid from the Population Council. By October 1967, over 64,000 IUD insertions had been reported by the government trained doctors, and an estimated 50,000 cycles of oral contraceptives were being sold monthly. The Ministry of Health and Social Welfare had given 473 gynaecologists and general practitioners training in family planning and plans were being formulated to train more doctors. These doctors would then be sent to the provinces; more recently 59 of the 67 provinces have been manned in this manner (AID brochure, FY, 1969; Guvenc, 1972). For a thorough discussion of the effectiveness of such programs, refer to Tuncer, M. (1971).

US on the relationship between parents and their children that comes close to the help-support relationship prevalent in Turkey is a study describing the help pattern that exists between elders and their child's family and its effect upon inter-generational family relationships (Sussman, 1953). Sussman found that a small sample of middle class, white, highly educated, urban Protestant families in New Haven, Connecticut was not "...as independent and isolated a unit as it is generally thought to be "(p.28). He found affectional and economic ties still linking the generational families and giving stability to their relationships.

Sussman did not go into the analysis of why such a relationship –which is really an exception to the kind of relationship that has been supposed to prevail between parents and married children in the United States– should exist in this particular community; however, the main point of interest for us here is that for the Turkish family this type of relationship is common, and not unique. For the middle class woman in Turkey, even though she may profess to want only two children, any unplanned or unexpected addition of more children is not perceived as a great hardship since she can hire outside help and/or count on the help of the extended family* while her children are growing up. In this sense, then, one can say that Turkey has a help-support culture. It is because of these factors that for the Turkish woman it may be less important to practice birth control conscientiously; she can afford the attitude of "que sera sera" that Thomlinson (1965) mentions.

The important question at that point is how best to study the effects of the aforementioned social psychological and immediate environmental factors on fertility behavior. In countries other than the US, there has been little research on the effects of social psychological and environmental variables. This lack of studies in other countries is a particularly troublesome deficiency because in any one country attitudes, concepts and values are confounded with environmental and demographic factors. If one can compare cultures where the attitudes are similar but environmental factors and values different, it is easier to pin down the source of the attitudes; and, one gets a greater variation of independent variables of interests. The case of Turkey versus the US presents an ideal opportunity to conduct this kind of comparative study.

HYPOTHESES

Cross-Cultural Hypotheses

1. Given the same socio-economic background and desired family size, women in

----

(*) The concept of the extended family has been defined and re-refined a variety of times by different researchers (Murdock, 1890; Parsons, 1945; Mindel, 1970; Sussman, 1963; Litwak, 1963; Adams, 1968; and others), taking sides and making the issue very complicated. Here, reference is made to the kin of orientation of the woman after she is married. Co-residence is not a point of concern. The general meaning, then, is closer to Litwak's (1963) "modified extended family" which "...operates without the need for geographical propinquity, occupational involvement, or nepotism in job placement of family members, or a rigid hierarchical authority structure (p. 477)."

Turkey will care less than women in the United States about birth control, find fewer birth control techniques acceptable, and have larger families.

2. Again holding background and family size desires constant, women in Turkey will believe there are more good reasons to have children (or, a large family) than women in the United States. The reason for this and for the previous hypothesis is the fact that, because of their immediate help-support environment, the negative aspects of having children are less salient in the former culture.

Intra-Cultural Hypotheses

1. Within Turkey, women who live near relatives or have comparable help-support from relatives will be less likely to care about birth control, find fewer birth control techniques acceptable, and have larger families.

2. Within Turkey, women who have hired child-care help will be less likely to care about contraception and limiting their family size than women who do not, at the time, employ such help.

METHOD

The United States sample of this study consisted of 177 married women selected as a subset of the 864 women studied previously. These were originally selected from randomly chosen groups of women living in Douglas Country (primarily, Lawrence, Kansas) whose husbands or who themselves were employed in skilled occupations or above. The 177 women were selected by a procedure which matched United States women with the sample of Turkish women in age, desired family size, and urban background.

In Turkey, a sample of 177 middle class* married Turkish women who had at least one child, who were 45 or under at the time of the interview and who had completed at least a partial high school education, was drawn from Ankara, the capital city of Turkey. Information from both samples was collected through a two-part survey.

Part one of the survey was a verbal interview covering (a) biographical information, (b) demographic information, (c) information on feelings about adoption, and (d) a section on the advice given by other people to the subject about the latter's job and/or schooling, the number of children she should have, and how she should raise

----

(*) Ankara is known to be a predominantly "middle class city." In choosing the sample (a) education (at least partial high school), (b) adequate housing in the "middle" or "upper middle class" regions of the city, and (c) employment (at least as a skilled worker) of the husband, were required. No other attempt was made to specify the concept of "middle class". The sample that was drawn, then, was a homogeneous group of women who were capable of understanding and intelligently responding to the questionnaire.

her children.

Part two was a confidential questionnaire on (a) attitudes toward birth control, family size, children and careers, (b) reasons for having four, three, two, one or no children, (c) opinions on population growth, the institution of marriage, working women, and the role the government should play with regard to these issues. In Part two, there were some questions unique to the Turkish study, which were used only in the intra-cultural analysis. The questions were on how far away the parents, or other relatives who might be called on for help with child care, lived; who the respondent would call if and when she needed help; whether or not she had, or anticipated having in the future, a maid.

RESULTS

The basic thesis of this paper has been that in Turkey the most influential reasons for having children are not simply family size desires and birth control non-availability, but also the help-support culture and the home environment conducive to having children.

For testing the hypotheses, analyses of variance and correlational techniques were employed for within Turkey analyses and for the cross-cultural analysis.

The reader should note that (a) only the significant results (but regardless of whether or not they are in line with the hypotheses) will be presented, in summary form, and data from the questionnaire considered irrelevant to this presentation will not be included; (b) the Turkish sample is drawn from the city of Ankara to ensure comparability across cultures; it may not represent Turkish women as a whole.

Cross-Cultural Results

1. The majority of Turkish women had help from servants, and more Turks than Americans had at least one relative living in the home.*

2. Turkish women considered adoption less acceptable, found fewer birth control methods acceptable and more of them unacceptable. There were also qualitative differences in that Turkish women approved more often of the less "safe" contraceptives (rhythm, condoms) than did US women.

3. Turkish women found more good reasons than their American counterparts

----

(*) In the Turkish sample, 54% of the women admitted to having a live-in maid, 26.5% to having a non-live-in cleaning woman, and 87% said that they anticipated either continuing on with the one they had or hiring one in the future. It is unfortunately not possible to compare the Turkish women with their American counterparts because the American study did not include questions on this subject, but an informal survey by colleagues of the author revealed that almost none had live-in help, and when help was employed it was a once-a-week cleaning woman.

to have four children while the latter group found more good reasons than the former for having two, one and no children. As a corollary to this, more Turkish women found "no good reasons" for having one and no children. In short, the Americans tended to accept a smaller family size more and a larger family size less than the Turkish women.

4. The results on the advice section of the questionnaire lead to certain insights about differential social pressure in the two cultures. (a) While American women got more advice on job and/or schooling, Turkish women got more advice on the number of children they should have and how they should raise their children. (b) Not only the quantity but the nature of advice differed. American women got advice much more than their Turkish counterparts to "get a job", to "continue their education", to not "quit" their jobs. Also, while American women were advised not to have any more children or were told that they had too many or just the right number, Turkish women reported being urged to have more children. (c) The source of such advice also differed for the women in the two cultures: Turkish women received advice from elder family members (extending to grandparents, aunts, great aunts, etc.), while American women were given advice mostly by friends.

Intra-Cultural Results

Live-in versus non-live in. One purpose of the intra-cultural analyses was to examine the effects of differences in help support within the Turkish culture.

One method of defining the amount of help support in the immediate environment in Turkey was to categorize married women according to whether or not they had hired child-care help. The assumption in doing this was that living with relatives, having relatives close by (physically or perception-wise), and/or employing servants is equal to greater help-support.

Results showed that the differences between the live-in and the non-live-in groups were all in the expected direction of help support effects: Live-in women, as compared to their non-live-in counterparts, (a) had more children than women living alone with their husbands and children, (b) found adoption less acceptable, (c) found more methods of birth control unacceptable, (d) found more good reasons to have three and four children, (e) received more advice from others to have no further schooling and/or jobs, to have more children, to raise their children this or that way. Also, the responses of all women in the sample to the question of whether or not they would ask someone from the family to look after their children revealed that regardless of how close they lived to them, they would not hesitate to do so.

Work versus no work. Another purpose of the intra-cultural analyses was to examine the effects of women's outside employment on the dependent variables discussed up to now.

Results were as follows:

a) More working women had live-in servants but there was no difference between working and non-working women in terms of having a live-in relative.

b) Working women expressed more good reasons for having no children, and more good reasons to have four children.

c) Working women found a greater number of birth control methods unacceptable.

d) Working women had intended fewer children than non-working women when they got married.

e) Working women received more advice to have more children.

Results (b), (c), and (d) above require comment:

The findings that working women, as compared to non-working women, found more good reasons to have four children and found a greater number of birth control methods unacceptable seem to be contradictory with research results mostly demonstrating outside employment to have a curbing effect on fertility.

These results, however, are in line with the general direction of the hypotheses. While these women had intended fewer children when they got married, and found more good reasons to have no children, once they have a child, the help-support culture becomes active to influence them in the direction of having more children, through either verbal advice or by becoming available to share the responsibility of child-care.

CONCLUDING COMMENTS

In general, then, this data supports the help-support culture effects between cultures and within Turkey. Also interesting are the effects of women's employment on the dependent variables.

Notwithstanding the fact that the Turkish sample was somewhat non-representative, the results of this study are suggestive in many ways and provide a good starting point for further research. Why is it, for example, that the mother and other relatives of a married women urge her to have more children knowing that it is they themselves who are or will be called upon to help with child care? Is it because these elder women consider it a duty such that it is a part of their mother role to help to take care of their children's children? Or is it because they don't have much to do and being around their grandchildren gives them a new energy and a "wish to live", so to speak? The answer to these questions may be the starting point of an effort to pin down the exact intervening variable at work in the relationship between the presence of an actual or a perceived extended family and fertility.

IMPACT OF FAMILY CROWDING
ON THE DESIGN PROBLEMS OF DWELLING UNITS WITHIN TURKEY

Nigan Bayazıt, Ahsen Özsoy, Mine Celesun

Istanbul Technical University, Faculty of Architecture-Faculty of Architecture and Engineering, Istanbul, Turkey

ABSTRACT: The concept of crowding is examined within the context of dwelling units. A review of the literature and the analysis of the surveys and case studies in some parts of Turkey are presented. A crowding model for dwellings has been developed in structuring the conceptual crowding problems and their effects on the physical setting. Influences on the sense of crowding and their effect on the adaptation or alteration processes in relation to values of density are discussed within the context of dwelling unit. Lastly the findings of the case studies from Turkey are matched with these subjects.

INTRODUCTION

This article aims to examine the consequences observed in the physical environment of dwelling units which are caused by the dense families' behaviours. The study discusses the probable results of this behaviour on the design and re-arrangement of physical settings. Dwelling density problems either in citiesor in the recently built up areas reached an important dimension because of the shortage of housing (İmar ve İskan Bakanlığı, 1974). The average density in Turkey is about 2.42 person/room (RD). This is very high in comparison to developed countries. Percentage of the dwellings with 1 or 2 rooms reaches 61.5% of all dwellings in the country (Irmak, 1976) while the size of family is still quite large.

Lack of privacy is obvious in these dwellings. The need for physical space reflects itself in or outside the buildings in making alterations. The composition of the family generally varies from house to house because of the economic dependence of some of the relatives and family generations. Material culture is in the process of major change. Within this context the crowding problem is discussed referring to the survey studies which were previously completed.

CONCEPTS AND DEFINITIONS

In relation to this study several concepts are used. Some of the concepts such as density and crowding are always used freely and interchangeably. The impression is given that these are completely different labels. Major concepts such as dwelling units, family, crowding and density within the dwelling units are defined in this section.

Dwelling Unit (DU)

Dwelling unit consists of a cluster of rooms within which there are sections reserved for the basic units (International Encylopedia of the Social Sciences, 1968) of the family members. The home is a menage for production as well as consumption and may therefore also reflect features of economic organization (International Encylopedia of the Social Sciences, 1968). So that the dwelling unit (DU) is inclusively the cluster of rooms which are occupied by the family for habitation purposes.

Family

The basic model of kinship and family ties is that of biological relatedness and sexual intercourse, so that kin and familial relations are thought of in terms of physical descent or sexual relations based on the socially recognized rules. "Nuclear Family" is the universal form of family relations, a universal human social grouping. It exists as a distinct and strongly functional group in every society and more complex familial forms are compounded.

The family structure is not well determined even in the urbanized areas in Turkey. It may show the character of either nuclear or a joint (or temporarily-large) family in any life cycle of the family (Kıray , 1977).

Crowding

The dictionary meaning of crowding is "a large number of persons congregated or collected into a close body without order. The terms population and overcrowding are generally used interchangeably" (Murray, 1974) as they are merely different words. The best description of over-crowding does not dwell at length on the shortage of space or restricted movement in an abstract sense as Sommer (Sommer, 1969) pointed out. There are many other definitions of crowding. If a person or a group of people try to hide themselves from others and fail to achieve it, this is defined as intrusion or crowding (Altman, 1974). In another way Choi defines as "the test level of psychological phenomenon which originates from the interrelations among spatial, social, individual and physical dimensions" (Choi, et al., 1976). People are influenced by properties of spaces during their activities so that we accepted the definition which was made by Proshansky (Proshansky, et al.,1970):crowding,"either for the member or observer of the crowd, is not simply a matter of the density of persons in a given space. For the crowded person, at least the experience of 'being crowded' depends also to some degree on the people crowding him, the activity going on and his previous experience involving numbers of people in similar situations."

Density

The studies on crowding in general take interest in interrelations among certain physiological, behavioural and social phenomena, and are followed by the studies onthe degrees of density to give a quantitative measure. The influence of density on people is less certain than on animals (Choi, et al, 1976). As it has been already indicated before many writers use the words density and overcrowding interchangeably which is not quite correct.

Density in this sense is the number of people per unit space (SD) or room (RD) or area per person (AD) so that these are unit indices while crowding is accepted as criterion that varies pertinent to objectives of the people, place and time. Inside or outside the dwelling person/area or person/space measures are most often utilized (Mercer, 1974; Murray, 1974; Wilner, et al., 1962). Also they indicate the other density measures, such as average number of people in a dwelling unit. Murray points out that in Britain at present any household where there are more than 1.5 persons per room is considered to be "over-crowded", many local authorities are tending to regard a level of 1.0 person per room as the cut-off point, a reflection of rising standards and expectation" (Murray, 1974). In Fairchild (Fairchild, 1964) also measures of crowding are defined with the values of density (RD) such as, one to one-and-one half persons, crowded; over one-and-one half but less than two persons overcrowded; two or more persons, gross overcrowding.

This kind of independent variable measures of density are used as parameters for the dependent variable of diseases and juvenile delinquency; they more easily countable and measurable than the other measures of crowding.

METHOD OF STUDY

The existence of the sense of crowding among the members of the family and its influences on the physical setting is the major subject of this paper. A model in Fig. 1. is designed to explain the hypothesis clearly. Interrelations are sought among the measurable parameters, such as density values (RD and AD), family composition, dwelling size and number of rooms, and dependent variables such as alterations on dwelling, distribution of activities among the rooms, arrangements of furniture inside the dwelling. The following conceptual model explaines the procedure and the results of crowding syndrome. Most of the research work on crowding is, on one hand, directed towards the relations among physiological, behavioural and social phenomena and, on the other, the degree of density (Choi, et al., 1976). Modification of dwelling spaces, services and the use of physical setting will be examined in relation to measurable influences. The case studies and surveys in the following sections are examined within the framework of this conceptual model.

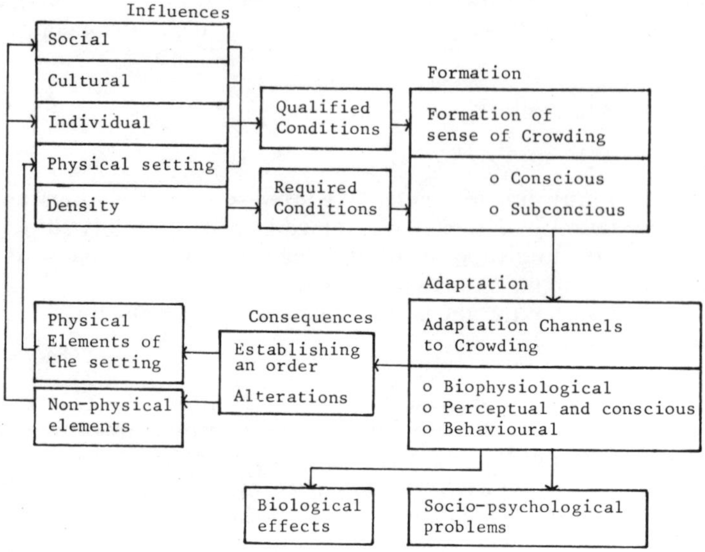

Fig.1. The Conceptual Model of Crowding in DU.

A SHORT REVIEW OF THE STUDIES ON CROWDING

This section is devoted to the review of the thoughts of other authors on the subject, in establishing a relation to the developed model.

The Proxemics Concept and Crowding

The experience of crowding in accordance with the proxemics concept (Hall, 1959) helps to clarify the thoughts. Cultural norms intervene in the perception and adjustment of the distance between people (Kıray, 1977). There are standards for the heights of buildings which are built to serve certain activities, as well as rules and constraints. But it is hard to give values of m2 or m3 per person exactly for every case. Distance and area values for different activities vary considerably from culture to culture so that as it is pointed out by Bird (Bird, 1972)crowding syndrome is not easy to measure universally. Hall's classification on the concept of psychological space, proxemics, can be admitted universally, and so the boundaries of the psychological distance differ widely from culture to culture. The highest density, as Bird explained (Bird, 1972), was in Manhattan , US, where people use more working space than anywhere else in the world. This leads one to the sense of crowding.

The Sense of Crowding

Architects, interior designers construct spaces using subjective and empirical principles, and the feelings of the people in relation to environment. Architects write and talk about it, and theories of architecture emerge to an extent from these suppositions. There are several studies on the sense of crowding relevant to the effects of physical environment which are tried to be grouped as follows:

If a room is used for more than two purposes that should give the feeling of crowding. A study-bedroom seems more crowded than a bedroom or a study room. Some density situations give different degrees of crowding according to the activities taking place in that space (Choi, et al., 1976).

The individual may also experience different degrees of crowding under the same situation according to his personal relations with others crowding him and the type of interaction with them. A room gives the sense of crowding when it is used in a deviation from its initial purpose and shared with a foreigner.Being crowded is reception of excessive social stimulation and not merely the lack of space (Mercer, 1974). If any one of the factors of social class, status and the age of the people in a room is a mixture, the room seems more crowded(Bird, 1972).

Properties of physical setting may cause the sense of crowding such as a room with a scenery looks more spacious than without scenery. Furniture density also proved to be a factor affecting the sense of crowding in Imamoğlu (1973).

Variables of physical environment seemed more relevant to the sense of crowding, whether or not high density was seen and felt as a crowded situation. This now receives empirical support in the work of Desor (Mercer, 1974). Her main thesis is that "being crowded is excessive perception of conspecifics" from which follows that if a high density residential area is to be rendered a non-crowded environment, it must be designed in such a way as to conceal people from each other for much of the time (Mercer, 1974). Crowding gives rise to the need for social regulations that limit the unwanted intimacy which would be likely to arise in the absence of physical barriers. Under crowded conditions, social norms for maintaining privacy partially substitute for the lack of physical devices (Sommer, 1969).

Physical variables such as lighting, temperature, humidity, shape of the room, number of doors and windows, density of furniture and so forth seem to be closely related to the perception of crowding. Desor fully devoted her study to the perception of crowding as a function of variation in physical environmental variables. On the other hand laboratory studies show that when physical components (such as temperature, shortage of air, area limitations, etc.) and the group size were kept constant, high density had negligible influence on work performance (Choi, et al., 1976; Mercer, 1974).

It was pointed out in Bagley (1974) that housing stress in relation to shared WC, shared bath and with the density of more than 1.5 (RD) caused psychiatric breakdown and incidence a short time after the subjects moved to that environment.

Effects of Crowding

The findings demonstrated however that crowding could affect in several ways. The experiments show that crowding could affect social and agressive behaviours (Smith, 1974). Positive correlation between high population density and the rate of socio-psychological problems such as crime, suicide, and mental illness were observed in some studies but not all (Choi, et al., 1976; Mercer, 1974). However it has been shown that people with mental illness were much more likely to experience overcrowded housing than people without such a diagnosis.

The results of the studies on the increase of crime in relation to population augmentation and density in Istanbul and Izmit showed that there was no correlation between the population increase and crime rate (Şehirleşmenin doğurduğu ceza adleti Sempozyumu, 1973). The influence of crowded dwellings has been argued by some writers. Home crowding, not residential density, is the main factor responsible for variations in the incidence of social and psychological problems of children within urban areas. The findings indicated that crowded children will tend to be more agressive, impulsive and extroverted (Murray, 1974). There should be a difference between spatially and socially dense households.

Privacy and Crowding

In this paper privacy is defined as "the selective control over access to and information about the self to be communicated to others". Privacy is a word opposed to crowding in respect to spatial distance, physical barriers and confinement. Different cultures developed alternative mechanisms and gained behaviours for the purpose of privacy.

The feeling of crowding emerges if desired privacy is greater than the achieved one according to Altman (1974). In most of the primitive societies homes spread into one another and the households intermingle in a communal life and without privacy, or the desire for it (Berardo, 1974). A study was made by Lewis on poor families in Mexico, where parents obscured their bed space in the kitchen using empty packing materials (Altman, 1974). Almost exactly the same situation was observed in the over-crowded housing in Osmaniye (İTÜ, Mimarlık Fakültesi, 1975) Fig. 2.b.

Some activities in a dwelling require confinement, not necessarily in association with crowding. Any factor or process which increases social or spatial distance and pushes the members apart will hasten confinement or separatedness (Murray, 1974).

Adaptation to Crowding

Adaptation to crowding is an important area of study in accustoming people to the closeness of others, or manipulating physical variables of the environmental enclosures so that crowding is less easily perceived. It is possible from the properly conducted scientific studies that are available to assert that man can adapt to conditions of high density living (Mercer, 1974).

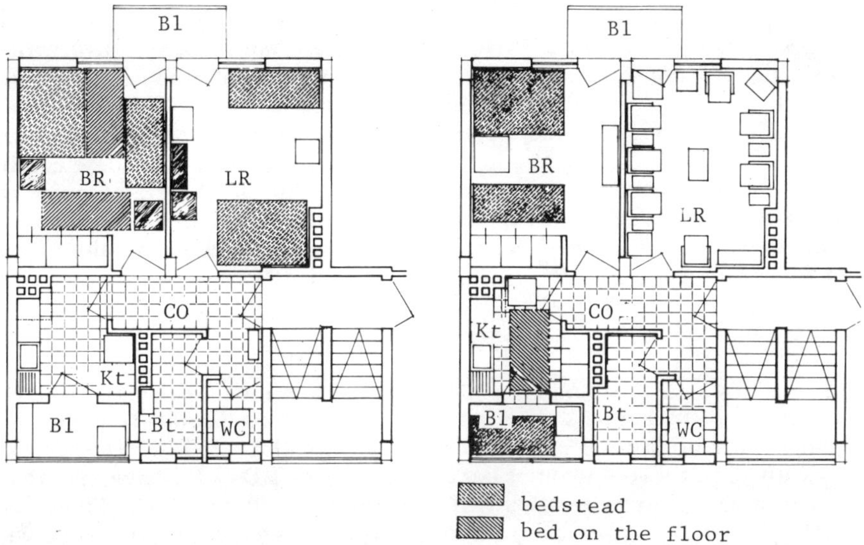

Fig. 2.a and b. Furniture arrangements in Osmaniye housing development (İTÜ, Mimarlık Fakültesi, 1975)

Adaptation may occur in three main ways. Firstly the individual cannot resist the socio-psychological pressures of the environment and suffers some psychiatric illnesses or frustrations with society. Secondly he may shows biological reactions to his environment. Thirdly he may try to solve the problem consciously and begin to manipulate and establish an order, and consequently make physical alterations in his dwelling.

STUDIES IN TURKEY

The sense of crowding in Turkish homes has not been properly studied before and one might say none of the discussed group of studies were directed to investigate the problem. Effects of crowding on the spatial and physical environment of DU deduced from the results of surveys and case studies, which were conducted in Istanbul and Izmit, are discussed were to give some indirect evidences. Studies are generally concentrated on the people who moved to cities from rural areas in the last decade and the second and third generations of the immigrants of the first half of the century. Dwellers live in either self-built (gecekondu) housing or in the dwellings which were built by the Ministry of Housing or subsidized co-operatives. The following table is designed to explain the place, subject of the cases or surveys, number of respondents, year of study and the type of housing.

Family Composition

Economic and social facilities ın Turkey are not still adequate at most of the income levels to live independently. This situation reflects the family composition, as in the case when people who move to urban areas are forced to live with their relatives or friends from their hometown. Until the newcomers find jobs, they have to live together, so that the family becomes temporarily large. This results in the physical insufficiency of space and increases the density within the DU. Irmak's study gave the evidence of 24% of the families around Istanbul living under these conditions in self-built (gecekondu) areas (Irmak, 1977). Similar situations were observed also in the Izmit area where newcomers were in general in the form of nuclear family but fairly soon they became a very crowded and temporarily-large or joint family in character for certain periods (Wilner, et al., 1962; Bird, 1972).

Density in DU

The majority of the families (91.4 %) in the Izmit area with 4-5 persons live in dwellings with 2,3 or 4 rooms. The modal DU is 3 rooms (Izmit Yeni Yerleşmeler Yapılabilirlik Araştırması, Istanbul, Birleşmiş Mimarlar, 1974). The density in dwellings with 2,3 or 4 rooms varies between 1.12-2.11 RD in this area. Another case study in the same region gave rather different results: in the dwellings with 3 rooms RD was 1.05 to 1.40, average population of a DU was 4.1 persons, while it was 3.88 - 4.28 persons in the previous study.

On the other hand the Osmaniye housing development in Istanbul is an extreme example of RD and AD figures, where RD varies between 1.72 - 3.15. The examples in this paper in Fig. 2 have 2.94 RD on the average

Distribution of Activities Among Spaces

Specialization of rooms is quite different from the European examples where living, sleeping, cooking, studying take place in sparate spaces. Kıray in her Western Anatolia study (Kıray, 1964) pointed out: "If number of rooms is increased, specialization of rooms may be possible such as separation of dining space from sleeping area, sleeping area from living place". "People who grew up in a dwelling where everybody slept all together in one room have no intention of sleeping in separate rooms from either hygienic or privacy habits viewpoints" (Kıray, 1964). Nearly the same facts had been observed by Kandiyoti (Kandiyoti, 1977) in Izmit", "As many as 65 % of house holds report that their family room is being used to sleep in, which is not surprizing given room densities. Apart from shortage of space, part of the tendency to concentrate a lot of different activities in this room lies in the fact that stoves are the only means of heating, and ....serve various functions such as warming water, or a meal or drying small laundry items aside from just heating". The pervasive characteristic of the houses visited is the multi-functional livingroom which serves as a focal point for all the families' activities (Kandiyoti, 1977).

These show the evidence of high density and crowding, but families do not complain about the multi functional use of living rooms and bedrooms, as it was in the old Turkish homes, in which rooms were used for many functions such as taking meals, sleeping and daily living (Kıray, 1977). In high density dwellings also specialization of rooms was not physically possible, consequently rooms reserved

Place	SUBJECT OF THE STUDY	Number of respondents	YEAR	TYPE OF HOUSİNG
İZMİT	Privacy Behavior in (DU) Survey	23	1976	Single storey Co-Operative Housing
İZMİT	Activity and Furniture Distribution Survey	40	1976	Co-Operative Block of Flats
İZMİT	Household Survey of all Neighbourhoods	1000	1974-75	. Rented . Self-built . Owner-occupied
İSTANBUL	Socio-Demographical Aspects of Urbanisation	500	1975	Self-built (Gecekondu) (Housing)
İSTANBUL (AKSARAY, LALELİ)	Survey on Bathing and Washing Activities	72	1976	2 Different Block of flats
İSTANBUL (OKMEYDANI)	Survey on Open Spaces	123	1976	. Blocks of flats . Detached H.
İSTANBUL (OSMANİYE)	Furniture Distribution Survey	14	1975	Housing site for Gecekondu Dwellers

Fig. 3. The list of survey and case studies from Turkey

for daily living and guest entertaining were used for sleeping purposes at night. Another consequence of over-crowding is seen in the number of people sleeping in each room, where it varies from 3-6 in the Osmaniye example and no bedroom is used for sleeping purposes of 2 people only (İTÜ, Mimarlık Fakültesi, 1975).

The other evidence for the lack of the sense of crowding among peasants and lower class families was in a dwelling with 3 rooms to keep one of the rooms only for the entertainment of guests even if the family had to sleep in the enclosed balcony or kitchen (İTÜ, Mimarlık Fakültesi, 1975) Fig. 2.

In the case study in subsidized co-operative flats (Özsoy, 1976) average RD is 1.05 in one group and 1.4 in another. Whenever the number of people increases, families prefered to divide the large guest room into two parts or began to use the second bedroom for many functions such as children's sleeping and studying as well as daily family activities, TV watching, sitting,chatting, etc.

Ostentatious display of the family's belongings in the guest room plays an important role in the use of rooms. This is a very important characteristic of Turkish people who like to appear richer, wealthier and more educated than they really are (Kıray 1977). This value is more important than mental and physical health and education of the children. There was no special room which was used for children in all the surveys and case studies.

Privacy

Privacy was important in the case of being seen by outsiders, or overlooking of other houses (Bayazit, et al., 1976). Only the privacy of young couples held importance in the temporarily-large or joint families. If there were more than one couple, they would not prefer to pass through a hall or a room while they had to reach the bathroom (Bayazıt, et al., 1976). Specially the distance between the bath-room and parents' bedroom become the focal point for the defence of privacy in crowding homes (Aytuğ, 1976). In these kinds of families the highest satisfaction correlation was found between the density and close or adjacent relations of bedroom and bathroom (Aytuğ, 1976).

Physical Results of Crowding and Density

Additional space need was recorded in all the case studies and surveys. To meet this need physical alterations were generally made on open spaces (balconies, terraces, porches , etc.) of flats and houses. Balconies were closed to enlarge the living room or kitchen when the AD exceeds 5-10 m2 per person. But this kind of alteration was also observed in lower densities as well (Celesun, 1976). The living rooms were generally enlarged through the balcony to gain space for an extra bed, and the kitchens were in general enlarged to gain more storage space and dining space (ITU, Mimarlık Fakültesi, 1975).

Whenever the population of the family exceeds 4 persons, need for a second WC or a WC separate from the bathroom played an important role in the satisfaction with DU (Aytuğ, 1976). Overlapping activities in WC-containing bathrooms caused several complaints and dissatisfaction. But there were no significant correlations between the family size and bathroom space need (Aytuğ, 1976). The same situation was also observed in Celesun's (1976) study.

CONCLUSIONS

Adaptation to the physical environment is seen as easier for Turkish families in a new dwelling environment, because the sense of crowding does not affect people in their crowded life too much. As in the past, the specialization of living and sleeping room is not certain yet. Relatives and kin may come in and stay with the family and sleep on the floor or bedstead not giving much consideration to the problems of crowding and personal privacy. The privacy of young couples may sometimes gain importance in the distribution of activities among the rooms. Lack of the sense of crowding and privacy generally affects the mental and physical health of the children negatively. Ostentatious display of the family is more important than the future of the children and the privacy.

CROWDING IN HOUSEHOLDS: CHICAGO, ILLINOIS VERSUS ISTANBUL, TURKEY

Mark Glazer

Pan American University

Edinburg, Texas, U.S.A.

ABSTRACT: The object of this paper is to do a cross-cultural analysis of household size and interaction, the goal being a better understanding of dense households as opposed to low density households, and the human interaction which takes place in these households. The households under scrutiny are in North Metropolitan Chicago and Istanbul, Turkey, and they are the households of the Jewish communities of these two areas. The results of the study show that the American community has smaller households and less human interaction than its Turkish counterpart, which seems to lead towards more personal stress.

1. Crowding: An Anthropological Perspective

It is, I believe, imperative to analyze the term "crowding"[1] from a social and anthropological perspective in addition to the social psychological, architectural, and other perspectives. Anthropology as a human science brings with it two major views which set it apart from other such sciences: 1) It is cross cultural, and this protects against the perspective that studying men in one culture helps us understand man in all societies, and 2) Anthropologists do field research in natural settings, thus collecting data based on daily human experience. It is only by such an approach that the pitfalls of research based on culture specific meaning of terms can be avoided. Does "crowding" mean as it does in American culture that one person is "crowding" another, or does it mean that we are dealing with cluttering as in the Turkish work "kalabalık"?

Research based on the American "crowding" is bound to yield very different results, especially when used in the isolation of laboratory conditions. The fact is that density or human crowds are not necessarily the setting for negative experiences,

----

(1) The term "crowding" is utilized to mean "density" in this paper, and will be used interchangeably with it.

even in the American experience. A great majority of Americans like to see their stadiums, arenas, and theaters full ("no-shows"are booed at professional football games). It must be added that a crowd is a must for a beer party. On the other hand, crowded buses, trains, and airplanes are not especially favored. It would seem then, that crowding in its actual day-to-day functions is not necessarily a stressful event; it may even be a pleasurable social and cultural experience in the American perspective.

It is, however, necessary to extend our view of crowding beyond the confines of American society towards a cross-cultural perspective. The singular and fundamental reason for this is that each culture structures its socio-psychological elements in a different way. What types of crowds are positive and what types of crowds are negative, when does an individual feel crowded and when does an individual feel in harmony with a large number of other people is a culturally grounded psychological relationship, a relationship which can only be deciphered in the context of a socio/cultural unit.

It must also be remembered that it is not only crowding which creates stress but that loneliness, its antithesis, creates its own share of human emotional problems. It must be added that the importance given to crowding is at least partially the result of North European and American normative values which regard crowding as being stressful (although this is not necessarily so, even in these cultures) while viewing privacy and personal autonomy as being positive life values. Although this may seem to be a paradox, such cultural contradictions are not unusual in any way or in any culture. On the other hand, the stress created by loneliness in the United States is often the result of the household organization of the nuclear family with the result that maintaining a marriage in the United States, where the nuclear family has so many functions to perform, has resulted in extremely high divorce rates.

I would like to make the point that neither crowding nor loneliness are absolute transcending values and that we cannot speak of one without the other.

What I propose to do in this paper is to look at social distance from the perspective of two types of household units: one in Chicago, where the nuclear family predominates with its values oriented toward privacy and personal space; the other in Istanbul, where the household is extended and where close family ties create a crowded household whose life style is based on mutual dependence rather than an individualistic world view.

2. Crowding in Households in Two Urban Communities

Culture patterns human life in multifaceted and multidimensional configurations. One of the most important of these configurations takes place in the realm of psycho-social distancing which deeply effects the quality of emotional relationships individuals have with each other. Research in urban settings in North Metropolitan Chicago in the USA and Istanbul in Turkey demonstrates different types of patterning for social distancing. While some of the differences can be attributed to cultural differences, others seem to be an attribute of different spacial relationships which result in different population densities in urban households.

Chicago and its northern suburbs are more spacious from the perspective of individual living units; in other words fewer people live together here in individual units than in Istanbul. Chicago is then less crowded per living unit than Istanbul. The main element involved in this is the difference in family structure; in Chicago nuclear families tend to live alone, without in-laws or other relatives being a part of the household. On the other hand, the Turkish household usually includes in-laws as well as unmarried siblings of the married couple. However, social distancing includes more than this, although cultural pressures exist for the persistence of these patterns in both cultures, the economic dimension is of importance here. In the USA the means to keep nuclear families in different households exists; this, however, is a situation which does not exist in Turkey. This indicates that the economic dimensions of the situation should be kept in mind as part of the socio-cultural determinants of household organization.

The communities under scrutiny here are both Jewish communities. One is the Jewish community of North Metropolitan Chicago, living in Rogers Park, Skokie, Evanston, and Glencoe; the other, the Sephardic community of Istanbul, Turkey. The reason for using two Jewish communities is that it was expected that the influence of the cultures and household sizes around them would be better understood. An expectation which turned out to be well founded. These two communities are as different as any two societies can be. They differ in their social organization, in the languages they speak, and in their value systems. As noted earlier, Jewish communities are generally affected by the patterns of the larger societies and cultures in which they live, e.g. Turkish Jews are centralized under a Grand Rabbinate; American Jews employ a highly individualized congregational system. Many of these differences stem from the fact that they are parts of two very different larger societies and cultures. The principal similarity between the groups is their common religious identity which on a second look is quite distinct.

Research into these communities was conducted on the basis of anthropological participant observation supplemented by a questionnaire in the Chicago community. The results of both types of research were the same. The questionnaire was, however, helpful in refining some of the classifications of social distancing. Five ranges of social distancing are utilized in this study.[2] These are organized in terms of social and psychological distance starting from the most distant and moving towards the most intimate[3] and meaningful to the individual. In this scheme, Range 1 is the most distant, and Range 4, the most intimate.

Range 1, Outer World, deals with people and cultural items which are most distant and therefore, least meaningful to the individual. This range basically includes people and items from other cultures which the individual is not familiar with or has little interest in. Range 2, Wider Society and Culture, includes aspect of one's own

----

(2) The ranges of social distancing utilized in this paper are an adoption of Francis L. K. Hsu's view on psychosocial intimacy. See Francis L.K.Hsu's, "Psychosocial Hemeostasis and Jen: Conceptual Tools for Advancing Psychological Anthropology," American Anthropologists, 1971, 73:23-24.

(3) The term 'intimate' has absolutely no sexual connotations in the context of this paper. It is used to indicate extremely close social and psychological relationships.

culture with which ego has no personal involvement, e.g., Edinburg, Texas for the typical New Yorker. Range 3, Operative Society and Culture, is the range in which the individual functions in daily life. It includes individuals with whom ego has daily contact in his or her occupation. Range 4, Intimate Society and Culture, includes human beings with whom ego has intimate social and psychological relationships. It is these individuals which make ego's life worthwhile or miserable and which provide him with a human context. This range is divided into two subranges: A, which is its outer context, and B, which includes the most important individuals to ego. The ranges in the two communities under study are stated below.

A. Ranges of Social Intimacy in the Jewish Community of North Metropolitan Chicago

Range 1, Outer World: This is the cognitive range which includes customs and artifacts belonging to other cultures and of which the individual has little or no knowledge. We must also include here categories in which the individual has no particular interest. This research reveals one of these as Europe, which a very large proportion of respondents found to be unimportant for them even though their parents were often European immigrants. In the interviews it was also obvious that the informants were not always at their knowledgable best when discussing other parts of the globe. I was, for example, once asked if the Shah of Turkey was in good health!

Range 2, Wider Society and Culture: This range consists of human beings, cultural roles and artifacts which are present in the larger society but which may not have any influence on ego. One category which I believe to fall into this niche is an individual's home town. Respondents generally feel no great loyalty to their home towns nor do they believe them to be of major importance in their lives.

There are four individual roles among those in the questionnaire which fall into this category: they are barmen, counselors, psychologists, and psychoanalysts. Very few respondents were willing to share good or bad news with barmen and even fewer are ready to share their secrets with them. This seems to be a role with which the respondents are familiar but with which they have nothing to do. Much as in the case of the barman, the individuals in the sample do not want to share either good or bad news with persons playing such roles as counselors, psychologists, or psychoanalysts. They can easily do without such relationships and needless to say, these relationships do not contribute to the happiness of the respondents in any way.

Range 3, Operative Society and Culture: The hallmark of this range is role relationships; it contains cultural roles and artifacts with which the individual deals without strong emotional attachment. Jewish culture and Jewish religion seem to fall into this range; though both respondents and informants find these important, most of them do not consider them essential.

As expected, business relationships fall within this category. Such relationships are found to be important and worth being loyal to, but not with the intensity given to some other relationships. Although respondents feel they would share good news with business associates they are reluctant to share bad news and extremely reluctant to share secrets with them. Other relationships which fall in this range are teachers, rabbis, neighbors, doctors, and cousins. Again, as in the case with business associates, good news is going to be shared with all of the above but neither bad news nor secrets are shared with them. Persons in this range are not confidants or intimates of ego.

Range 4, Intimate Society and Culture: This range contains human beings with whom ego is intimate as well as cultural items which are of great importance to the individual. Affect is of prime importance in this range. All indications in the research show that there are two ranges of intimacy involved here, one which is extremely close to the individual and a second one which is less close. An individual's father and 2nd best friend as well as, for example, citizenship, fall into the second category while his spouse and children fall in the first. There are no cultural artifacts in the first category which, as a rule, seems to be inhabited by only a very few human beings. This second range will be called intimate society and culture sub-range A. It is, however, sub-range B which is of prime importance to the individual.

Intimate Society and Culture Sub-Range A: This sub-range includes from most distant to closest, in-laws, brother, sister, second best friend, best friend, father, and mother. These are individuals with whom ego has a great deal of personal contact and to whom he feels quite close. Often, these individuals are ego's confidants, with best friend, father, and mother having the best chances of being so. Good news will be shared quite openly with all the individuals here and bad news will be shared with all except in-laws. When it comes to sharing secrets, many are ready to share them with all these individuals with the exception of, again, in-laws. It must, however, be noted that in the sharing of secrets best friend is second only to spouse; he seems to be more important as a confidant than either ego's children or his mother.

Intimate Society and Culture Sub-Range B: For ego, from age 25 on, there are only two individuals in this category; his children and his spouse, who, in all the different aspects of the questionnaires emerge as considerably closer to ego than any other individual or individuals.

Ego's children seem to rank behind his spouse in the following categories; most close to ego, not being able to do without, sharing good news and would miss the most. Ego's children are also behind best friend in the sharing of secrets and of good news. It must, however, be noted that ego (from age 25 on) generally believes his children contribute more to his personal happiness, more than anyone else.

Ego's spouse, on the other hand, rates as being the closest to the individual and also as being his closest confidant. Ego also finds that he cannot do without his spouse and the spouse would be the person whom he would miss most. Ego's spouse is also the second greatest contributor to his personal happiness. For some reason, unknown to the author, ego often prefers to share good news with his best friend rather than with his spouse!

Although ego's spouse and children seem to be the most intimate relationships he has, it must be noted that, from the point of view of confidant, ego's best friend is extremely important. Although he may not be loved as much as, for example, ego's parents, his importance to ego is considerable, especially in the realm of self-disclosure. As a result, it may be said that from the point of view of "love" a best friend's place is in intimate society and culture sub-layer 2, but as ego's confidant he belongs in intimate society and culture sub-layer 1. Best friend, in other words, bridges the two sub-layers.

B. Ranges of Social Intimacy in the Sephardic Community of Istanbul, Turkey

The Jews of Istanbul live in a small, tightly-knit community where the individual has continual contact with others. This community has an ideal of mutual help within the family and among friends. This is a community where privacy is secondary, if not almost totally non-existent, and where it is nearly impossible to be alone for a long time. The person here is judged not only on his own merit but also on the merits of his family and friends. He is a part of a highly interacting and mutually dependent small group.

Range 1, Outer World: This cognitive range includes peoples, customs and artifacts belonging to other cultures and of which the individual is not knowledgable. The Jews of Istanbul have an astounding awareness of foreign lands and people. Because the Jews of Turkey have been open to Western culture for a very long time, their awareness of Israel, European countries, and the United States is quite accurate. On the other hand, the awareness of this community of places such as Africa or the Far East is more than negligible.

Range 2, Wider Society and Culture: This range consists of human beings, cultural roles, and artifacts which are present in the larger society and culture. To the Turkish Jew, this means not only a large part of Turkey but also large parts of the city in which he lives. There are numerous Jews in Istanbul who do not know the poorer sections of the city at all. On the other hand, these people love the city of Istanbul dearly. Certain individuals and roles are also quite definitely ignored.

Range 3, Operative Society and Culture: This range contains human beings, cultural roles and artifacts with which the individual deals without strong emotional involvement. The situation of the Jew in Istanbul is such that relationships with rabbis, teachers, bureaucrats, doctors would fall within such a category. The most important role relationship Istanbul Jewish males have is with their customers. In this community of businessmen, most seller-buyer relationships are of the ordinary sort, those with steady customers taking precedence over others. The more important relationships here are with steady customers who do a good deal of business with an individual businessman. Often a businessman socializes with and entertains his better customers and though this tends to enrich their relationship, it does not add emotional depth to it. The feelings of a businessman towards his clients are of a strictly business nature and part of role relationships.

There is a certain ambiguity in a man's relationships with his business partners. If business partners are relatives or close friends, their relationships are going to be more than strictly business. If the partners are all Jewish, it will be a semi-close relationship. If, however, a partnership includes a Jew and e.g., a Greek, theirs will, in most cases, be a role relationship.

Women have by far fewer role relationships than men. These would include the personnel of the stores in which they shop, servants (in wealthy families), and the spouses of their husbands' business associates.

Range 4, Intimate Society and Culture: This range contains human beings with whom ego is intimate as well as cultural items which are of great importance to him. Affect is of prime importance in this range. I believe that two ranges of intimacy are operant among the Jews of Istanbul as for the Jews of Northern Metropolitan Chicago. The first range of intimacy is inhabited by an individual's best friend, his parents, his wife, his children and his siblings. The first range of intimacy is inhabited by an individual's closest friends, his first and sometimes second cousins and often by his in-laws. The second range is inhabited by an individual's best friend, his parents, his wife, his children, and his siblings. It is this range which is of great importance to the individual.

Intimate Society and Culture Sub-Range A: Every person in the community has a circle of close friends. He can call upon these for help and, if they are knowledgable, for problems. An individual will have approximately five close friends whom he can call upon for both problems and good times. A person will spend long hours in their company during the week. These friends represent relationships which are close to the individual and in which he is emotionally involved. First, second, and sometimes third cousins are part of this range of psychosocial intimacy. A person's circle of close relationships will always include a few cousins of different distances the parents of these cousins, uncles and aunts, are also part of this range. An individual will also visit his in-laws a few times a week and these visits will be returned as often. The Turkish Jew has a feeling of responsibility towards all these relationships; the feeling is, of course, mutual. The feeling of responsibility that the members of any self-defined group have for each other is one of the most important aspects of interpersonal relationships among the Jews of Istanbul.

Intimate Society and Culture Sub-Range B: There is a Turkish message which says, "Come and help me even if both your hands are hurt." In other words, "I need help and I am relying on you to help me whatever else you may be doing." The plea is very rarely, if ever, left unanswered. This is the kind of relationship which reigns among individuals of this range. It is a range full of mutual help and understanding, it is one where individuals care for and feel responsible for each other to a high degree. The inhabitants of this range are always ready to listen, help, and act on behalf of the others. For example, if an accident happens to one of the members of this circle, the others will flock to his side to comfort him and advise him on a variety of subjects. One will advise on the best doctors available, another will try to see what can be done to help his business out, and a third will try to get him what he would like to have. These same individuals will be right there at every important event in a person's life and try to help. In a situation like this an individual does not need a psychoanalyst or a social worker; rich or poor, every man is so involved with others that his own problems become partially submerged.

The individual will usually include among his intimates his parents, spouse, children(especially after they are 18), brothers and sisters, and finally his best friend. The relationships between these individuals is as described above.

C. Comparisons Between the Two Communities

While the Turkish Jew includes a number of close friends, cousins of different closeness and in-laws in his sub-range A, the American includes only in-laws, brother, sister, second best friend, father, and mother. In the former case, this sub-layer includes people outside the individual's primary group and outsiders, while in the American case it includes members of the primary group only.

This contrast would seem to suggest that the distinction between relatives and non-relatives is less clear among Turkish Jews than among American Jews.

The above interpretation is commensurate with the next finding. Sub-layer 3(b), namely the most intimate society and culture, is in the case of the Jews of North Metropolitan Chicago very small, including only three relationships; one's spouse, one's children and one's best friend. There are no other members of one's primary group here. But the same sub-range 4(b) among the Jews of Istanbul includes all of the above, plus the individual's initial natal group, namely father, mother, and siblings. The Turkish group is obviously more inclusive than its American counterpart. It, therefore, enjoys a more massive source of close psychosocial intimacy than the latter. Every individual, Turkish or American, begins life with parents and often siblings. In the early years, he naturally resorts to these individuals as sources for his social dependency. For the American Jew, growing up means some sort of psychic break with them.Forthe Turkish Jew, the same process means incorporating more and other individuals without giving up those who nurtured him.

This difference is psychosocial intimacy between the two communities is basic to many behavioral dissimilarities between them. The Chicago Jew lives in a culture which encourages self-reliance, so he tries to become self-reliant; the Istanbul Jew lives in a culture that encourages mutuality and cooperation, so he considers the network of mutual obligations important. The Turkish Jew, therefore, has more automatic, permanent, and continuous sources of psychosocial intimacy than his American counterpart. As a result, the Turkish Jew's achievement motivation is oriented toward the group rather than the individual. The Chicago Jew draws his sources of greatest psychosocial intimacy from his spouse and children for this is what his society demands of him. In so doing, he pushes parents and siblings farther away. The meagerness of sources of intimacy compels the Chicago Jew to seek compensation by becoming active in a variety of organizations. Although there are non-kinship Jewish organizations in Turkey, their importance to the individual does not seem to be great. Changes which effect the inner range 4B are more difficult to bring about than those which effect only the outer range. Therefore, changes in sub-range 4(a) are more difficult to implement than those in sub-range 4(b). From this, I would deduce that it would be very difficult to change the patterns of the husband-wife dyad in the American community, just as it would be difficult to alter the network of mutual obligations in the Turkish community. In the Turkish community, the individual is locked into his social responsibilities (Hypothesis F(a)), thus restricting his extension towards other ranges.

The differences in the makeup of their respective and social distancing fits in with the larger picture of differences between Turkish and American societies. The American Jew's pattern of social distance quite closely resembles that of Western societies in the hypothesis outlined above, and the Istanbul Jew's pattern resembles that of the larger Turkish society.

In summary, in the American situation in Chicago, the individual has far fewer opportunities for human interaction. This state of loneliness becomes especially acute in later life. The consequence of this loneliness is increased in pressure on the mental health of the individual. In Turkey, the situation is reversed; the individual has larger or more crowded networks of social interaction in both youth and old age. As the result of this situation, there seems to be fewer pressures on the mental health of the individual in comparison to the individual in the USA.

3. Tentative Conclusions

The questions of density or crowding are numerous. Many of these are problems which have to be translated into solutions for urban planners or architects, while others are problems of human interaction under stress. Both are social and cultural problems. The psychological or mental health problems are consequences of these social realities. It is, therefore, essential in problems of density to treat the social and cultural as independent variables and the psychological as a dependent variable.

It is essential that before we plan, build, or treat, we understand the social configurations and their semiotics in the cultural context. We further have to understand that before we act upon human beings with different cultural, subcultural, and class properties, it is essential to understand each group before we plan for it.

What are some suggestions that would be possible as the result of the study above? If we are to plan for the American group, given its preferences, it would be possible to build for nuclear families which are relatively isolated in their social interaction. This, however, should be done with care; the fact that old people are very alone in American society should be taken into account. The interaction of old people with children should also be taken into account. In this situation the cultural preference for people to be alone should not be confused with the existen al situation of the need for human interaction and the balance between the two would be of major importance.

In Turkey, on the other hand, a high interaction and higher density situation suggest that planning should revolve around larger units and greater emphasis on walking distances between relatives, the important point being that it would be essential to maintain human interaction here, as the loss of it would be potentially negative.

Much as any other problem of the human reality, crowding is a relative reality which has to be understood and acted upon in social and cultural terms; to lose sight of this is an invitation of misunderstanding crowding, as it indicates that we have lost sight of the human condition.

CLUSTER LAYOUTS IN HIGH DENSITY URBAN SITUATIONS

Reinhart Goethert
Research Associate, Urban Settlement Design Program
MIT, Cambridge, Mass. USA

Mark Butler, Nedret Butler
Architects, housing consultants
Istanbul, Turkey

Bülent Tokman
Architect, housing consultant
Ankara, Turkey

ABSTRACT: The paper is focused on physical planning aspects in the context of developing countries, the low income majority and rapidly urbanizing areas. Specific aspects of developing country problems are explored, particularly squatter areas and the need for government intervention for provision of utilities in the high density areas. Various physical planning indicators are identified and trade-offs are discussed in the context of Boston, urban areas of Latin America and Nairobi, Kenya. Two cases studies in Turkey are compared clarifying the physical determinants of clusters. Brief mention is made of housing policies and the need to encourage higher densities to avoid costly sprawl.

A cluster design which groups individually owned lots around a shared common court is proposed as answering many of the problems. Cost, social, cultural and administrative advantages are outlined for the clusters. A model development is presented illustrating various densities in a flexible development.

CONTEXT: LOW INCOME PEOPLE, DEVELOPING COUNTRIES, RAPID URBANIZATION

The focus of this paper is on physical planning aspects in the context of developing countries, the low income majority, and rapidly urbanizing areas. Although the comments are particularly appropriate to the many urban areas suffering from the effects of the rapid population growth typical in the developing areas, many of the principles may directly apply to other areas as well.

Since the 1930's and 1940's Latin America, and later in other countries, developing countries have been faced with an exponential population growth with one of its results rapid urbanization focused on the central cities. The immigrants are poor, relatively unskilled, and culturally and socially not adapted to the urban way of life. The governments had and still do not have the technical, financial and administrative resources to cope with this sudden influx and as a result tenement areas in the center of the cities have become overcrowded with these immigrants and large areas of land on the periphery were overrun with illegal or "squatter" developments. Studies indicate that the high density center city tenements –1 room family units grouped around a shared kitchen, bathroom facilities– are the reception areas for the migrants who later move into illegal squatter areas on the fringe of the urban areas.

In the squatter settlements, land is occupied, subdivided and very soon covered by all kinds of shelters and dwellings built by the occupants solely by their own efforts and resources. If the provision of shelter is a relatively simple matter, the provision of utilities and services for a community is not; it demands more than individual effort. It demands a collective effort both from the community and from the government to plan and mobilize political, economical and technical resources.

In squatter or illegal settlements, utilities and services are neither provided nor anticipated initially. Very soon, this becomes a critical matter and invariably, the development and improvement of the settlement depends entirely on the community's ability to procure water, sewage, electricity and other services from the city. More often than not, unplanned sprawl and inefficient or inadequate use of the land makes the provision of utilities and services both difficult and uneconomical. The resulting alternatives are either to pay a high price for the development and improvement, or not to pay and let the community stagnate.

Most governments have adopted an attitude of "benign neglect" towards squatting and illegal settlements. Under political pressure only sporadic actions are taken. No consideration is given to the urgent issues: squatting should be anticipated and controlled through land utilization policies. The existing housing stock should incorporate slums and squatter settlements realistically appraised and upgraded. Popular participation (initiative, responsibility, "sweat equity") should be fully utilized. The lowest income sectors should be reached.

PHYSICAL PLANNING INDICATORS

Overcrowding and excessively high densities are often used in describing these squatter settlements, particilarly for the reception areas in the center of the cities.

Crowding, as defined by Webster, implies "a massing together and often suggests a loss of individuality of the ...member." Indicators of crowding are perhaps not too difficult to define: "net density," "people per unit," and "area per person," are three of direct interest to architects and planners. Defining the acceptable limits more difficult, particularly when social/cultural factors are taken into consideration. For example, it is clear from observation that in Asian countries there exists a higher tolerance to "crowding" (high density-higher ratio of people per unit) than in Latin

America. Generalizing (and with all its dangers) it could be said that Asia, Africa, and Latin America, in that order, tend to tolerate progressively more crowding.

Acceptable/desirable gross density ranges (including areas for streets, in an urban residential setting, up to 4 story walk-ups) of 300-600 people per hectare appear appropriate. Defining people per unit is more difficult, dependent on family structure (nuclear vs. extended). Area per person in a dwelling is generally considered in the range of 8 to $12m^2$ per person as desirable (approximately 80 to 120 ft^2 per person).

TABLE OF TYPICAL DENSITIES

Number of floors	Gross Density p/Ha	Net Density p/Ha
1	147	225
2	294	450
4	588	900

LAND USE: Circulation area 20%; semi-public area 15% (open 12%, buildings 3%); Private area 65% (open 43%, covered 22%);

DWELLINGS: Land coverage 1/3 of private land; dwelling area per person $12m^2$; Shops and misc. area per person $3m^2$; open area per person $10m^2$. Types: tenements, apartments, houses, walk-ups.

Physical factors affect densities in different degrees. It can be assumed that higher densities are associated with specific values: higher number of floors; higher land coverage; smaller lot and dwelling unit areas; smaller dwelling area per person; rooms and shanties as opposed to houses. The table following shows that in reality some values can offset others, as in the case in Latin America and Nairobi, where people are crowded into one-story shelters. In Boston, the high and medium densities are the result of apartment dwelling units and also result from the number of floors: 3-5 in walk-ups, 7 in the only case of high rise. High densities are reached despite low land coverage and large dwelling area per person. The cases in Latin America are of one story houses, but still densities are high and medium because of high land coverage, small lots, small dwelling units and small dwelling area per person. In the cases of Nairobi, Kenya, the extreme living conditions in some of the cases have people packed in one-story shanties or rooms, covering the land and leaving narrow passages that are the only spaces for circulation and outdoor activities.

The high densities in general are viewed as negative, but determining what specifically is inappropriately high is difficult. The lower is the density, the larger is the land area required for a given population, which results in higher costs per capita in land and infrastructure. At the opposite extreme, very high densities not only may put an excessive load on the services, but what is more serious, it could create negative and destructive social condition. It is always possible to determine a range of density limits compatible with adequate services and infrastructure. But very little can be done to determine similar indices to anticipate or forecast social and behavioral implications of population densities in a given physical environment.

LOCALITY	DWELLING TYPE AND UNIT	DWELLING DEVELOPER	DWELLING FLOORS	LOT AREA	LOT COVERAGE	DWELLING UNIT AREA	PERSONS/ DWELLING	DWELLING UNIT AREA/ PERSON	GROSS DENSITY	NET DENSITY
			No.	m^2	%	m^2	No.	m^2	No./Ha	No./Ha
BOSTON AREA, U.S.A.										
Columbia Point	High Rise, Apartment	Public	7	NA	-	78	4	20.0	747	1449
North End	Walk-up, Apartment, Row	Private	4-5	168	65	47	2	24.0	708	1040
Charlestown	Walk-up, Apartment	Public	3-4	NA	-	70	4	18.0	395	570
East Boston	Walk-up, Apartment	Public	3-4	NA	-	65	4	16.0	414	510
South End	Walk-Up, Apartment, Row	Private	4-5	144	50	58	3	19.0	287	480
Washington Park	Walk-up, Apartment, Detached	Private	3-4	243	50	119	4	30.0	232	317
Cambridge Port	Walk-up, Apartment, Detached	Private	3-4	319	38	114	3	38.0	112	148
Lincoln	Detached, House	Private	1-2	17500	1	195	4	49.0	2	2
LATIN AMERICA										
El Agostino (flat), Lima, Peru	Row House	Popular	1	36	100	36	6	6.0	525	664
Villa Socorro, Medellin, Colombia	Grouped House	Private	1	96	45	43	6	7.0	279	574
El Agostino (hill), Lima, Peru	Row House	Popular	1	62	69	43	6	7.0	403	552
Cuevas, Lima, Peru	Row House Shop	Popular	1	158	89	140	6	14.0	275	410
El Ermitaño, Lima, Peru	Row House	Popular	1	160	69	110	10	11.0	208	341
Mendocita, Lima, Peru	Row House	Popular	1	46	100	46	8	6.0	166	238
El Gallo, Ciudad Guayana, Venezuela	Detached House	Public	1	300	22	65	6	11.0	124	186
Mariano Melgar, Arequipa, Peru	Row House Shop	Private	1	240	92	221	10	18.0	87	126
NAIROBI, KENYA										
Mathare Valley	Grouped, Rooms, Tenements	Private	1	-	100	12	4	3.0	1600	3333
Kariobangi	Grouped, Rooms, Site & Services	Public	1	-	100	14	4	3.5	532	2660
Karura Village	Grouped, Shanties, Temporary	Popular	1	-	100	41	11	3.7	720	2400
Kirinyaga Village	Grouped, Shanties, Temporary	Popular	1	-	100	13	2	6.5	450	2250
River Road	Row, Rooms, Tenement	Private	2-3	255	87	10	3	3.3	768	1280
Uhuru Phase 4	Row, Houses, Subsidized	Public	2	-	-	71	6	12.0	312	780
Kawangware	Grouped, Rooms, Tenement	Private	1	-	-	14	4	3.5	552	699
Eastleigh	Row, Rooms, Tenement	Private	1	300	41	10	15	0.7	480	666
Pumwani	Walk-up, Apartments, Subsidized	Public	3	-	-	52	6	9.0	313	549
Bahati	Row, Rooms, Subsidized	Public	1	-	-	17	11	1.5	320	405
Kariobangi South	Row, Houses, Subsidized	Public	1	133	47	63	4	16.0	270	380
Woodley-Kibera	Row, Houses, Subsidized	Public	2	222	47	166	7	24.0	217	310
Quarry Road	Detached, Houses, Subsidized	Public	1	150	32	48	6	8.0	114	173
Quarry Road	Semidetached, Houses, Subsidized	Public	1	250	40	100	9	11.0	72	83
Woodley I	Detached, Houses, Subsidized	Public	1	1800	6	100	7	14.0	35	41
Dagoretti	Grouped, Rooms, Traditional	Private	1	-	-	28	19	1.5	36	41

TABLE OF PHYSICAL CHARACTERISTICS AND DENSITIES IN BOSTON, U.S.A.: LATIN AMERICA: AND NAIROBI, KENYA. (Taken from "Urbanization Primer, for design of site and services projects," Horacio Caminos, Reinhard Goethert; MIT PRESS, to be published in Spring 1978)

LAND USE

Streets/Walkways	89%
Playgrounds	2%
Cluster Courts	—
Dwellings/Lots	9%

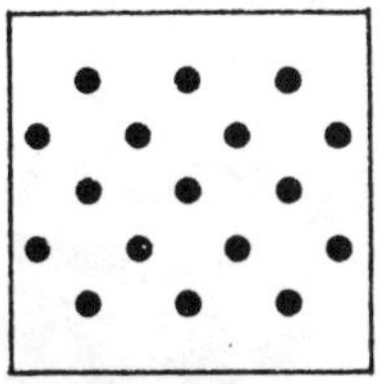

DENSITY

Persons/Hectare 333

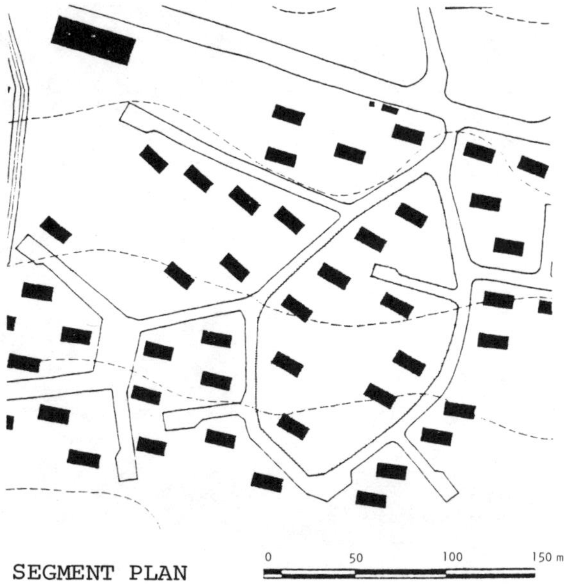

SEGMENT PLAN

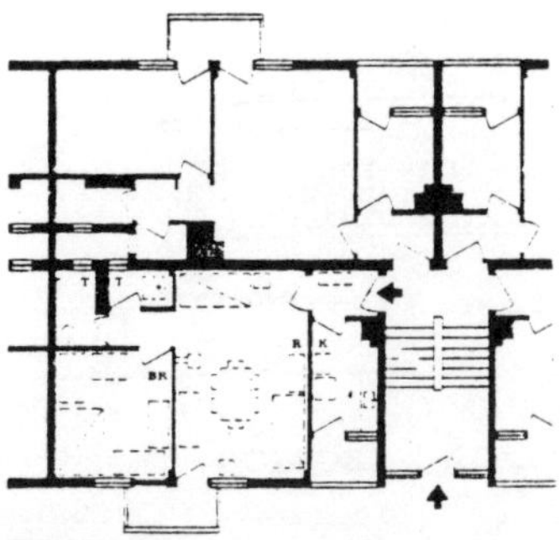

DWELLING PLAN

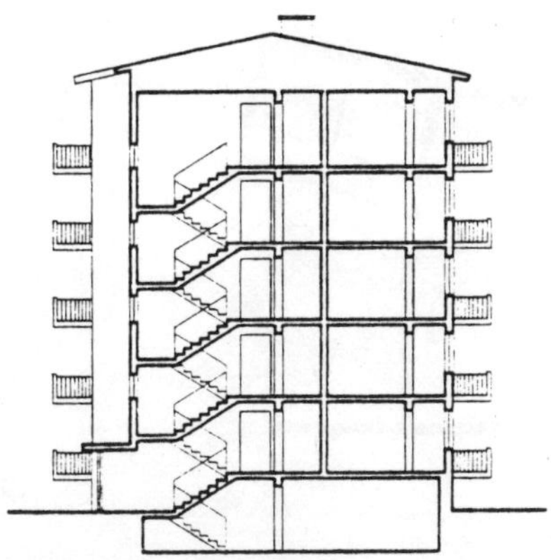

DWELLING SECTION

OSMANIYE: Public, low income, apartments
Istanbul, Turkey
Density - 333 people/Hectare
Private area - 9%
Dwelling area - 45 m²
Height of dwellings - 7 stories

The high density is achieved at the expense of small apartment units and concentration of people in seven story units in only 9% of the land. The remainder of the land is a costly problem of maintenance and responsibility of the government. The layout allows concentrating of utilities in servicing the buildings, which allows a simpler network and fewer service connections.

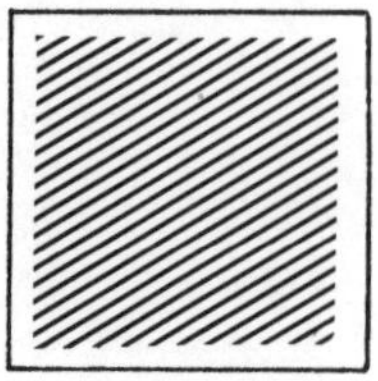

LAND USE

Streets/Walkways	21%
Playgrounds	—
Cluster Courts	—
Dwellings/Lots	79%

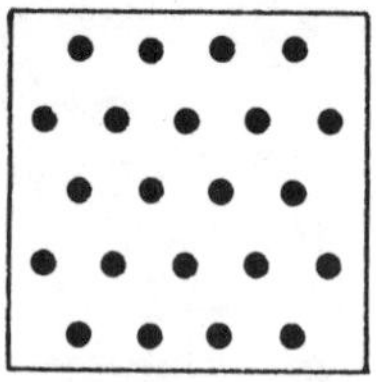

DENSITY

Persons/Hectare 458

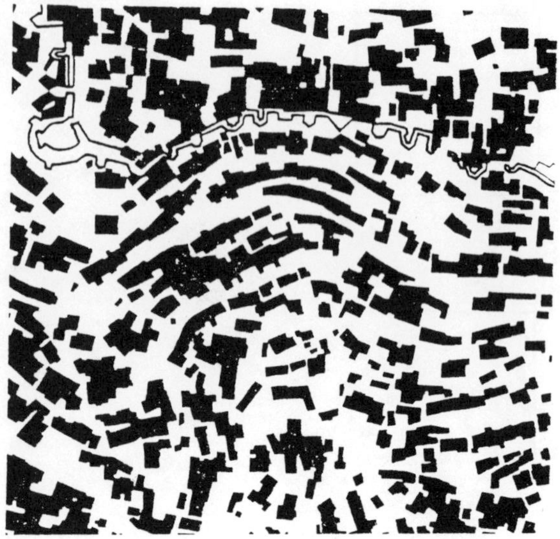

SEGMENT PLAN 0 50 100 150 m

DWELLING PLAN

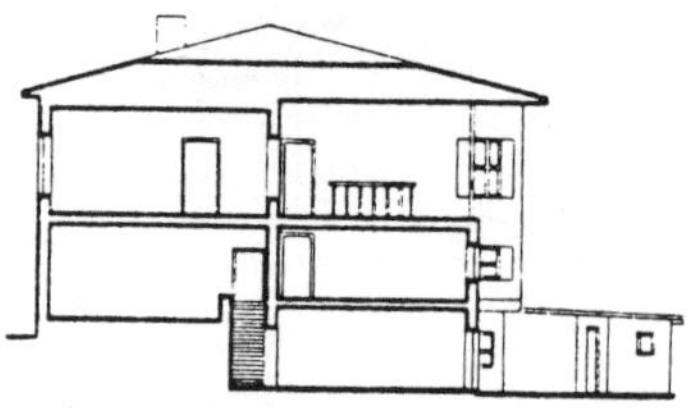

DWELLING SECTION

KALE: popular, low income, traditional urban houses
Ankara, Turkey
Density - 458 people/Hectare
Private area - 79 %
Dwelling area - 160 m² (2 stories)
Height of dwellings - 2-3 stories)

The even higher density is better distributed over the area. Notice that the individual dwelling unit area is 3 times greater than the case of Osmaniye. The land is essentially distributed among the private lot owners allowing a more manageable distribution of population as well as removing land from the responsibility/maintenance of the government. Expensive utility lines are required in all of the streets. Service connections, one per lot, are impossible to concentrate to save costs.

However, there are many positive factors to higher densities: a) cost advantages: lower per unit costs in the provisions of services: water supply, sewer, electiricity, paved streets, street lighting, storm dainage, etc. The utilities are shared among a greater number of people. In addition, sewage perticularly requires a minimum density for proper functioning with less maintenance problems. b) commercial advantages: a greater variety of stores can thrive in a higher density area. There would be better chance of success in maintaining a store. c) social advantages: (although no guarantee!) there exists the potential for less alienation, loneliness; there can be mutual support, encouragement.

HOUSING POLICIES

Many approaches taken by public housing agencies have only aggravated the negative aspects of high densities. Everyone is quite familiar with the huge complexes of 4-5 story walk-up apartments intended for the low income which invariably results in disaster. Although density perhaps is not the prime villain, it at least provides the catalyst for failure. As seen from the examples in Turkey, the high density in Osmaniye is achieved through high rise structures, and small dwelling unit area. Only 9% of the land is under the direct control/responsibility of the private sector. Kale, on the other hand, has even higher densities (450 p/Ha vs. 333 p/Ha) but the land is better distributed, the dwelling unit area is three times larger ($160m^2$ vs. $45m^2$) and the average private lot area is $254m^2$. No private lots exist as such in the public housing project of Osmaniye.

High densities must still be encouraged, particularly for lower income groups in developing countries. It is more appropriate to achieve these higher densities through dwellings of up to 4 stories with clear definition of responsibility in the land. The high density policies avoid costly sprawl developments as found in both developing as well as in developed countries. The current codes require unreasonably large lots wich have a costly effect: many urban areas are forced to unnecessarily extend beyond existing transportation, water, sewer, and electrical networks. The developments force costly expansion to the relatively sparsely populated areas. The lower the density, the greater is the amount of land needed for a given population and the greater the cost per capita for land, utilities, and services.

In developing countries, the provision of utilities and services and the control of land subdivision are the most important tasks of the hard pressed governments. Efficient and economical provision of infrastructure is imperative. In addition, for the vast number of squatters areas, secure tenure is prerequisite to stability and consolidation of the areas.

CLUSTERS

Designs grouping individually owned lots around a jointly owned court have proven successful as a means of coping with many of the problems. Small lots with high densities may be used minimizing the adverse effects of "regular gridiron layouts." These designs are often called "clusters" but should not be confused with the developments now being advocated in the United States. Clusters as meant here

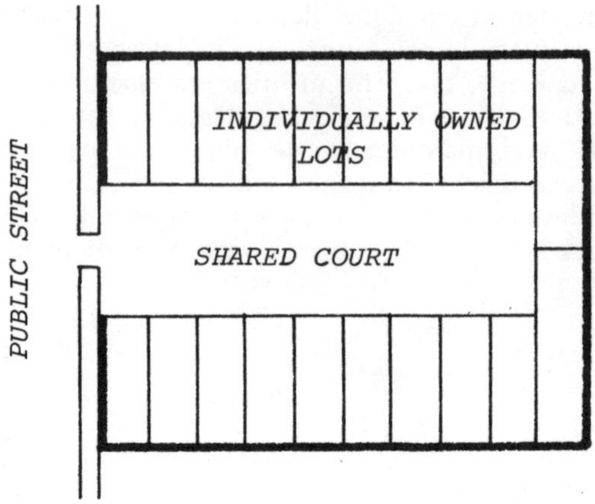

TYPICAL CLUSTER LAYOUT

have specific requirements: primarily, the physical arrangement and land ownership is such that the group of families has complete Use, Responsibility, and Control. The physical design appropriate to this is a dead-end solution, with each lot with individual title and the central area held in condominium. Another name of the cluster could be "horizontal condominium".

There are many areas of advantages: a) Cost: it allows aggregation of people/ units for more efficient utility services, it allows concentrating the main utility networks along primary streets to facilitate maintenance, installation, and to minimize the total length of networks. It provides the potential for lowering the costs in developing countries by offering staged or progressive services levels. For example, provide a single utility unit (water, sewer, electricity, laundry, shower, etc.) for each cluster, with the option of future individual lot connections as the users can afford it. In addition, it takes a large percentage of land and street areas out of the government's responsibility and assigns it to smaller groups of people. b) Social: it defines smaller community groups, increasing the prospect of the sense of "belonging" and identity, it encourages cooperation of the families which can lead to other endeavors (commercial ventures, shared dwelling construction, etc.) and encourages responsibility. c) Cultural: the arrangement is related to many traditional communal groupings dating from early aborigines in the Americas as well as in Indonesia, China, and India. Specific planned dwelling groups have been documented as early as c. 2670 B.C. in Egypt. d) Administrative: grouping allows simplified administrative function in title allocation, tax collecting, utility meter reading and billing, centralized refuse collection, etc. e) Flexibility: future change can more readily be accommodated in the larger condominium structure. As often common in urban areas, changes for commercial uses which require larger properties are difficult because of the many diverse property sizes and owners. Various lot sizes and dwelling types may readily be used. More "square" lots are even feasible without penalty of added infrastructure costs.

On a larger neighborhood/community scale, the cluster also has many advantages. It minimizes through traffic in the residential areas. It focuses bus routes and defines more clearly streets with shopping potential, and can help in defining/ identifying neighborhood centers and the focus of activities.

COMPARISON OF GRIDIRON AND CLUSTER LAYOUTS

GRIDIRON LAYOUT

Lot area	100%
Private area (lots)	52%
Schools, community facilities	16%
Public area(streets)	31%
Street length	4253m
Unit length	266m/Ha
Net density	600p/Ha

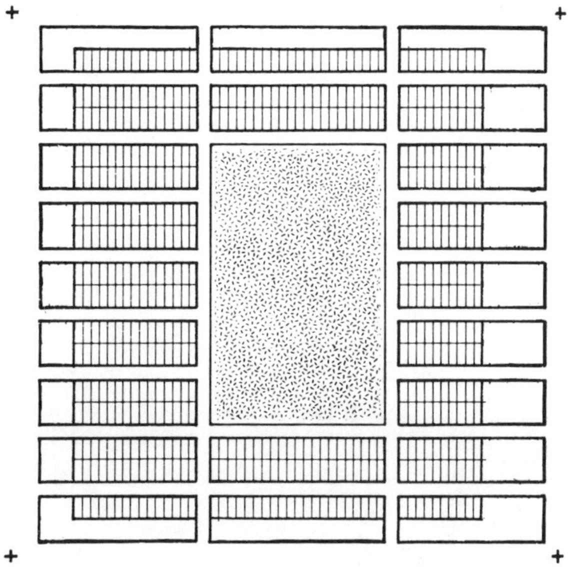

CLUSTER LAYOUT

Lot area	100 m^2
Private area(lots)	62%
Schools, community facilities	16%
Public area(streets)	21%
Street length	2400m
Unit length	150m/Ha
Net density	600p/Ha

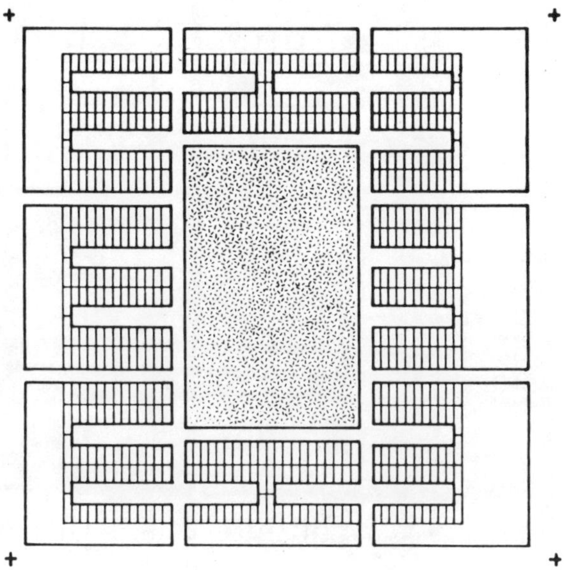

From a physical standpoint (which can be translated into costs), there is an obvious advantage to the cluster layout: private land is 20% more, public land is 32% less, total length and unit length of circulation is 44% less.

From a social standpoint there is a potential for better social aspects despite the high net densities.

CLUSTER MODEL ILLUSTRATING VARIOUS DENSITIES. The model of a development composed of clusters illustrates various densities of population according to various lot sizes, and various dwelling types (walk-ups, row, detached and site and services). The layout allows flexibility and adjusts to a variety of densities.

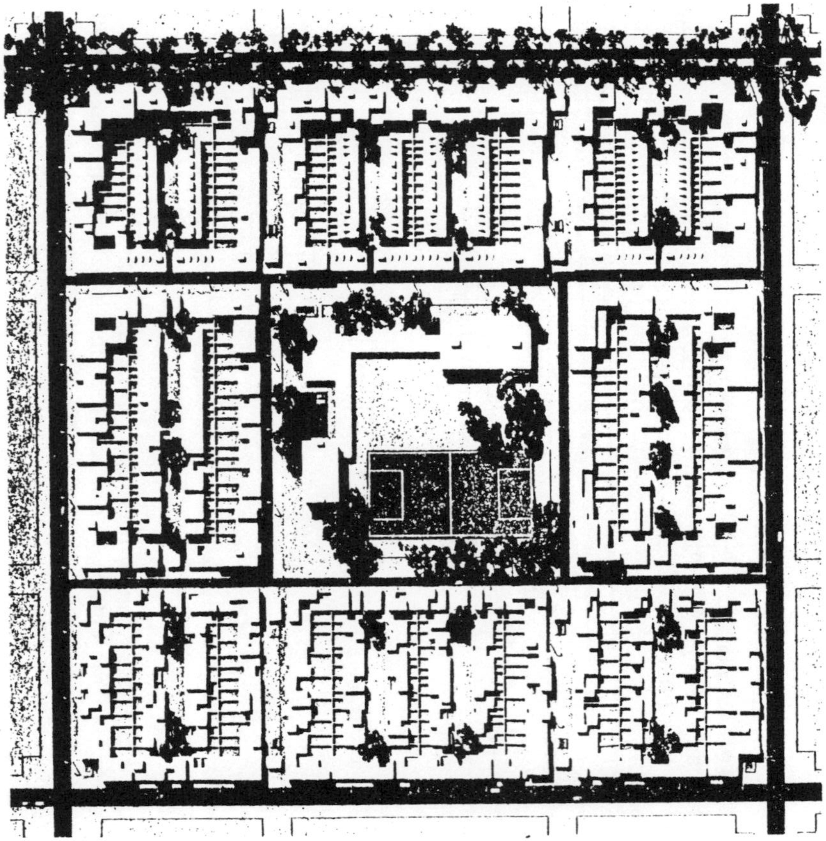

800 PEOPLE/HECTARE
$12 m^2$/person
Average number of floors: 2.4
Dwelling types: walk-ups, row

600 PEOPLE/HECTARE
$12 m^2$/person
Average number of floors: 1.8
Dwelling types: walk-ups, row

400 PEOPLE/HECTARE
$12 m^2$/person
Average number of floors: 1.2
Dwelling types: row, site and services

In the case of small lots as found in low income urban areas of developing areas (70m^2 to 120m^2) coverage approaches 100% with little or no light and ventilation. Despite codes, it is difficult and perhaps impossible to control the coverage in the lots because of administration, personnel, and financial limitations. Gross densities in some cases have reached 1600 people/Hectare. The shared court areas in cluster layouts provides an open space which assures at least a minimum degree of open space requirements. Encroachments on this area would probably be low because of 'social pressure' from the joint owners.

The size of the cluster in terms of number of lots and area has not been precisely defined. From a technical standpoint, a maximum depth of 100 m would be dictated by manhole spacing for eventual sewage installation. Also, fire codes set limits for maximum length of dead end streets. A 100 m depth could accommodate from 20 (10 m lot frontages) to 24 (6 m lot frantages) families/lots. A smaller number of families would be potentially more successful. Densities could be quite high, reaching 500-600 people/hectare with only an average of 1.8 floors as seen in the cluster model.

ACTUAL EXPERIENCE WITH CLUSTER LAYOUTS

Although the concept is old and various physical examples can be seen down through the ages, little large scale implementation of the cluster has been carried out. However, recent World Bank funded projects in Central and South America have incorporated the cluster design into several layouts. The projects are of the "site and services" type which requires the user to construct his own dwelling when given a piece of land. The projects are too recent to draw any conclusions as to the success of the clusters. The projects are intended to be monitored and data should be available within a few more years.

CONCLUSION

There are many advantages to higher densities: more efficient use of utilities and services, lower cost to government and users, and various social advantages. These become particularly advantageous in the context of developing countries of limited resources and an overwhelming demand for services. Clusters perhaps are a physical planning device which answer many of the problems. Particularly where there are smaller lots and high densities as in the majority of low income areas, the clusters provide open areas for light and ventilation and allow smaller social groupings for providing a sense of "belonging." In addition, the government is saved from costly over-extension of utility and services lines and maximum land is in private hands and out of the responsibility of the hard-pressed government.

REFERENCES

Adams, B.N. Kinship in an urban setting. Chicago: Markham, 1968.

Agency for International Development, U.S. Dept. of State. FY 1969 AID objectives. (Washington, D.C., Author, 1969).

Aiello, J.R., and Cooper, R.E. The use of personal space as a function of social effect. Proceedings of the 80th Annual Convention of the American Psychological Association, 1972, 7, 207-208.

Aiello, J.R., Epstein, Y.M., and Karlin, R.A. The effects of crowding on electrodermal activity. Sociological Symposium, 1975, 14, 43-57.

Albert, S., and Dabbs, J.M., Jr. Physical distance and persuasion. Journal of Personality and Social Psychology, 1970, 15, 265-270.

Altman, I. The environment and social behavior: Privacy, personal space, territory, crowding. Monterey,Cal.: Brooks/Cole, 1975.

Altman, J. Privacy: A conceptual analysis. In D.H. Carson (Ed.), EDRA 5, 1974.

Appleyard, D., Gerson, S., and Lintell, M. Liveable urban streets: Managing auto traffic in neighborhoods. Washington, D.C.: Government Printing Office, 1976.

Argyle, M., and Dean, J. Eye contact, distance and affiliation. Sociometry, 1965, 28, 289-304.

Aytuğ, A. Orta nitelikli şehir konutlarında yıkanma ve yıkama mekanları konusunda bir araştırma. Unpublished M.S. Thesis, Istanbul Teknik Üniversitesi, Mimarlık Fakültesi, 1976.

Bagley C. The built environment as an influence on personality and social behavior. In D.Canter (Ed.), Psychology and built environment. The Architectural Press, 1974.

Baker, J., and Young, M. The Hornsey plan. 3rd/4th eds., 1971/1973.

Baldassare, M., and Fischer, C.S. The relevance of crowding experiments to urban studies. In D. Stokols (Ed.), Perspectives on environment and behavior. New York: Plenum, 1977.

Bandura, A. Social learning theory. New York: McCaleb-Seiler, 1971.

Baril, L. L'image urbaine. Recherches Sociographiques, 1972, 12, 227-237.

Barker, R.G. Ecological Psychology. Stanford: Stanford University Press, 1968.

Barker, R.G. Ecology and motivation. In M.R. Jones (Ed.), Nebraska Symposium on Motivation, 1960. Lincoln, Neb.: University of Nebraska Press, 1960.

Barker, R.G. Explorations in ecological psychology. The American Psychologist, 1965, 20, 1-14.

Barker, R. and Gump, P. Big school, small school: High school size and student behavior. Stanford: Stanford University Press, 1964.

Baron, R., Byrne, D., and Griffitt, W. Social psychology: Understanding human interaction. Boston: Allyn and Bacon, 1974.

Baron, R.M., Mandel, D.R., and Adams, C.A. Effects of social density on university residential environments. Journal of Personality and Social Psychology, 1976, 34 (3), 434-446.

Baron, R., and Rodin, S. Personal control and crowding stress: Processes mediating the impact of spatial and social density. In B. Baum and Y. Epstein, Human responses to crowding. Hillsdale, N.J.: Lawrence Earlbaum, 1977.

Bass, M.H., and Weinstein, M.S. Early development of interpersonal distance in children. Canadian Journal of Behavior Science, 1971, 3, 368-376.

Baum, A., and Greenberg, C. Waiting for a crowd: Behavioral and perceptual effects of anticipating crowding. Journal of Personality and Social Psychology, 1975, 32, 671-679.

Baum, A., and Koman, S. Differential response to anticipated crowding: Psychological effects of social and spatial density. Journal of Personality and Social Psychology, 1976, 34, 526-536.

Baum, A., and Valins, S. Architecture and social behavior: psychological studies of social density. Hillsdale, N.J.: Lawrence Earlbaum, 1977.

Baum, A., and Valins, S. Residential environments, group size and crowding. Proceedings of the American Psychological Association, 1973, 211-212.

Baum, A., Valins, S., and Harpin, R. The role of group phenomena in the experience of crowding. In S. Saegert (Ed.), Crowding in real environments. Beverly Hills, Cal.: Sage, 1975.

Bayazit, N., Yönder, A., and Bakir (Özsoy), A. Three levels of privacy behaviour in the appropriation of dwelling spaces in Turkish homes. In 3. IAPC, Strasbourg, France, 1976.

Bechtel, R.B. Enclosing behavior. Stroudsberg, Pa.: Dowden, Hutchinson and Ross, 1971.

Bechtel, R.B. The undermanned environment: A universal theory? Paper presented at meetings of the Environmental Design and Research Association, Milwaukee, Wisc., 1974.

Behar, L., and Springfield, S. Behavior rating scale for the preschool child. Developmental Psychology, 1974, 10, 601-610.

Berardo, M.F. Marital invisibility and family privacy. In D. Carson (Ed.), EDRA 5, 1974.

Bergman, B.S. The effects of group size, personal space and success-failure on physiological arousal, test performance, and questionnaire response. Unpublished doctoral dissertation, Temple University (Pennsylvania), 1971. Abstract, Dissertation Abstracts International, 1971, 2319-3420-A.

Berlyne, D.E. Aesthetics and psychobiology, New York: Appleton Century Crofts, 1971.

Bernstein, B. Some sociological determinants of perception. British Journal of Sociology, 1958, 10, 153-174.

Berry, B. Geography of market centers and retail distribution. Englewood Cliffs, N.J.: Prentice-Hall, 1967.

Berry, B., and Garrison, W. Alternative explanations of urban rank-size relationships. Annals of the Association of American Geographers, 1958, 48, 83-91.

Berry, B., Goheen, P.G., and Goldstein,. Metropolitan area definition: A re-evaluation of concept and statistical practice. In L.S. Bourne (Ed.), Internal structure of the city. New York: Oxford University Press, 1971.

Berry, B., and Horton, F. Geographical perspectives on urban systems. Englewood Cliffs, N.J.: Prentice-Hall, 1967.

Berry, B., and Pred, A. Central place studies: A bibliography of theory and applications. Philadelphia: Regional Science Research Institute, 1961.

Berry, B., Simmonds, J.W., and Tennant, R.I. Urban population densities. The Geographical Review, 1963, 53 (3), 299-405.

Best, H.R. Extent of urban growth and agricultural displacement in post-war Britain. Urban Studies, 1968, 5(Feb.).

Bickman, L., Teger, A., Gabriele, T., McLaughlin, L, Berger, M., and Sunaday, E. Dormitory density and helping behavior. Environment and Behavior, 1973, 5, 465-490.

Bharucha-Reid, R. Environmental psychology, NEPA and the challenge of the real world. Invited address, American Psychological Association, Chicago, 1975.

Bharucha-Reid, R. Organized behavior under stress. Unpublished doctoral dissertation, Wayne State University (Detroit), 1972.

Bharucha-Reid, R., and Kiyak, A. Density, noise and personality. Paper presented at meetings of the American Psychological Association, Washington, D.C., 1976.

Bharucha-Reid, R., and Kiyak A. Human and non-human components of crowding. Paper presented at the Symposium on Human Consequences of Crowding, Antalya, Turkey, November, 1977.

Bird, C. The crowding syndrome: Learning to live with too much and too many. New York: David McKay, 1972.

Blake, J. Demographic science and the redirection of population policy. In M.C. Sheps and J.C. Ridley (Eds.), Public health and population change. Pittsburgh: University of Pittsburgh Press, 1965.

Blau, P.M. Exchange and power in social life. New York: Wiley, 1964.

Blumenfeld, H. The modern metropolis. Cambridge, Mass.: MIT Press, 1967.

Boaden, N., et al. Planning and participation in practice. (Forthcoming, /1977/)

Boaden, N., et al. Public participation in planning within a representative local democracy. Political Quarterly, in press (1977).

Boal, F.W. Territoriality on the Shankill-Falls Divide. Irish Geography (Belfast), 1969, 6, 30-50.

Booth, A. Final report: Urban crowding project. Ottawa: Ministry for Urban Affairs, 1975.

Booth, A., and Edwards, J.N. Crowding and family relations. American Sociological Review, 1976, 41, 308-321.

Booth, A., and Johnson, D.R. The effects of crowding on child health and development. American Behavioral Scientist, 1975, 18, 736-749.

Brehm, J. A theory of psychological reactance. London: Academic Press, 1966.

Broadment, D.E. Decision and stress. London: Academic Press, 1971.

Broady, M. Planning for people? Bedford Square Press, 1968.

Brower, S. The design of neighborhood parks. Baltimore: Department of Planning, 1977.

Brower, S. and Williams, P. Outdoor recreation as a function of the urban housing environment. Environment and Behavior, 1974, 6 (3), 295-345.

Brush, J.E. Some dimensions of urban population pressure in India. In W. Zelinsky, et al., (Eds.), Geography and crowding world. New York: Oxford University Press, 1970.

Bryant, D., and Hall, D. Neighbourhood councils in Brentwood. London: Polytechnic of North London, Department of Geography, 1971.

Butler, M., and Butler, N. Urban dwelling environments: Istanbul, Turkey, Unpublished M.Arch. A.S. thesis, Urban Settlement Design Program, MIT. Cambridge, Mass.: MIT Press, 1976.

Byrne, D., Baskett, G.O., and Hodges, L. Behavioral indicators of interpersonal attraction. Journal of Applied Social Psychology, 1971, 1, 137-149.

Calhoun, J. Death squared: The explosive growth and demise of a mouse population. Proceedings of the Royal Society of Medicine, 1973, 66, 80-89.

Calhoun, J.B. Population. In A. Allison (Ed.), Population control. Hardmondsworth: Penguin, 1970.

Calhoun, J.B. Population, density, and social pathology. Scientific American, 1962, 206, 139-140.

Calhoun, J.B. Space and the strategy of life. In A.H. Esser (Ed.), Behavior and environment: The use of space by animals and men. New York: Plenum, 1971.

Calhoun, J.B. What sort of box? Man-Environment-Systems, 1973, 3, 3-30.

Caminos, H., and Goethert, R. Urbanization primer: For design of site and service projects. Cambridge, Mass.: MIT Press, 1978.

Campbell, D.T., and Fiske, D.W. Convergent and discriminant validation by the multitrait-multimethod matrix. Psychological Bulletin, 1959, 56, 81-105.

Campbell, D.T., and Stanly, J.C. Experimental and quasi-experimental designs for research. Chicago: Rand McNally, 1966.

Canter, D., and Canter, S. Closer together in Tokyo. Design and Environment, 1971, 2, 60-63.

Carnahan, D., Gove, W., and Galle, D. Urbanization, population density and overcrowding: Trends in the quality of life in urban America. Social Forces, 1974, (53), 62-72.

Carroll, J.D. Individual differences and multidimensional scaling. In R.N. Shepard, A.K. Romney, and S.B. Nerlove (Eds.), Multidimensional scaling: Theory and applications in the behavioral sciences, Vol. 1. New York: Academic Press, 1972.

Carroll, J.D., and Chang, J.J. Analysis of individual differences in multidimensional scaling via an n-way generalization of 'Eckhart-Young' decomposition. Psychometrika, 1970, 35, 283-319.

Casetti, E. Alternate urban population density models: An analytical comparison of their validity range. In A.J. Scott (Ed.), Studies in regional science, London papers. In Regional Science, No. 1, Pion Ltd., 1969.

Cattell, R.B. Personality: A systematic, theoretical and factual study. New York: McGraw-Hill, 1950.

Celesun, M. Istanbul'da kent konutlarında açık mekanlar konusunda bir araştırma. Unpublished Master's thesis. Istanbul: Istanbul Teknik Üniversitesi, Mimarlık Fakültesi, 1976.

Chadwick, G. A systems view of planning. Oxford: Pergamon Press, 1971.

Chevan, A. Family growth, household density and moving. Demography, 1971, 8, 451-458.

Choi, S.C., Mirjafari, A., and Weaver, H.B. The concept of crowding. Environment and Behaviour, 1976, 8 (3).

Chombart de Lauwe, P. Famille et habitation. Paris: Edition Centre National de la Recherche Scientifique, 1959.

Christian, J. The pathology of overpopulation. Military Medicine, 1963, 128, 571-603.

Christian, J.J., Flyger, V., and Davis, D.C. Factors in mass mortality of a herd of sika deer. Chesapeake Science, 1, 79-95.

Clark, C. Urban densities compared. Town and Country Planning, 1967, 35 (Jan.), 32-33.

Clark, C. Urban population densities. Journal of the Royal Statistical Society, 1951, 114, 490-496.

Clough, G.C. Social behavior and ecology of Norwegian lemmings during a population peak and crash. Papers of the Norwegian State Game Research Institute, 1968, 2, 328.

Coale, A.J. Should the U.S. start a campaign for fewer births? Population Index, 1968, 34 (4), 47.

Cohen, J., Sladen, B., and Bennett, B. The effects of situational variables on judgments in crowding. Sociometry, 1975, 38, 273-281.

Cohen, S. Environmental load and the allocation of attention. In A. Baum and S. Valins, Advances in environmental psychology. Hillsdale, N.J.: Lawrence Earlbaum, 1977.

Cohen, S. Glass, D., and Phillips, S. Environment and health. In H.E. Freeman, S. Levine, and L.G. Reeder (Eds.), Handbook of medical sociology. Englewood Cliffs, N.J.: Prentice-Hall, 1977.

Cohen, S., Glass, D.C., and Singer, J.E. Apartment house, auditory discrimination and reading ability in children. Journal of Experimental Social Psychology, 1973, 9, 407-427.

Cole, R.L. Citizen participation and the urban policy process. Lexington, Mass.: D.C. Heath, 1973.

Commission on the Constitution. Devolution and other aspects of government: An attitudes survey. London: Her Majesty's Stationery Office, 1973.

[Committee of Inquiry on Industrial Democracy.] Report on the committee of inquiry on industrial democracy. London: Her Majesty's Stationery Office, 1977.

Committee on the Management of Local Government. The local government elector. London: Her Majesty's Stationery Office, 1967.

Coppock, J.T., and Sewell, D. (Eds.). Public participation in planning. London: Wiley, 1977.

Cozby, P.G. Effects of density, activity and personality on environmental preferences. Journal of Research in Personality, 1973, 7, 45-60.

D'Atri, D.A. Psycho-physiological responses to crowding. Environment and Behavior, 1975, 7, 237-252.

Delos Report. Ekistics, 1965, 21; 1966, 22 (July); 1967, 24 (Nov.), 144.

Department of the Environment. Children at play. Design Bulletin No. 27. London: Her Majesty's Stationery Office, 1973.

Desor, J.A. Toward a psychological theory of crowding. Journal of Personality and Social Psychology, 1972, 21, 79-95.

Dixey, B. A guide to neighbourhood councils. Halstead, Essex: The Association for Neighbourhood Councils, 1975.

Doherty, J.M. Developments in behavioural geography. London: London School of Economics, Graduate Geography Department, 1969.

Dooley, B.B. Crowding stress: The effects of social density on men with "close" or "far" personal space. Unbuplished doctoral dissertation, University of California, Los Angeles, 1974.

Dooley, B. Effects of social density on men with "close" or "far" personal space. Journal of Population: Behavioral, Social and Environmental Issues, in press (1977).

Dubos, R. The social environment. In H.M. Proshansky, W.H. Ittelson, and C.G. Rivlin (Eds.), Environmental psychology. New York: Holt, Rinehart and Winston, 1970.

Duke, M.P., and Nowicki, S. A new measure and social learning model for interpersonal distance. Journal of Experimental Research in Personality, 1972, 6, 1-16.

Ebel, R.L. Estimation of reliability of ratings. Psychometrika, 1951, 16, 407-424.

Ehrlich, P., and Ehrlich, A. Population resources, environment. San Francisco: Freeman, 1970.

Elliott, D. Policy and participation. London: Milton Keynes, Open University, 1975.

Epstein, Y., and Karlin, R. Effects of acute experimental crowding. Journal of Applied Social Psychology, 1975, 5 (1), 34-53.

Esmer, Ö. Elements of residential density. Unpublished paper, Middle East Technical University, Ankara, Turkey, 1970.

Esser, A.H. A biosocial perspective on crowding. In J. Wohlwill and D. Carson (Eds.), Environment and the social sciences: Perspectives and applications. [Washington, D.C.]: American Psychological Association, 1972.

Esser, A. Experiences of crowding: Illustration of a paradigm for man-environment relations. Representative Research in Social Psychology, 1973, 4, 207-218.

Esser, A.H. The psychopathology of crowding in institutions for the mentally ill and retarded. Paper presented at the 5th World Congress of Psychiatry, Mexico City, 1971.

Esser, A.H. Toward a definition of crowding. The Sciences, 1971, 11, 6.

Evans, G.W. Behavioral and physiological consequences of crowding in humans. Unpublished doctoral dissertation, University of Massachusetts, 1975.

Evans, G.W., and Eichelman, W. Preliminary models of conceptual linkages among proxemic variables. Environment and Behavior, 1976, 8, 87-116.

Everitt, J.C. Community and propinquity in a city. Annals of the Association of American Geographers, 1976, 66, 104-116.

Fairchild, H.P. (Ed.). Dictionary of sociology and related sciences. Patterson, N.J: Littlefield and Adams, 1964.

Faris, R., and Dunham, H. Mental disorders in urban areas. Chicago: University of Chicago Press, 1939.

Farley, J. Graduate women: Career aspirations and desired family size. American Psychologist, 1970, 25 (12), 1099-1100.

Fawcett, J.T. Psychology and population. The Population Council, 1970.

Feather, N.T. The relationship of persistence at a task to expectation of success and achievement relative to motive. Journal of Abnormal and Social Psychology, 1961, 63, 552-561.

Fienberg, S. The analysis of multidimensional contingency tables. Ecology, 1970, 51, 419-433.

Fischer, C.S. The urban experience. New York: Harcourt Brace Jovanovich, 1976.

Fischer, C.S., Baldassare, M., and Olshe, R.J. Crowding studies and urban life: A critical review. Working Paper 24, Berkely Institute of Urban and Rural Development, [n.d.]. In American Institute of Planners Journal, November, 1975, 406-418.

Fishbein, M., and Ajzen, I. Belief, attitude, intention and behavior. New York: Addison-Wesley, 1975.

Fisher, T. Situation-specific variables as determinants of perceived environmental aesthetic quality and perceived crowdedness. Unpublished manuscript, Department of Psychological Sciences, Purdue University, 1973.

Freedman, J.L. Crowding and Behavior. New York: Viking, 1975. (Paperback published by Freeman of San Francisco, 1975.)

Freedman, J.L., Heshka, S., and Levy, A. Crowding as an intensifier of the effect of success and failure. In J.L. Freedman, Crowding and Behavior. New York: Viking/San Francisco: Freeman, 1975.

Freedman, J.L. Heshka, S., and Levy, A. Population density and pathology: Is there a relationship? Journal of Experimental Social Psychology, 1975, 11, 539-552.

Freedman, J.L., Klevansky, S., Ehrlich, P., and Price, J. The effects of crowding on human task performance. Journal of Applied Social Psychology, 1971, 1, 7-25.

Freedman, J.L., Levy, A.S., Buchanan, R.W. and Price, J. Crowding and human aggressiveness. Journal of Experimental Social Psychology, 1972, 8, 528-548.

Gale, A., Lucas, B., Nissim, R., and Harpman, B. Some EEG correlates of face-to-face contact. British Journal of Social and Clinical Psychology, 1972, 11, 326-332.

Galle, O.R., Gove, W., and MacPherson, J. Population density and pathology: What are the relations for man? Science, 1972, 176, 23-30.

Galle, O.R., McCarthy, J.D., and Gove, W. Population density and pathology. Paper presented at the Annual Meeting of the Population Associates of America, New York, 1974.

Gallis, R. Population density and social pathology: The case of building type, social allowance and juvenile delinquency. Social Forces, 1973, 53, 306-314.

Garner, B.J. Models of urban geography and settlement location. In R. Charley and P. Haggett, Models in geography. London: Methuen, 1967.

Gibbs, J. (Ed.). Urban research methods. New York: Von Nostrand, 1961.

Glass, D.C., and Singer, J.E. Urban stress. New York: Academic Press, 1972.

Globig, L. Effects of crowding, sensitivity to crowding and set on task performance. Unpublished doctoral dissertation, University of Nevada, 1976.

Goldstein, S. Facets of redistribution: Research challenges and opportunities. Demography, 1976, 13, 423-434.

Goodchild, B. Class differences in environmental perception. Urban Studies, 1974, 11, 157-169.

Gordon, A.I. Jews in suburbia. Boston: Beacon Press, 1959.

Greenberg, C.I., and Baum, A. Compensatory response to anticipated densities. Journal of Applied Social Psychology, in press (1977).

Greenberg, C.I., and Firestone, I.J. Compensatory responses to crowding: Effects of personal space intrusion and privacy reduction. Journal of Personality and Social Psychology, 1977, 35, 656-663.

Griffith, N., and Veitch, R. Hot and crowded: Influences of population density and temperature on interpersonal affective behavior. Journal of Personality and Social Psychology, 1971, 17, 92-98.

Guardo, C. Personal space in children. Child Development, 1969, 40, 143-151.

Guardo, J., and Meisels, M. Child-parent spatial patterns under praise and reproof. Developmental Psychology, 1975, 5, 365.

Gurin, P., Gurin, G., Lao, R., and Beattie, M. Internal-external control in the motivational dynamics of negro youth. Journal of Social Issues, 1969, 25, 29-54.

Güvenç, B. An experimental family planning program in the Etimesgut health region. In F.C. Shorter and B. Guvenc (Eds.), Turkish Demography: Proceedings of a Conference. Ankara: Hacettepe University Publications, 1972.

Hafaez, E. (Ed.). The behavior of domestic animals. Baltimore: Williams and Wilkins, 1962.

Hall, E.T. The hidden dimension. New York: Doubleday, 1966.

Hall, E.T. The madding crowd: Space and its organization as a factor in mental health. Landscape, 1962, 12 (1).

Hall, E.T. The silent language. New York: Doubleday, 1959.

Hampton, W. Democracy and community. Oxford: Oxford University Press, 1970.

Hausser, D. Agency size and organizational functioning in the federal sector. Paper presented at the American Psychological Association Convention Symposium, "Is Small Beautiful?", 1977.

Hayes, W.L. Statistics for the social sciences. New York: Holt, Rinehart and Winston, 1973.

Heider. The psychology of interpersonal relations. New York: Wiley and Sons, 1958.

Heller, J., Groff, B., and Solomon, S. Toward an understanding of crowding: The role of physical interaction. Journal of Personality and Social Psychology, 1977, 35, 183-190.

Helson, H. Adaptation-level theory. New York: Harper and Row, 1964.

Heraud, B.J. The new towns and London's housing problems. Urban Studies, 1966, 3 (1).

Heraud, B.J. New towns: The end of a planner's dream. New Society, 1968, July 11.

Heraud, B.J. Social class and the new towns. Urban Studies, 1968, 5 (1).

Horowitz, M.J., Duff, D.F., and Stratton, L.O. Human spatial behavior. American Journal of Psychotherapy, 1965, 19, 20-28.

Horowitz, M., Duff, D., and Stratton, L. Personal space and the body buffer zone. Archives of General Psychology, 1964, 11, 651-656.

Hsu, Francis L. K. Clan, caste and club. New York: Von Nostrand, 1963.

Hsu, Francis, L. K. The effect of dominant kinship on kin and kin behavior: A hypothesis. American Anthropologist, 1965, 67, 638-660.

Hsu, Francis, L.K. Psychosocial hemeostasis and Jen: Conceptual tools for advancing psychological anthropology. American Anthropologist, 1971, 73, 23-44.

Hutt, C., and Vaizey, M.J. Differential effects of group density on social behavior. Nature, 1966, 209, 1371-1372.

International Encyclopedia of the Social Sciences. New York: Macmillan, 1968.

Irmak, Y. Kentleşmenin Sosyo-Demografik Yönleri. Istanbul: Habilitation Thesis, 1976.

Isard, W., et al. Methods of regional analysis. Cambridge, Mass. MIT Press, 1960.

Israel, J. Stipulations and constructions in the social sciences. In L. Israel and H. Tajfel (Eds.), The context of social psychology: A critical assessment. London: Academic Press, 1972.

Ittelson, W.H., Proshansky, H.M., and Rivlin, L.G. A study of bedroom use on two psychiatric wards. Hospital and Community Psychiatry, 1970, 21 (6), 177-180.

Ittelson, W. Proschansky, H., Rivlin, L., and Winkel, G. An introduction to environmental psychology. New York: Holt, Rinehart and Winston, 1972.

Iwata, O. Factors in the perception of crowding. Japanese Psychological Research, 1974, 16, 65-70.

Izmit yeni yerleşmeler yapılabilirlik araştırması. Istanbul: Birleşmiş Mimarlar, 1974.

İmamoğlu, V. The effect of furniture density on the subjective evalution of spaciousness and estimation of size of rooms. In 2. IAPC (Ed. by R. Küller), Strasbourg: Dowden, Hutchinson and Ross, 1973.

İmar ve İskan Bakanlığı. İstatistik verilere göre Türkiye'de konut nitelikleri. Ankara: İmar ve İskan Bakanlığı, M.G. M., 1974.

Jacobs, Jane. The death and life of great American cities. New York: Random House, 1961.

James, J.R. Residential densities and housing layouts. Town and Country Planning, 1967, 35, 552-561.

Janowitz, M. The community press in an urban setting. [n.p.]: Free Press, 1952.

Johnson, R. (Ed.). Space settlements. Washington, D.C.: National Aeronautics and Space Administration, 1977.

Jourard, S.M. The transparent self. New York: Van Nostrand-Reinhold, 1964.

Kandiyoti, D. Some implications of social change for housing design in human consequences of crowding. Paper presented at the Conference on Human Consequences of Crowding, Antalya, Turkey, November, 1977.

Kaplan, K.J., and Greenberg, C.I. Regulation of interaction through architecture, travel and telecommunication: A distance equilibrium approach to environmental planning. Environmental Psychology and Nonverbal Behavior, 1976, 1, 17-29.

Kasperson, R.E., and Minghi, J.V. The structure of political geography. London: University of London Press, 1969.

Keleş, R. Eski Ankara'da bir şehir tipolojisi. Yayınları No. 314. Ankara: Ankara University, Faculty of Political Sciences, 1971.

Kelvin, P.A. A social psychological examination of privacy. British Journal of Social and Clinical Psychology, 1973, 12, 248-261.

Kiray, M.B. Personal communication, 1977.

Kiray, M.B. Yedi yerleşme noktasında turizmle ilgili sosyal yapı analizi. Ankara: T.T.B., 1964.

Kinze, A. Body-buffer zones in violent prisoners. New Society, 1971, 28, 148.

Korte, C., Ypma, I., and Toppen, A. Helpfulness in Dutch society as a function of urbanization and environmental input level. Journal of Personality and Social Psychology, 1975, 32, 996-1003.

Küller, R. Beyond semantic measurement. In R. Küller (Ed.), Architectural psychology: Proceedings of the Lund Conference. Stroudsberg, Pa.: Dowden, Hutchinson and Ross, 1973.

Küller, R. Environment and activation. Report to be published by the National Swedish Institute for Building Research, Stockholm, 1978.

Küller, R. Psycho-physiological conditions in theatre construction. In Theatre Space: Proceedings of the 8th World Congress of the International Federation for Theatre Research, Münich, September 18-25, 1977. Munich: Prestel Verlag, 1977.

Küller, R. A semantic model for describing perceived environment. Document D 12. National Swedish Institute for Building Research, Stockholm, 1972.

Küller, R. The use of space: Some physiological and philosophical aspects. Paper presented at the 3rd International Architectural Psychology Conference at the Louis Pasteur University, Strasbourg, June 21-25, 1976.

Langer, E., and Saegert, S. Crowding and cognitive control. Journal of Personality and Social Psychology, 1977, 35, 175-182.

Lansing, J.B., and Hendricks, G. Automobile ownership and residential density. Ann Arbor: Institute for Social Research, 1967.

Laufler, R.S., Proshansky, H.M., and Wolfe, M. Some analytic dimensions of privacy. In H.M. Proshansky, W.H. Ittelson, and L.G. Rivlin (Eds.), Environmental Psychology, New York: Holt Rinehart Winston, 1970.

LeBon, G. The crowd: A study of the popular mind. New York: Ballentine, 1964.

Lee, T.R. Urban neighbourhood as a socio-spatial schema. Human Relations, 1968, 21, 241-268.

Levy, L., and Herzog, A.N. Effects of population density and crowding on health and social adaptation in the Netherlands. Journal of Health and Social Behavior, 1974, 15, 228-240.

Lewis, G.L. Turkey. New York: Praeger, 1965.

Linder, D.E. Personal space. In J.W. Thibaut, J.T. Spence, and R.C. Carson (Eds.), Contemporary topics in social psychology. Morristown, N.J.: General Learning Press, 1976.

Lipset, S.M. The American Jewish community in a comparative context. In P.I. Rose, (Ed.), The Ghetto and beyond. New York: Random House, 1969.

Little, K.B. Cultural variation in social schemata. Journal of Personality and Social Psychology, 1968, 10, 1-7.

Little, K. Personal space. Journal of Experimental Social Psychology, 1965, 1, 237-247.

Litwak, E. The use of extended family groups in the achievement of social goals. In M.B. Sussman (Ed.), Sourcebook in marriage and the family. Boston:Houghton-Mifflin, 1963.

Loo, C. The differential effects of spatial density on low and high scorers on behavioral problem indices. Environment and Behavior, in press (1977).

Loo, C.M. The effects of spatial density on the social behavior of children. Journal of Applied Social Psychology, 1972, 4, 372-381.

Loo, C.M. Ethical issues in research: Considerations for laboratory and survey settings and for minority participants. In J. Sieber (Ed.), Ethics of decision-making in social science research. In press (1977).

Loo, C.M., and Smetana, J. The effects of crowding on the behaviors and perceptions of 10-year-old boys. Environmental Psychology and Nonverbal Behavior, in press (1977).

Lott, D.F., and Sommer, R. Seating arrangements and status. Journal of Personality and Social Psychology, 1967, 7, 90-95.

Luria, A.R. The working brain: An introduction to neuro-psychology, Harmondsworth : Penguin, 1973.

Lynch, K. Site planning. Cambridge, Mass.: MIT Press, 1962.

Mabry, J.H. Toward the concept of housing adequacy. Sociology and Social Research, 1959-1960, 44, 86-90.

Maccoby, E., Dowley, E., Hagen, J., and Dagerman, R. Activity level and intellectual functioning in normal preschool children. Child Development, 1965, 36, 761-770.

Maccoby, E., and Jacklin, C. The psychology of sex differences. Stanford: Stanford University Press, 1974.

March, L., and Steadman, P. The geometry of environment. London: RIBA Publications, 1971.

Marsden, H. Crowding and animal behavior. In J. Wohlwill and D. Carson (Eds.) Environment and the social sciences: Perspectives and applications. Washington, D.C.: American Psychological Association, 1972.

Marsella, A. Escudero, M., and Gordon, P. The effects of dwelling density on mental disorders in Filipino men. Journal of Health and Social Behavior, 1970, 11, 288-294.

Marshall, J.E., and Heslin, R. Boys and girls together: Sexual composition and the effect of density and group size on cohesiveness. Journal of Personality and Social Psychology, 1975, 31, 952-961.

Marshall, N.J. Orientations toward privacy: Environmental and personality components. Unpublished doctoral dissertation, University of California, Berkeley, 1971.

Marshall, N. Privacy and environment. Human Ecology, 1972, 1, 93-110.

Martin, L., and March, L. (Eds.). Urban space and structures. Cambridge (England): Cambridge University Press, 1972.

Maslow, A.H., and Mintz, N.L. Effects of esthetic surroundings: I. Initial short-term effects of three esthetic conditions upon perceiving "energy" and "well-being" in faces. Journal of Psychology, 1956, 41, 247-254.

Mauldin, W.P. Population surveys: An essential tool. In B. Berelson (Ed.), Family planning programs. New York: Basic Books, 1969.

McBride, G., King, M.G., and James, J.W. Social proximity effects on galvanic skin response in adult humans. Journal of Psychology, 1965, 61, 153-157.

McCain, G., Cox, V., and Paulus, D. The relationship between illness complaints and degree of crowding in a prison environment. Environment and Behavior, 1976, 8, 289-291.

McCarthy, D., and Saegert, S. Residential density, social overload and social withdrawal. In J. Aiello and A. Baum, Crowding in residential environments. New York: Plenum, 1977.

McClelland, L.A. Crowding and social stress. Unpublished doctoral dissertation, University of Michigan, 1974.

McClelland, L., and Auslander, N. Determinants of perceived crowding and pleasantness in public settings. In P. Suidfeld and J. Russell (Eds.), The behavioral basis of design, Book I: Selected papers. Stroudsberg, Pa.: Dowden, Hutchinson and Ross, 1976.

McClelland, L, and Auslander, N. Perceptions of crowding and pleasantness in public settings. Environment and Behavior, in press (1977).

Mclenahan, The changing urban neighbourhood. University of California Press, 1929.

McClenahan. The communality: The urban substitute for the tradition community. Sociology and Social Research, 1945, 30, 264-274.

McGrew, P.L. Social and spatial density effects on spacing behavior in pre-school children. Journal of Child Psychology and Psychiatry, 1970, 11, 197-205.

McHarg, I. The ecology of the city. American Institute of Architects Journal, 1962, 38, 103.

McLoughlin, B.J. Urban and regional planning--A systems approach. London: Faber and Faber, 1969.

McNemar, Q. Psychological statistics. New York: Wiley, 1969.

McPherson, J.M. Population density and social pathology: A re-examination. Sociological Symposium, 1975, 141, 76-90.

Mead, G.H. The social psychology of G.H. Mead. (A.Strauss, Ed.) Chicago: University of Chicago Press, 1956.

Mead, M., and Mofraux, R. The study of culture at a distance. Chicago: University of Chicago Press, 1953.

Mehrabain, A. Relationship of attitude to seated posture, orientation and distance. Journal of Personality and Social Psychology, 1968, 10, 26-30.

Meier, R. A communication theory of urban growth. Cambridge, Mass, 1962.

Meisels, M., and Guardo, C.J. Development of personal space schemata. Child Development, 1969, 40, 1167-1178.

Mercer, J.C. Towards standing room only. In Psychology and Built Environment, 1974.

Michelsen, W., and Garland, R. The differential role of crowded homes and dense residential areas in the incidence of selected symptoms of human pathology. Research paper No. 67. Toronto: Centre for Urban and Community Studies, University of Toronto, 1974.

Miles, R.E., Jr. Ways of stopping U.S. population growth. Princeton Alumni Weekly, September 29, 1970.

Midgram, S. The experience of living in cities. Science, 1970, 167, 1461-1468.

Mindel, C. Issues and controversies in kinship research: Some critical observations. Paper presented at the 1970 meeting of the Midwest Sociological Society. St. Louis, Mo., April 16-18.

Mirels, H. Dimensions of internal vs. external control. Journal of Consulting and Clinical Psychology, 1970, 34 (2), 226-228.

Mitchell, R.E. Personal, family and social consequences arising from high density housing in Hong Kong and other major cities in Southeast Asia. Mimeo of the American Publich Health Association.

Mitchell, R. Some social implications of high density housing. American Sociological Review, 1971, 36, 18-29.

Moos, R.H. The human context: Environmental determinants of behavior. New York: Wiley, 1976.

Murdock, G.P. Social structure. New York: Macmillan, 1949.

Murgatroyd, S., Rees, B., and Reynolds, D. Taking local decisions. Leeds: ILP Square One, 1977.

Murray, R. The influence of crowding on children's behavior. In D. Canter (Ed.), Psychology and built environment. The Architectural Press, 1974.

Murray, R. Overcrowding and aggression in primary schoolchildren. In C.M. Morrison (Ed.), Educational priority, EPA, a Scottish study. Edinburgh: Her Majesty's Stationery Office, 1974.

Muth, R. The spatial structure of the housing market. PPRSA, 1961, 7, 207-220. Also in R. Muth, Cities and Housing, University of Chicago Press, 1969.

Newling, B. The spatial variation of urban population densities. Geographical Review, 1969, 59, 242-252.

Newling, B. Urban growth and spatial structure: Mathematical models and empirical evidence. Geographical Review, 1966, 56, 213-225.

Newman, Oscar. Defensible space. Chicago: University of Chicago Press, 1972.

Newsweek. The American Jew. Newsweek, 1971, (March 1).

Nisbett, R.E., and Wilson, T.D. Telling more than we can know: Verbal reports on mental processes. Psychological Review: 1977, 84, 231-299.

Nygren, T.E., and Jones, L.E. Individual differences in perceptions and preferences for political candidates. Journal of Experimental Social Psychology, 1977, 13, 182-197.

Osgood, C.F., May, W.H., and Miron, M.S. Cross-cultural universals of affective meaning. Urbana: University of Illinois Press, 1975.

Osmaniye gecekondu önleme bölgesi tespit çalışması. Istanbul: Istanbul Technical University, Faculty of Architecture, 1975.

Overmeier, J.B., and Seligman, M.E.P. Effects of inescapable shock upon subsequent escape and avoidance learning. Journal of Comparative and Physiological Psychology, 1967, 63, 23-33.

Özsoy, A. Kooperatif konutlarında kullanıcı mimar ilişkisinin davranış farklarının değerlendirilmesi. Unpublished Master's Thesis, Istanbul: Istanbul Technical University, Faculty of Architecture, 1976.

Parsons, T. Essays in sociological theory: Pure and applied. Free Press, 1949.

Patai, R. Tents of Jacob. Englewood Cliffs, N.J.: Prentice-Hall, 1971.

Patterson, A. Crowding: It ain't necessarily so. Contemporary Psychology, 1976, 21, 530-531.

Patterson, A. Methodological developments in environmental behavioral research. In D. Stokols (Ed.), Perspectives on environment and behavior. New York: Plenum, 1977.

Patterson, M.L. Compensation in nonverbal immediacy behaviors: A review. Sociometry, 1973, 36, 237-252.

Paulus, P.B., Aunis, A.B., Seta, J.J., Schkade, J.K., and Matthews, R.W. Density does affect task performance. Journal of Personality and Social Psychology, 1976, 34, 248-253.

Petty, R.M. Experimental investigation of undermanning theory. In D.H. Carson (Ed.), Man environment interactions: Evaluations and applications. Milwaukee: Environmental Design Research Assoc., 1974.

Petty, R.M., and Wicker, A.W. Degree of manning and degree of success of a group as determinants of members' objective experiences and their acceptance of a new group member: A laboratory study of Barker's theory. Unpublished paper, Department of Psychology, University of Illinois, 1971.

Preiser, W.F.E. Behavior of nursery school children under different spatial densities. Man-Environment-Systems, 1972, 2, 247-250.

Prescott, J.R.V. Political geography. Hutchinson, 1972.

Price, J. The effects of crowding on the social behavior of children. Unpublished doctoral dissertation, Columbia University, 1971.

Proshansky, H.M., and Rivlin, L.G. (Eds.), Environmental psychology: Man in his physical setting. New York: Holt, Rinehart and Winston, 1970.

Prohansky, H.N., Ittelson, W.H. and Rivlin, L.G. Freedom of choice and behavior in physical settings. In J.F. Wohlwill, and D.H. Carson (Eds.), Environment and the social sciences: Perspectives and applications. Washington, D.C.: American Psychological Association, 1972.

Proshansky, H.M., Ittelson, W.H., and Rivlin, L.G. Freedom of choice and behavior in a physical setting. In H.M. Proshansky and L.G. Rivlin (Eds.), Environmental psychology: Man in his physical setting. New York: Holt, Rinehart and Winston, 1970.

Putney, S., and Putney, G.T. The adjusted American. New York: Harper and Row, 1964.

Rainwater, L. Fear and the house-as-haven in the lower class. Journal of the American Institute of Planners, 1966, 32, 23-31.

Rapoport, A. Toward a redefinition of density. Environment and Behavior, 1975, 7, 133-158. Also in S. Saegert (Ed.), Crowding in real environments. Beverly Hills, Cal.: Sage Publications, 1976.

Rees, P.H. The axioms of intra-urban structure and growth. In B. Berry and F. Horton, Geographic perspectives on urban systems. Englewood Cliffs, N.J.: Prentice-Hall, 1970.

Research Services, Ltd. Community attitudes survey: England. [London]: Her Majesty's Stationery Office, 1969.

Research Services, Ltd. Community attitudes survey: Scotland. [London]: Her Majesty's Stationery Office, 1969.

Richardson, H. Optimality in city size, systems of cities and urban policy. Urban Studies, 1972, 9(1).

Rodin, J. Crowding, perceived choice and response to controllable and uncontrollable outcomes. Journal of Experimental Social Psychology, 1976, 12.

Rodin, J., and Baum, A. Crowding and helplessness: Potential consequences of density and loss of control. In A. Baum and Y. Epstein (Eds.), Human response to crowding. Hillsdale, N.J.: Lawrence Earlbaum, 1977.

Roncek, D.W. Density and crime. American Behavioral Scientist, 1975, 18, 843-860.

Ross, M., Layton, B., Erickson, B., and Schopler, J. Affect, facial regard and reactions to crowding. Journal of Personality and Social Psychology, 1973, 28, 69-76.

Ross, R.T. Optimal orders in the method of paired comparisons. Journal of Experimental Psychology, 1939, 25, 417-421.

Roth, C. A short history of the Jewish people. London: East and West Library, 1953.

Rotter, J.B. Generalized expectancies for internal-external control of reinforcement. Psychological Monographs, 1966, 80 (1).

Royal Commission on Local Government in England. Community attitudes survey. London: Her Majesty's Stationery Office, 1969.

Saegert, S. Crowding: Cognitive overload and behavioral constraint. In W. Preiser (Ed.), Environmental design research, Vol. II. Stroudsberg, Pa.: Dowden, Hutchinson and Ross, 1973.

Saegert, S. (Ed.), Crowding in real environments. Environment and Behavior, 1975, 7 (2), 131-252. Also published separately by Sage Publications, 1976.

Saegert, S.C. Effects of spatial and social density on arousal, mood and social organization. Unpublished doctoral dissertation, University of Michigan, 1974.

Saegert, S. High-density environments: Their personal and social consequences. In A. Baum and Y. Epstein (Eds.), Human responses to crowding. Hillsdale, N.J.: Lawrence Earlbaum, 1977.

Saegert, S. Stress inducing and stress reducing qualities of environments. In H. Proshansky. W. Ittelson, and L. Rivlin, Environmental Psychology. New York: Holt, Rinehart and Winston, 1975.

Saegert, S., Mackintosh, E., and West, S. Two studies of crowding in urban public spaces. Environment and Behavior, 1975, 7, 159-184.

Sarbin, T.R., and Allen, V.L. Role theory. In M.G. Lindzey and E. Aronson (Eds.), The handbook of social psychology, Vol. I. Reading, Mass.: Addison–Wesley, 1969.

Schachar, A. Patterns of population densities in the Tel-Aviv metropolitan area. Environment and Planning, 1975, 7, 279-291.

Schettino, A., and Borden, R. Sex differences in response to naturalistic crowding: Affective reactions to group size and group density. Personality and Social Psychology Bulletin, 1976, 2, 67-70.

Schmitt, R. Density, delinquency and crime in Honolulu. Sociology and Social Research, 1957, 41, 274-276.

Schmitt, R.C. Density, health and social disorganization. Journal of the American Institute of Planners, 1966, 32, 38-40.

Schmitt, R.C. Implications of density in Hong kong. Journal of the American Institute of Planners, 1968, 24, 210-217.

Schopler, J, Langmeyer, D., Stokols, D., and Reisman, S. The North Carolina internal-external scale: Validation of the short form. Research Previews, 1973, 20, 3-12.

Schopler, J, McCallum, R., and Rusbult, C.E. Behavioral interference and internality-externality as determinants of subjective crowding. Unpublished paper, University of North Carolina, Chapel Hill, 1977.

Schopler, J., Rusbult, C.E., and McCallum, R. Conceptual dimensions of crowding: A miltidimensional scaling analysis. Paper presented at the Symposium on the Human Consequences of Crowding, Antalya, Turkey, November, 1977.

Schopler, J., and Stockdale, J.E. An interference analysis of crowding. Journal of Environmental Psychology and Nonverbal Behavior, 1977, 1, 81-88.

Schopler, J., and Stokols, D. A psychological approach to human crowding. Morristown: General Learning Press, 1976.

Schopler, J., and Walton, M. The effects of expected structure, expected enjoyment and participants' internality-externality upon feelings of being crowded. Unpublished Manuscript, University of North Carolina, Chapel Hill, 1974.

Schumacher, E.F. Small Is Beautiful. New York: Harper and Row, 1973.

Seligman, M.E.P. Helplessness: On depression, development and death. San Fransisco: Freeman, 1975.

Seligman, M.E.P., and Maier, S.F. Failure to escape traumatic shock. Journal of Experimental Psychology, 1967, 74, 1-9.

Selltiz, C., Jahoda, M., Deutsch, M, and Cook, S.W. Research methods in social relations. New York: Holt, Rinehart and Winston, 1963.

Sensing, J., Reed, T.E., and Miller, J.S. Cooperation in the prisoner's dilemma game as a function of inter-personal distance. Psychonomic Science, 1972, 26, 105-106.

Seta, J.J., Paulus, P.B., and Schkade, J.K. Effects of group size and proximity under cooperative and competitive conditions. Journal of Personality and Social Psychology, 1976, 34 (1), 47-53.

Severy, L.J. Ecosystems: Cognitive style differences regarding population-environment phenomena. In J.B. Calhoun (Ed.), Perspectives on adaptation, environment and population. In press (1977).

Severy, L.J., and Atkins, T. A social-psychological conceptualization of the fertiltymigratory relationship: Spacing and ecosystems. Paper presented at the Southeastern Psychological Association Meetings, New Orleans, 1976.

Severy, L.J., Brigham, J.C., and Schlenker, B.R. A contemporary introduction to social psychology. New York: McGraw-Hill, 1976.

Sharf, Al. Byzantine Jewry. New York: Schocken Books, 1971.

Shepard, R.N., Romney, A.K., and Nerlove, S.B. (Eds.). Multidimensional scaling: Theory and applications in the behavioral sciences, Vol. 1. New York: Academic Press, 1972.

Sherrod, D.R. crowding, perceived control, and behavioral aftereffects. Journal of Applied Social Psychology, 1974, 4, 171-186.

Simmel, G. The metropolis and mental life. In P.K. Hatt, and A.J. Ross, Jr. (Eds.), Cities and society. New York: Free Press, 1957.

Simmie, J. Citizens in conflict. London:Hutchinson, 1974.

Skalre, M, and Greenblum, T. Jewish identity on the suburban frontier. New York: W.W. Norton, 1967.

Smith, P.K. Aspects of the play group environment. In D.Canter (Ed.), Psychology and Built Environment. The Architectural Press, 1974.

Smith, P.K. and Connelly, K. Patterns of play and social interaction in preschool children. In N.B. Jones (Ed.), Ethological studies of child behavior. Cambridge, England: Cambridge University Press, 1972.

Smith, R.W. A theoretical basis for participatory planning. Policy Sciences, 1975, 4, 275-295.

Solomon, S. Room at the bottom: Public housing projects are alive and well. The Sciences, 1972, 12, 25-29.

Sommer, R. Further studies in small group ecology. Sociometry, 1965, 28, 337-348.

Sommer, R. Personal space: The behavioral basis for design. London: Prentice-Hall, 1969.

Southwick, C.H. Peronycus Leucopus: An interesting subject for studies of socially induced stress responses. Science, 1964, 143, 55-56.

Speare, A., Goldstein, S., and Frey, W.H. Residential mobility, migration, and metropolitan change. Cambridge, Mass.: Ballinger, 1975.

Spencer, D. An evaluation of cognitive mapping in neighbourhood perception. CURS Research Memo 23. [Birmingham, England]: University of Birmingham, 1973.

Srivastava, R.K. Undermanning theory in the context of mental health care environments. In D.H. Carson (Ed.), Man-environment interactions, Part II. Stroudsberg, Pa.: Dowden, Hutchinson and Ross, 1974.

Stewart, J.D. The responsive local authority. London: Charles Knight, 1974.

Stewart, J.Q. and Warntz, W. Physics of population distribution. Journal of Regional Science, 1958, 1, 99-123. Also in G. Dimko, et al. (Eds.),Population geography: A reader. New York: McGraw-Hill, 1970.

Stockdale, J.E. Crowding: Determinants and effects. In L. Berkowitz (Ed.), Advances in experimental social psychology. New York: Academic Press, In Press (1977).

Stockdale, J.E., Jones, L., and Wittman, L. The perception of crowding: A multidimensional approach. London School of Economics, in preparation (1977).

Stockdale, J.E., and Schopler, J. The effects of room size and resource availability on subjective crowding. Unpublished manuscript, London School of Economics, 1976.

Stokols, D. The experience of crowding in primary and secondary environments. Paper presented at the sumposium on Theoretical and Empirical Issues with Regard to Privacy, Territorality, Personal Space and Crowding, at the 82nd Annual Convention of the American Psychological Association, New Orleans, August, 1974.

Stokols, D. The experience of crowding in primary and secondary environments. Environment and Behaviour, 1976, 8 (1), 49-86.

Stokols, D. On the distinction between density and crowding: Some implications for future research. Psychological Review, 1972, 79, 275-277.

Stokols, D. A social psychological model of human crowding phenomena. Journal of the American Institute of Planners, 1972, 38, 72-84.

Stokols, D. Toward a psychological theory of aienation. Psychological Review, 1975, 82, 26-44.

Stokols, D., Rall, M., Pinner, B., and Schopler, J. Physical, social and personal determinants of the perception of crowding. Environment and Behaviour, 1973, 5, 87 - 115.

Stokols, D., Smith, T.E., and Prostor, J.J. Partitioning and perceived crowding in a public space. American Behavioral Scientist, 1975, 18, 792-814.

Strahan, R.F., Todd, J.B., and Inglis, G.B. A palmar sweat measure particularly suited for naturalistic research. Psychophysiology, 1974, 11, 715-720.

Stringer, P. Attitudes, decision-making and environmental participation. London: Wiley, forthcoming (1977).

Stringer, P. Participation in personal construct theory. In D. Bannister (Ed.), New perspectives in personal construct theory. London: Academic Press, 1977.

Stringer, P., and Plumridge, G. Consultation with organisations on the North East Lancashire Advisory Plan. Interim Research Paper 3: Public participation in structure planning. Sheffield, England: University of Sheffield, 1974.

Stringer, P., and Plumridge, G. Review of the development control system. London: Her Majesty's Stationery Office, 1975.

Stycos, J.M. Family and fertility in Puerto Rico. New York: Columbia University Press, 1955.

Styles, B.J. Public participation: A reconstruction. Journal of the Town Planning Institute, 1971, 57.

Sundstrom, E. Crowding as a sequential process: Review of research on the effects of population density in humans. In A. Baum, and Y. Epstein (Eds.), Human response to crowding. Hillsdale, N.J.: Lawrence Earlbaum, 1977.

Sundstrom, E. An experimental study of crowding: Effects of room-size, intrusion, and goal-blocking on nonverbal behaviors, self-disclosure, and self-reported stress. Journal of Personality and Social Psychology, 1975, 32, 645-654.

Sundstrom, E. Toward an interpersonal model of crowding. Sociological Symposium, 1975, 14, 129-144.

Sussman, M.B. The help pattern in the middle class family. American Sociological Review, 1953, 18 (Feb.), 22-28.

Sussman, M.B. The isolated nuclear family: fact or fiction. In M.B. Sussman (Ed.), Sourcebook in marriage and the family. Boston: Houghton-Mifflin, 1963.

Takane, Y, Young, F.W., and de Leeuw, J. Nonmetric individual differences multidimensional scaling: An alternative least squares method with optimal scaling features. Psychometrica, 1977, 42, 7-67.

Tanaka-Matsumi, J., and Marsella, A. Cross-cultural variation in the phenomenological experience of depression. Journal of Cross-Cultural Psychology, 1976, 7, 379-413.

Tangri, S. Policies that affect the status of women and fertility. JSAS (Catalog of Documents in Psychology), 1972, 2, 107.

Tangri, S. Role innovation in occupational choice among college women. Unpublished doctoral dissertation, University of Michigan, 1969.

Thibault, J.W. and Kelley, H.H. The social psychology of groups. New York: Wiley, 1959.

Thomlinson, R. Population dynamics. New York: Random House, 1965.

Timur, S. Fertility and related attitudes among two social classes in Ankara, Turkey. Unpublished master's thesis, Hacettepe University, Ankara, 1955.

Toffler, A. Future shock. London: Bodley Head, 1970.

Tokman, K.B. Urban dwelling environments: Ankara, Turkey, Unpublished M.Arch. A.S. Thesis, Urban Settlement Design Program, MIT, Cambridge, Mass, 1975.

Tuncer, B. The impact of population growth on the Turkish economy. Ankara: Hacettepe University Publications, No. 3, 1968.

Tuncer, M. Ankara doğum evinde postpartum aile planlaması çalışması sonuçları. Unpublished paper (mimeo), January 25, 1971.

T.C. Başbakanlık, Devlet Planlama Teşkilatı, Dördüncü Beş Yıllık Kalkınma Planı (1977 - 1982)-- Özel İhtisas Komisyon Raporu. D.P.T. Publications, No. 1536, February, 1977.

U.S. Department of Housing and Urban Development, Federal Housing Administration. Land use intensity. Washington, D.C., 1966.

Valins, S., and Baum, A. Residential group size, social interaction and crowding. Environment and Behavior, 1973, 5, 421-439.

Vucinich, W.W. The Ottoman Empire. New York: Van Nostrand, 1965.

Waite, L.J., and Stolzenberg, R.M. Intended childbearing and labor force participation of young women: Insights from nonrecursive models. American Sociological Review, 1976, (41), 235-252.

Webber, M. Culture, territoriality and the elastic mile. Papers and Proceedings, Regional Science Association, 1964, 11, 59-69.

Webber, M. Urban space and the non-place urban realm. In M. Webber, et al., Explorations in urban structure. University Park, Pa.: University of Pennsylvania Press, 1964.

Welch, S. and Booth, A. The effect of crowding on aggression. Sociological Symposium, 1975, 14, 105-128.

Westin, A. Privacy and freedom. New York: Atheneum, 1970.

Whyte, W. The last landscape. New York: Doubleday, 1968.

Wicker, A. From church to laboratory to national park: A program of research on excess and insufficient populations in behavior settings. In D. Stokols (Ed.), Perspectives on environment and behavior: Theory, research and applications. New York: Plenum, 1977.

Wicker, A. Size of church membership and members' support of church behavior settings. Journal of Personality and Social Psychology, 1969, 13, 278-288.

Wicker, A. Too many, too few: Effects of overmanning and undermanning on human behavior. Colloquium presented at the University of California, Irving, May, 1974.

Wicker, A. Undermanning, performances and students' subjective experiences in behavior settings of large and small high schools. Journal of Personality and Social Psychology, 1968, 10, 255-261.

Wicker, A.W. Undermanning theory and research: Implications for the study of psychological and behavioural effects of excess populations. Representative Research in Social Psychology, 1973, 4, 265-277.

Wicker, A., and Kauma, C. Effects of a merger of a small and a large organization on members' behaviors and experiences. Journal of Applied Psychology, 1974, 59, 24-30.

Wicker, A., McGrath, J., and Armstrong, G. Organization size and behavior setting capacity as determinants of member participation. Behavioral Science, 1972, 17, 499-513.

Wilner, D.M., et al. The housing environment and family life. Baltimore: The Johns Hopkins Press, 1962.

Wilson, A.G. Entropy in urban and regional modelling. London: Pion, 1970.

Wilson, A.G. A statistical theory of spatial distribution models. Transportation Research, 1967, 1, 253-269.

Winer, R.J. Statistical principles in experimental design. New York: McGraw-Hill, 1962.

Winkel, G.H. The role of ecological validity in environmental research. In Proceedings of the Wisconsin Conference on Behavior-Environment Research Methods, sponsored by the Institute for Environmental Studies, University of Wisconsin, October, 1977.

Winsborough, H. The social consequences of high population density. Law and Contemporary Problems, 1965, 30 (1), 120-126.

Wirth, L. Urbanism as a way of life. American Journal of Sociology, 1938, 44. Also in P. Hatt, and A. Reiss (Eds.), Cities and society. The Free Press of Glencee, 1957.

Wolfe, M. Room size, group size and density. Environment and Behavior, 1975, 7, 199-224.

Wolpert, J. Migration as an adjustment to environmental stress, Journal of Social Issues, 1966, 22, 92-102.

Worchel, S., and Teddlie, C. Factors affecting the experience of crowding. A two-factor theory. Journal of Personality and Social Psychology, 1976, 34, 30-40.

Yaffe, J. The American Jews. New York: Paperback Library, 1968.

Yancey, W.L. Architecture, interaction and social control: The case of a large-scale public housing project. Environment and Behavior, 1971, 3 (1), 3-20.

Young, F.W. Scaling replicated conditional rank-order data. In D.R. Heise (Ed.), Sociological methodology. San Francisco: Josey-Bass, 1975.

Zelinsky, W. The geographer and his crowding world: Cautionary notes toward the study of population pressure in developing lands. In W. Zelinsky, et al. (Eds.), Geography and a crowding world. New York: Oxford University Press, 1970.

Zimmels, H.T. Ashkenazim and Sephardim. London: Oxford University Press, 1958.

Zlutnick, S., and Altman, I. Crowding and human behavior. In J. Wohlwill and D. Carson (Eds.), Environment and the social sciences: Perspectives and applications. Washington, D.C.: American Psychological Association, 1971.

INDEX OF NAMES

SUBJECT INDEX